LONDON MATHEMATICAL SOCIETY LECTURE NOTE SERIES

Managing Editor: Professor M. Reid, Mathematics Institute,
University of Warwick, Coventry, CV4 7AL, United Kingdom

The titles below are available from booksellers, or from Cambridge University Press at www.cambridge.org/mathematics

194 Independent random variables and rearrangement invariant spaces, M. BRAVERMAN
195 Arithmetic of blowup algebras, W. VASCONCELOS
196 Microlocal analysis for differential operators, A. GRIGIS & J. SJÖSTRAND
197 Two-dimensional homotopy and combinatorial group theory, C. HOG-ANGELONI et al.
198 The algebraic characterization of geometric 4-manifolds, J. A. HILLMAN
199 Invariant potential theory in the unit ball of C^n, M. STOLL
200 The Grothendieck theory of dessins d'enfants, L. SCHNEPS (ed.)
201 Singularities, J.-P. BRASSELET (ed.)
202 The technique of pseudodifferential operators, H. O. CORDES
203 Hochschild cohomology of von Neumann algebras, A. SINCLAIR & R. SMITH
204 Combinatorial and geometric group theory, A. J. DUNCAN, N. D. GILBERT & J. HOWIE (eds)
205 Ergodic theory and its connections with harmonic analysis, K. PETERSEN & I. SALAMA (eds)
207 Groups of Lie type and their geometries, W. M. KANTOR & L. DI MARTINO (eds)
208 Vector bundles in algebraic geometry, N. J. HITCHIN, P. NEWSTEAD & W. M. OXBURY (eds)
209 Arithmetic of diagonal hypersurfaces over infite fields, F. Q. GOUVÉA & N. YUI
210 Hilbert C*-modules, E. C. LANCE
211 Groups 93 Galway / St Andrews I, C. M. CAMPBELL et al. (eds)
212 Groups 93 Galway / St Andrews II, C. M. CAMPBELL et al. (eds)
214 Generalised Euler–Jacobi inversion formula and asymptotics beyond all orders, V. KOWALENKO et al.
215 Number theory 1992–93, S. DAVID (ed.)
216 Stochastic partial differential equations, A. ETHERIDGE (ed.)
217 Quadratic forms with applications to algebraic geometry and topology, A. PFISTER
218 Surveys in combinatorics, 1995, P. ROWLINSON (ed.)
220 Algebraic set theory, A. JOYAL & I. MOERDIJK
221 Harmonic approximation, S. J. GARDINER
222 Advances in linear logic, J.-Y. GIRARD, Y. LAFONT & L. REGNIER (eds)
223 Analytic semigroups and semilinear initial boundary value problems, KAZUAKI TAIRA
224 Computability, enumerability, unsolvability, S. B. COOPER, T. A. SLAMAN & S. S. WAINER (eds)
225 A mathematical introduction to string theory, S. ALBEVERIO et al.
226 Novikov conjectures, index theorems and rigidity I, S. FERRY, A. RANICKI & J. ROSENBERG (eds)
227 Novikov conjectures, index theorems and rigidity II, S. FERRY, A. RANICKI & J. ROSENBERG (eds)
228 Ergodic theory of Z^d actions, M. POLLICOTT & K. SCHMIDT (eds)
229 Ergodicity for infinite dimensional systems, G. DA PRATO & J. ZABCZYK
230 Prolegomena to a middlebrow arithmetic of curves of genus 2, J. W. S. CASSELS & E. V. FLYNN
231 Semigroup theory and its applications, K. H. HOFMANN & M. W. MISLOVE (eds)
232 The descriptive set theory of Polish group actions, H. BECKER & A. S. KECHRIS
233 Finite fields and applications, S. COHEN & H. NIEDERREITER (eds)
234 Introduction to subfactors, V. JONES & V. S. SUNDER
235 Number theory 1993–94, S. DAVID (ed.)
236 The James forest, H. FETTER & B. G. DE BUEN
237 Sieve methods, exponential sums, and their applications in number theory, G. R. H. GREAVES et al.
238 Representation theory and algebraic geometry, A. MARTSINKOVSKY & G. TODOROV (eds)
240 Stable groups, F. O. WAGNER
241 Surveys in combinatorics, 1997, R. A. BAILEY (ed.)
242 Geometric Galois actions I, L. SCHNEPS & P. LOCHAK (eds)
243 Geometric Galois actions II, L. SCHNEPS & P. LOCHAK (eds)
244 Model theory of groups and automorphism groups, D. EVANS (ed.)
245 Geometry, combinatorial designs and related structures, J. W. P. HIRSCHFELD et al.
246 p-Automorphisms of finite p-groups, E. I. KHUKHRO
247 Analytic number theory, Y. MOTOHASHI (ed.)
248 Tame topology and o-minimal structures, L. VAN DEN DRIES
249 The atlas of finite groups: ten years on, R. CURTIS & R. WILSON (eds)
250 Characters and blocks of finite groups, G. NAVARRO
251 Gröbner bases and applications, B. BUCHBERGER & F. WINKLER (eds)
252 Geometry and cohomology in group theory, P. KROPHOLLER, G. NIBLO, R. STÖHR (eds)
253 The q-Schur algebra, S. DONKIN
254 Galois representations in arithmetic algebraic geometry, A. J. SCHOLL & R. L. TAYLOR (eds)
255 Symmetries and integrability of difference equations, P. A. CLARKSON & F. W. NIJHOFF (eds)
256 Aspects of Galois theory, H. VÖLKLEIN et al.
257 An introduction to noncommutative differential geometry and its physical applications 2ed, J. MADORE
258 Sets and proofs, S. B. COOPER & J. TRUSS (eds)
259 Models and computability, S. B. COOPER & J. TRUSS (eds)
260 Groups St Andrews 1997 in Bath, I, C. M. CAMPBELL et al.
261 Groups St Andrews 1997 in Bath, II, C. M. CAMPBELL et al.
262 Analysis and logic, C. W. HENSON, J. IOVINO, A. S. KECHRIS & E. ODELL
263 Singularity theory, B. BRUCE & D. MOND (eds)
264 New trends in algebraic geometry, K. HULEK, F. CATANESE, C. PETERS & M. REID (eds)
265 Elliptic curves in cryptography, I. BLAKE, G. SEROUSSI & N. SMART
267 Surveys in combinatorics, 1999, J. D. LAMB & D. A. PREECE (eds)
268 Spectral asymptotics in the semi-classical limit, M. DIMASSI & J. SJÖSTRAND
269 Ergodic theory and topological dynamics, M. B. BEKKA & M. MAYER
270 Analysis on Lie Groups, N. T. VAROPOULOS & S. MUSTAPHA
271 Singular perturbations of differential operators, S. ALBEVERIO & P. KURASOV
272 Character theory for the odd order theorem, T. PETERFALVI
273 Spectral theory and geometry, E. B. DAVIES & Y. SAFAROV (eds)
274 The Mandelbrot set, theme and variations, TAN LEI (ed.)
275 Descriptive set theory and dynamical systems, M. FOREMAN et al.

276 Singularities of plane curves, E. CASAS-ALVERO
277 Computational and geometric aspects of modern algebra, M. D. ATKINSON *et al.*
278 Global attractors in abstract parabolic problems, J. W. CHOLEWA & T. DLOTKO
279 Topics in symbolic dynamics and applications, F. BLANCHARD, A. MAASS & A. NOGUEIRA (eds)
280 Characters and automorphism groups of compact Riemann surfaces, T. BREUER
281 Explicit birational geometry of 3-folds, A. CORTI & M. REID (eds)
282 Auslander–Buchweitz approximations of equivariant modules, M. HASHIMOTO
283 Nonlinear elasticity, Y. FU & R. OGDEN (eds)
284 Foundations of computational mathematics, R. DEVORE, A. ISERLES & E. SÜLI (eds)
285 Rational points on curves over finite fields, H. NIEDERREITER & C. XING
286 Clifford algebras and spinors 2ed, P. LOUNESTO
287 Topics on Riemann surfaces and Fuchsian groups, E. BUJALANCE *et al.*
288 Surveys in combinatorics, 2001, J. HIRSCHFELD (ed.)
289 Aspects of Sobolev-type inequalities, L. SALOFF-COSTE
290 Quantum groups and Lie theory, A. PRESSLEY (ed.)
291 Tits buildings and the model theory of groups, K. TENT (ed.)
292 A quantum groups primer, S. MAJID
293 Second order partial differential equations in Hilbert spaces, G. DA PRATO & J. ZABCZYK
294 Introduction to operator space theory, G. PISIER
295 Geometry and Integrability, L. MASON & YAVUZ NUTKU (eds)
296 Lectures on invariant theory, I. DOLGACHEV
297 The homotopy category of simply connected 4-manifolds, H.-J. BAUES
298 Higher operands, higher categories, T. LEINSTER
299 Kleinian Groups and Hyperbolic 3-Manifolds, Y. KOMORI, V. MARKOVIC & C. SERIES (eds)
300 Introduction to Möbius Differential Geometry, U. HERTRICH-JEROMIN
301 Stable Modules and the D(2)-Problem, F. E. A. JOHNSON
302 Discrete and Continuous Nonlinear Schrödinger Systems, M. J. ABLOWITZ, B. PRINARI & A. D. TRUBATCH
303 Number Theory and Algebraic Geometry, M. REID & A. SKOROBOGATOV (eds)
304 Groups St Andrews 2001 in Oxford Vol. 1, C. M. CAMPBELL, E. F. ROBERTSON & G. C. SMITH (eds)
305 Groups St Andrews 2001 in Oxford Vol. 2, C. M. CAMPBELL, E. F. ROBERTSON & G. C. SMITH (eds)
306 Peyresq lectures on geometric mechanics and symmetry, J. MONTALDI & T. RATIU (eds)
307 Surveys in Combinatorics 2003, C. D. WENSLEY (ed.)
308 Topology, geometry and quantum field theory, U. L. TILLMANN (ed.)
309 Corings and Comdules, T. BRZEZINSKI & R. WISBAUER
310 Topics in Dynamics and Ergodic Theory, S. BEZUGLYI & S. KOLYADA (eds)
311 Groups: topological, combinatorial and arithmetic aspects, T. W. MÜLLER (ed.)
312 Foundations of Computational Mathematics, Minneapolis 2002, FELIPE CUCKER *et al.* (eds)
313 Transcendatal aspects of algebraic cycles, S. MÜLLER-STACH & C. PETERS (eds)
314 Spectral generalizations of line graphs, D. CVETKOVIC, P. ROWLINSON & S. SIMIC
315 Structured ring spectra, A. BAKER & B. RICHTER (eds)
316 Linear Logic in Computer Science, T. EHRHARD *et al.* (eds)
317 Advances in elliptic curve cryptography, I. F. BLAKE, G, SEROUSSI & N. SMART
318 Perturbation of the boundary in boundary-value problems of Partial Differential Equations, DAN HENRY
319 Double Affine Hecke Algebras, I. CHEREDNIK
320 L-Functions and Galois Representations, D. BURNS, K. BUZZARD & J. NEKOVÁŘ (eds)
321 Surveys in Modern Mathematics, V. PRASOLOV & Y. ILYASHENKO (eds)
322 Recent perspectives in random matrix theory and number theory, F. MEZZADRI, N. C. SNAITH (eds)
323 Poisson geometry, deformation quantisation and group representations, S. GUTT *et al.* (eds)
324 Singularities and Computer Algebra, C. LOSSEN & G. PFISTER (eds)
325 Lectures on the Ricci Flow, P. TOPPING
326 Modular Representations of Finite Groups of Lie Type, J. E. HUMPHREYS
328 Fundamentals of Hyperbolic Manifolds, R. D. CANARY, A. MARDEN & D. B. A. EPSTEIN (eds)
329 Spaces of Kleinian Groups, Y. MINSKY, M. SAKUMA & C. SERIES (eds)
330 Noncommutative Localization in Algebra and Topology, A. RANICKI (ed.)
331 Foundations of Computational Mathematics, Santander 2005, L. PARDO, A. PINKUS, E. SULI & M. TODD (eds)
332 Handbook of Tilting Theory, L. ANGELERI HÜGEL, D. HAPPEL & H. KRAUSE (eds)
333 Synthetic Differential Geometry 2ed, A. KOCK
334 The Navier–Stokes Equations, P. G. DRAZIN & N. RILEY
335 Lectures on the Combinatorics of Free Probability, A. NICA & R. SPEICHER
336 Integral Closure of Ideals, Rings, and Modules, I. SWANSON & C. HUNEKE
337 Methods in Banach Space Theory, J. M. F. CASTILLO & W. B. JOHNSON (eds)
338 Surveys in Geometry and Number Theory, N. YOUNG (ed.)
339 Groups St Andrews 2005 Vol. 1, C. M. CAMPBELL, M. R. QUICK, E. F. ROBERTSON & G. C. SMITH (eds)
340 Groups St Andrews 2005 Vol. 2, C. M. CAMPBELL, M. R. QUICK, E. F. ROBERTSON & G. C. SMITH (eds)
341 Ranks of Elliptic Curves and Random Matrix Theory, J. B. CONREY, D. W. FARMER, F. MEZZADRI & N. C. SNAITH (eds)
342 Elliptic Cohomology, H. R. MILLER & D. C. RAVENEL (eds)
343 Algebraic Cycles and Motives Vol. 1, J. NAGEL & C. PETERS (eds)
344 Algebraic Cycles and Motives Vol. 2, J. NAGEL & C. PETERS (eds)
345 Algebraic and Analytic Geometry, A. NEEMAN
346 Surveys in Combinatorics, 2007, A. HILTON & J. TALBOT (eds)
347 Surveys in Contemporary Mathematics, N. YOUNG & Y. CHOI (eds)
348 Transcendental Dynamics and Complex Analysis, P. J. RIPPON & G. M. STALLARD (eds)
349 Model Theory with Applications to Algebra and Analysis Vol 1, Z. CHATZIDAKIS, D. MACPHERSON, A. PILLAY & A. WILKIE (eds)
350 Model Theory with Applications to Algebra and Analysis Vol 2, Z. CHATZIDAKIS, D. MACPHERSON, A. PILLAY & A. WILKIE (eds)
351 Finite von Neumann Algebras and Masas, A. SINCLAIR & R. SMITH
352 Number Theory and Polynomials, J. MCKEE & C. SMYTH (eds)
353 Trends in Stochastic Analysis, J. BLATH, P. MÖRTERS & M. SCHEUTZOW (eds)
354 Groups and Analysis, K. TENT (ed)
355 Non-equilibrium Statistical Mechanics and Turbulence, J. CARDY, G. FALKOVICH & K. GAWEDZKI, with S. NAZARENKO & O. V. ZABORONSKI (eds)
356 Elliptic Curves and Big Galois Representations, D. DELBOURGO

London Mathematical Society Lecture Note Series: 357

Algebraic Theory of Differential Equations

Edited by

MALCOLM A. H. MACCALLUM
Queen Mary, University of London

ALEXANDER V. MIKHAILOV
University of Leeds

CAMBRIDGE
UNIVERSITY PRESS

CAMBRIDGE
UNIVERSITY PRESS

University Printing House, Cambridge CB2 8BS, United Kingdom

One Liberty Plaza, 20th Floor, New York, NY 10006, USA

477 Williamstown Road, Port Melbourne, VIC 3207, Australia

314-321, 3rd Floor, Plot 3, Splendor Forum, Jasola District Centre, New Delhi - 110025, India

79 Anson Road, #06-04/06, Singapore 079906

Cambridge University Press is part of the University of Cambridge.

It furthers the University's mission by disseminating knowledge in the pursuit of education, learning and research at the highest international levels of excellence.

www.cambridge.org
Information on this title: www.cambridge.org/9780521720083

First published 2009

A catalogue record for this publication is available from the British Library

ISBN 978-0-521-72008-3 Paperback

Contents

Preface   *page* vi

1 Galois Theory of Linear Differential Equations   *Michael
   F. Singer*   1

2 Solving in closed form   *Felix Ulmer and Jacques-Arthur
   Weil*   83

3 Factorization of Linear Systems   *Sergey P. Tsarev*   111

4 Introduction to D-modules   *Anton Leykin*   132

5 Symbolic representation and classification of integrable
   systems   *A.V. Mikhailov, V.S. Novikov and Jing Ping
   Wang*   156

6 Searching for integrable (P)DEs   *Jarmo Hietarinta*   217

7 Around Differential Galois Theory   *Anand Pillay*   232

v

Preface

This book presents lectures given during a school and workshop organized at Heriot-Watt University in July and August 2006, with funding from the London Mathematical Society (LMS), which supported the Invited Lectures of Michael Singer, and the International Centre for Mathematical Sciences (ICMS) in Edinburgh, which supported the Workshop.

The origin of these events lies in a suggestion made by one of us (AVM) to the other that we should make a proposal for an Isaac Newton Institute programme. After some thought we concluded that a smaller scale proposal would be more appropriate to the present state of the field and the wishes of likely participants, as well as our capacity to write proposals.

We therefore, with his agreement, proposed Michael Singer as the speaker for the 2006 LMS Invited Lecture series, and simultaneously make a proposal to ICMS for a workshop to follow the Invited Lectures: happily both proposals were accepted. We were very glad that in making the workshop proposal we had the assistance of Michael himself and of Sergei Tsarev, who were very helpful in suggesting topics for inclusion and people to invite, whether or not they became speakers.

Between the initial proposal and carrying it out we were fortunate to recruit Chris Eilbeck of Heriot-Watt as an additional organizer. His local knowledge and connections with ICMS were invaluable in actually mounting the event, and it is hard to express our thanks forcefully enough!

Michael Singer is an excellent lecturer of great clarity: from personal experience we also knew that he is generous with his time and ideas, especially to new workers in the field. He has made outstanding contributions in Differential Algebra and Differential (and Difference) Galois Theory, and naturally it was that which we asked him to lecture on. We also had in mind from the start that we should complement Michael's lectures with some shorter series of lectures on other topics in the general field, which, together with Michael's longer series, would enable people new to the field to profit from the workshop. In a way, we were lucky that because he had recently published books in the field, Michael did not wish to write up the lectures at book length, as many LMS Invited Lecturers have, and we are therefore able to present the work of other lecturers as well.

We aimed, in both the school and the workshop, to cover the full range of algebraic approaches to differential equations: differential algebra, D-modules, model-theoretical aspects of the theory of differential equations, the Inverse Transform method for nonlinear partial differential equations, and the algebraic theory of factorization of differential equations and systems of differential equations. The only major approach which may have been under-represented was Lie symmetry methods and their generalizations, on which there are already excellent texts. We were also particularly keen to encourage and facilitate interactions between the various approaches, which had not really been brought together before.

As it turned out we were able to mount a good many of the subsidiary shorter lecture series we hoped for, with speakers distinguished for both content and presentation. As in the school itself, the major contribution in this book is Singer's and is an expanded written version of his lectures, but we also have written versions of most of the others. Anand Pillay (and others working in model theory and related areas) had written a number of other expository and research papers about the same time, and so did not wish to make a full writeup, but Anand nevertheless kindly agreed to write a short account referring to those other articles, while the general area covered by Vladimir Sokolov's lectures is discussed in the contribution of Mikhailov, Wang and Novikov.

In general we felt that the workshop talks, although covering a lot of very interesting and novel work, were too specialized for a lecture notes volume, but we invited some of those whose talks had more of an overview character, or had included a section of that kind, to contribute. Felix Ulmer and Jacques-Arthur Weil provided a combined contribution, Wang and Novikov joined theirs with Mikhailov's, and Hietarinta also kindly agreed to contribute.

We heartily thank our co-organizers, without whom the event would have been poorer both intellectually and practically, our sponsors, LMS and ICMS, the lecturers whose work is presented here, and the many other contributors to the workshop. We also gratefully acknowledge additional support from the Edinburgh Mathematical Society and the Royal Society of Edinburgh.

Finally, we would like to thank the ICMS very much for its organizational help, as well as the funding, in particular its Director, John Toland, and the administrative staff, Tracey Dart, Morag Burton and Audrey Brown, who made all the practical arrangements including coping with the knock-on effects of the sudden change in airline baggage regulations that happened at the end of the second week, and to thank the Conference staff of Heriot-Watt for enabling such a pleasant and stimulating two weeks. In recognition of the work of ICMS, any royalties from this book will be given to them to add to student support at future events.

Malcolm MacCallum
Alexander Mikhailov
January 2008

1

Introduction to the Galois Theory of Linear Differential Equations

Michael F. Singer

Department of Mathematics
North Carolina State University
Raleigh, NC 27695-8205
singer@math.ncsu.edu

1.1 Introduction

This paper is an expanded version of the 10 lectures I gave as the 2006 London Mathematical Society Invited Lecture Series at the Heriot-Watt University, 31 July - 4 August 2006†. My goal was to give the audience an introduction to the algebraic, analytic and algorithmic aspects of the Galois theory of linear differential equations by focusing on some of the main ideas and philosophies and on examples. There are several texts ([1, 2, 3, 4, 5] to name a few) that give detailed expositions and I hope that the taste offered here will encourage the reader to dine more fully with one of these.

The rest of the paper is organized as follows. In Section 1.2, *What is a Linear Differential Equation?*, I discuss three ways to think about linear differential equations: scalar equations, linear systems and differential modules. Just as it is useful to think of linear maps in terms of linear equations, matrices and associated modules, it will be helpful in future sections to go back and forth between the different ways of presenting linear differential equations.

In Section 1.3, *Basic Galois Theory and Applications*, I will give the basic definitions and describe the Galois correspondence. In addition I will describe the notion of monodromy and its relation to the Galois

† I would like to thank the London Mathematical Society for inviting me to present these lectures, the Heriot-Watt University for hosting them and the International Centre for the Mathematical Sciences for sponsoring a mini-programme on the Algebraic Theory of Differential Equations to complement these talks. Thanks also go to the Edinburgh Mathematical Society and the Royal Society of Edinburgh, who provided support for some of the participants and to Chris Eilbeck, Malcolm MacCallum, and Alexandre Mikhailov for organizing these events. This material is based upon work supported by the National Science Foundation under Grant No. 0096842 and 0634123

theory. I will end by giving several applications and ramifications, one of which will be to answer the question *Although $y = \cos x$ satisfies $y'' + y = 0$, why doesn't $\sec x$ satisfy a linear differential equation?*

In Section 1.4, *Local Galois Theory*, I will discuss the formal solution of a linear differential equation at a singular point and describe the asymptotics which allow one to associate with it an analytic solution in a small enough sector at that point. This will involve a discussion of Gevrey asymptotics, the Stokes phenomenon and its relation to the Galois theory. A motivating question (which we will answer in this section) is *In what sense does the function*

$$f(x) = \int_0^\infty \frac{1}{1+\zeta} e^{-\frac{\zeta}{x}} d\zeta$$

represent the divergent series

$$\sum_{n \geq 0} (-1)^n n! x^{n+1} \ ?$$

In Section 1.5, *Algorithms*, I turn to a discussion of algorithms that allow us to determine properties of a linear differential equation and its Galois group. I will show how category theory, and in particular the tannakian formalism, points us in a direction that naturally leads to algorithms. I will also discuss algorithms to find "closed form solutions" of linear differential equations.

In Section 1.6, *Inverse Problems*, I will consider the problem of which groups can appear as Galois groups of linear differential equations. I will show how monodromy and the ideas in Section 1.4 play a role as well as ideas from Lie theory.

In Section 1.7, *Families of Linear Differential Equations*, I will look at linear differential equations that contain parameters and ask *How does the Galois group change as we vary the parameters?*. This will lead to a discussion of a a generalization of the above theory to a Galois theory of parameterized linear differential equations.

1.2 What is a Linear Differential Equation?

I will develop an algebraic setting for the study of linear differential equations. Although there are many interesting questions concerning differential equations in characteristic p [6, 7, 8, 9], we will restrict ourselves throughout this paper, without further mention, to fields of characteristic 0. I begin with some basic definitions.

Definition 1.2.1 *1) A differential ring* (R, Δ) *is a ring* R *with a set* $\Delta = \{\partial_1, \ldots, \partial_m\}$ *of maps* (derivations) $\partial_i : R \to R$, *such that*

(i) $\partial_i(a+b) = \partial_i(a) + \partial_i(b)$, $\partial_i(ab) = \partial_i(a)b + a\partial_i(b)$ *for all* $a, b \in R$, *and*

(ii) $\partial_i \partial_j = \partial_j \partial_i$ *for all* i, j.

2) The ring $C_R = \{c \in R \mid \partial(c) = 0 \ \forall \ \partial \in \Delta\}$ *is called the* ring of constants *of* R.

When $m = 1$, we say R is an *ordinary differential ring* (R, ∂). We frequently use the notation a' to denote $\partial(a)$ for $a \in R$. A differential ring that is also a field is called a *differential field*. If k is a differential field, then C_k is also a field.

Examples 1.2.2 1) $(C^\infty(\mathbb{R}^m), \Delta = \{\frac{\partial}{\partial x_1}, \ldots, \frac{\partial}{\partial x_m}\})$ = infinitely differentiable functions on $\mathbb{R}^m$.

2) $(\mathbb{C}(x_1, \ldots, x_m), \Delta = \{\frac{\partial}{\partial x_1}, \ldots, \frac{\partial}{\partial x_m}\})$ = field of rational functions

3) $(\mathbb{C}[[x]], \frac{\partial}{\partial x})$ = ring of formal power series

$\mathbb{C}((x))$ = quotient field of $\mathbb{C}[[x]] = \mathbb{C}[[x]][\frac{1}{x}]$

4) $(\mathbb{C}\{\{x\}\}, \frac{\partial}{\partial x})$ = ring of germs of convergent series

$\mathbb{C}(\{x\})$ = quotient field of $\mathbb{C}\{\{x\}\} = \mathbb{C}\{\{x\}\}[\frac{1}{x}]$

5) $(\mathcal{M}_\mathcal{O}, \Delta = \{\frac{\partial}{\partial x_1}, \ldots, \frac{\partial}{\partial x_m}\})$ = field of functions meromorphic on $\mathcal{O}^{\text{open,connected}} \subset \mathbb{C}^m$

The following result of Seidenberg [10, 11] shows that many examples reduce to Example 5) above:

Theorem 1.2.3 *Any differential field* k, *finitely generated over* $\mathbb{Q}$, *is isomorphic to a differential subfield of some* $\mathcal{M}_\mathcal{O}$.

We wish to consider and compare three different versions of the notion of a linear differential equation.

Definition 1.2.4 *Let* (k, ∂) *be a differential field.*

(i) *A* scalar linear differential equation *is an equation of the form*

$$L(y) = a_n y^{(n)} + \ldots + a_0 y = 0, \quad a_i \in k.$$

(ii) *A matrix linear differential equation is an equation of the form*

$$Y' = AY, \; A \in \mathrm{gl}_n(k)$$

where $\mathrm{gl}_n(k)$ *denotes the ring of* $n \times n$ *matrices with entries in* k.

(iii) *A* differential module *of dimension* n *is an* n-*dimensional* k-*vector space* M *with a map* $\partial : M \to M$ *satisfying*

$$\partial(fm) = f'm + f\partial m \text{ for all } f \in k, m \in M.$$

We shall show that these three notions are equivalent and give some further properties.

From scalar to matrix equations: Let $L(y) = y^{(n)} + a_{n-1}y^{(n-1)} + \ldots + a_0 y = 0$. If we let $y_1 = y, y_2 = y', \ldots y_n = y^{(n-1)}$, then we have

$$
\begin{pmatrix} y_1 \\ y_2 \\ \vdots \\ y_{n-1} \\ y_n \end{pmatrix}'
=
\begin{pmatrix}
0 & 1 & 0 & \cdots & 0 \\
0 & 0 & 1 & \cdots & 0 \\
\vdots & \vdots & \vdots & \vdots & \vdots \\
0 & 0 & 0 & \cdots & 1 \\
-a_0 & -a_1 & -a_2 & \cdots & -a_{n-1}
\end{pmatrix}
\begin{pmatrix} y_1 \\ y_2 \\ \vdots \\ y_{n-1} \\ y_n \end{pmatrix}
$$

We shall write this last equation as $Y' = A_L Y$ and refer to A_L as the companion matrix of the scalar equation and the matrix equation as the companion equation. Clearly any solution of the scalar equation yields a solution of the companion equation and *vice versa*.

From matrix equations to differential modules (and back):

Given $Y' = AY$, $A \in \mathrm{gl}_n(k)$, we construct a differential module in the following way: Let $M = k^n$, $e_1, \ldots, e_n$ the usual basis. Define $\partial e_i = -\sum_j a_{j,i} e_j$, i.e., $\partial e = -A^t e$. Note that if $m = \sum_i f_i e_i$ then $\partial m = \sum_i (f_i' - \sum_j a_{i,j} f_j) \bar{e}_i$. In particular, we have that $\partial m = 0$ if and only if

$$
\begin{pmatrix} f_1 \\ \vdots \\ f_n \end{pmatrix}'
= A
\begin{pmatrix} f_1 \\ \vdots \\ f_n \end{pmatrix}
$$

It is this latter fact that motivates the seemingly strange definition of this differential module, which we denote by (M_A, ∂).

Conversely, given a differential module (M, ∂), <u>select</u> a basis $e = (e_1, \ldots, e_n)$. Define $A_M \in \mathrm{gl}_n(k)$ by $\partial e_i = \sum_j a_{j,i} e_j$. This yields a matrix equation $Y' = AY$. If $\bar{e} = (\bar{e}_1, \ldots, \bar{e}_n)$ is another basis, we get another equation $Y' = \bar{A}Y$. If f and $\bar{f}$ are vectors with respect to these two bases and $f = B\bar{f}$, $B \in \mathrm{GL}_n(k)$, then

$$\bar{A} = B^{-1}AB - B^{-1}B' \ .$$

Definition 1.2.5 *Let* (M_1, ∂_1) *and* (M_2, ∂_2) *be differential modules.*
1) A differential module homomorphism $\phi : M_1 \to M_2$ *is a* k-*linear map* ϕ *such that* $\phi(\partial_1(m)) = \partial_2(\phi(m))$ *for all* $m \in M_1$.
2) The differential modules (M_1, ∂_1) *and* (M_2, ∂_1) *are* isomorphic *if there exists a bijective differential homomorphism* $\phi : M_1 \to M_2$.
3) Two differential equations $Y' = A_1Y$ *and* $Y' = A_2Y$ *are* equivalent *if the differential modules* M_{A_1} *and* M_{A_2} *are isomorphic*

Instead of equivalent, some authors use the term "gauge equivalent" or "of the same type".

Differential modules offer us an opportunity to study linear differential equations in a basis-free environment. Besides being of theoretical interest, it is important in computations to know that a concept is independent of bases, since this allows one to then select a convenient basis in which to compute.

Before we show how one can recover a scalar equation from a differential module, we show that the standard constructions of linear algebra can be carried over to the context of differential modules. Let (M_1, ∂_1) and (M_2, ∂_2) be differential modules and assume that for certain bases these correspond to the equations $Y' = A_1Y$ and $Y' = A_2Y$.

In Table 1.1 we list some standard linear algebra constructions and how they generalize to differential modules. In this table, I_{n_i} represents the $n_i \times n_i$ identity matrix and for two matrices $A = (a_{i,j})$ and B, $A \otimes B$ is the matrix where the i, j-entry of A is replaced by $a_{i,j}B$. Also note that if $f \in \mathrm{Hom}_k(M_1, M_2)$ then $\partial(f) = 0$ if and only if $f(\partial m_1) = \partial_2(f(m_1))$, that is, if and only if f is a differential module homomorphism.

Referring to the table, it is not hard to show that $\mathrm{Hom}_k(M_1, M_2) \simeq M_1 \otimes M_2^*$ as differential modules. Furthermore, given (M, ∂) with $\dim_k(M) = n$, corresponding to a differential equation $Y' = AY$, we have that $M \simeq \oplus_{i=1}^n 1_k$ if and only if there exist $y_1, \ldots, y_n$ in k^n, linearly independent over k such that $y_i' = Ay_i$.

Table 1.1.

Construction	∂	Matrix Equation
$(M_1 \oplus M_2, \partial)$	$\partial(m_1 \oplus m_2) = \partial_1 m_1 \oplus \partial_2 m_2$	$Y' = \begin{pmatrix} A_1 & 0 \\ 0 & A_2 \end{pmatrix} Y$
$(M_1 \otimes M_2, \partial)$	$\partial(m_1 \otimes m_2) = \partial_1 m_1 \otimes m_2 + m_1 \otimes \partial_2 m_2$	$Y' = (A_1 \otimes I_{n_2} + I_{n_1} \otimes A_2)Y$
$(\mathrm{Hom}_k(M_1, M_2), \partial)$	$\partial(f)(m) = f(\partial_1 m) - \partial_2(f(m))$	$Y' = YA_2^T - A_1^T Y$
$\mathbf{1}_k = (k, \partial) = $ trivial differential module	$\partial \equiv 0$	$Y' = 0$
$(M^*, \partial) = \mathrm{Hom}_k(M, \mathbf{1}_k)$	$\partial(f)(m) = f(\partial(m))$	$Y' = -A^T Y$

From matrix to scalar linear differential equations: Before I discuss the relationship between matrix and scalar linear differential equations, I need one more concept.

Definition 1.2.6 *Let k be a differential field. The ring of differential operators over $k = k[\partial]$ is the ring of noncommutative polynomials $\{a_n \partial^n + \ldots + a_1 \partial + a_0 \mid a_i \in k\}$ with coefficients in k, where multiplication is determined by $\partial \cdot a = a' + a\partial$ for all $a \in k$.*

We shall refer to the degree of an element $L \in k[\partial]$ as its order $\mathrm{ord}\, L$. The following properties are not hard to check ([5], Chapter 2.1):

Lemma 1.2.7 *Let $L_1 \neq 0, L_2 \in k[\partial]$.*

1) There exist unique $Q, R \in k[\partial]$ with $\mathrm{ord}\, R < \mathrm{ord}\, L_1$ such that $L_2 = QL_1 + R$.

2) Every left ideal of $k[\partial]$ is of the form $k[\partial]L$ for some $L \in k[\partial]$.

We note that any differential module M can be considered a left $k[\partial]$-module and conversely, any left $k[\partial]$-module that is finite dimensional as a k-vector space is a differential module. In fact we have a stronger statement:

Theorem 1.2.8 *(Cyclic Vector Theorem) Assume there exists an $a \in k$ such that $a' \neq 0$. Every differential module is of the form $k[\partial]/k[\partial]L$ for some $L \in k[\partial]$.*

This result has many proofs ([12, 13, 14, 15, 16, 17] to name a few), two of which can be found in Chapter 2.1 of [5].

Corollary 1.2.9 *(Systems to Scalar equations) Let k be as above. Every system $Y' = AY$ is equivalent to a scalar equation $L(y) = 0$.*

Proof Let $A^* = -A^t$. Apply the Cyclic Vector Theorem to M_{A^*} to find an $L \in k[\partial]$ such that $M_{A^*} = k[\partial]/k[\partial]L$. If A_L is the companion matrix of L, a calculation shows that $M_A \simeq M_{A_L}$. $\qquad\qquad\square$

We note that the hypothesis that k contain an element a with $a' \neq 0$ is necessary. If the derivation is trivial on k, then two matrix equations $Y' = A_1Y$ and $Y' = A_2Y$ are equivalent if and only if the matrices are similar. There exist many examples of matrices not similar to a companion matrix.

Before we leave (for now) our discussion of the ring $k[\partial]$, I wish to make two remarks.

First, we can define a map $i : k[\partial] \to k[\partial]$ by $i(\sum a_j\partial^j) = \sum(-1)^j\partial^j a_j$. This map is is an involution ($i^2 = id$). Denoting $i(L) = L^*$, we have that $(L_1L_2)^* = L_2^*L_1^*$. The operator L^* is referred to as the *adjoint* of L. Using the adjoint one sees that there is right euclidean division as well and that every right ideal of $k[\partial]$ is also principal.

Second, Lemma 1.2.7.2 allows us to define

Definition 1.2.10 *Let $L_1, L_2 \in k[\partial]$.*

1) The least common left multiple $\mathrm{LCLM}(L_1, L_2)$ *of L_1 and L_2 is the monic generator of $k[\partial]L_1 \cap k[\partial]L_2$.*
2) The greatest common right divisor $\mathrm{GCRD}(L_1, L_2)$ *of L_1 and L_2 is the monic generator of $k[\partial]L_1 + k[\partial]L_2$.*

A simple modification of the usual euclidean algorithm allows one to find these objects. One can also define least common right multiples and greatest common left divisors using right ideals.

Solutions I will give properties of solutions of scalar and matrix linear differential equations and define the notion of the solutions of a differential module. Let (k, ∂) be a differential field with constants C_k (see Definition 1.2.1).

Lemma 1.2.11 *Let $v_1, \ldots v_r \in k^n$ satisfy $v_i' = Av_i$, $A \in \mathrm{gl}_n(k)$. If $v_1, \ldots, v_r$ are k-linearly dependent, then they are C_k-linearly dependent.*

Proof Assume, by induction, that $v_2, \ldots, v_r$ are k-linearly independent and $v_1 = \sum_{i=2}^{r} a_i v_i, a_i \in k$. We then have

$$0 = v_1' - A v_1 = \sum_{i=2}^{r} a_i' v_i + \sum_{i=2}^{r} a_i (v_i' - A v_i) = \sum_{i=2}^{r} a_i' v_i$$

so by assumption, each $a_i' = 0$. $\square$

Corollary 1.2.12 *Let (k, ∂) be a differential field with constants C_k, $A \in \mathrm{gl}_n(k)$ and $L \in k[\partial]$.*

1) The solution space $\mathrm{Soln}_k(Y' = AY)$ of $Y' = AY$ in k^n is a C_k-vector space of dimension at most n.

2) The elements $y_1, \ldots, y_r \in k$ are linearly dependent over C_k if and only if the wronskian determinant

$$yr(y_1, \ldots, y_r) = \det \begin{pmatrix} y_1 & \cdots & y_r \\ \vdots & \vdots & \vdots \\ y_1^{(r-1)} & \cdots & y_r^{(r-1)} \end{pmatrix}$$

is zero.

3) The solution space $\mathrm{Soln}_k(L(y) = 0) = \{y \in k \mid L(y) = 0\}$ of $L(y) = 0$ in k is a C_k-vector space of dimension at most n.

Proof 1) This follows immediately from Lemma 1.2.11.

2) If $\sum_{i=1}^{r} c_i y_i = 0$ for some $c_i \in C_k$, not all zero, then $\sum_{i=1}^{r} c_i y_i^{(j)} = 0$ for all j. Therefore, $wr(y_1, \ldots, y_r) = 0$. Conversely if $wr(y_1, \ldots, y_r) = 0$, then there exists a nonzero vector $(a_0, \ldots, a_{r-1})$ such that

$$(a_0, \ldots, a_{r-1}) \begin{pmatrix} y_1 & \cdots & y_r \\ \vdots & \vdots & \vdots \\ y_1^{(r-1)} & \cdots & y_r^{(r-1)} \end{pmatrix} = 0$$

Therefore each y_i satisfies the scalar linear differential equation $L(y) = a_{r-1} y^{(r-1)} + \ldots + a_0 y = 0$ so each vector $v_i = (y_i, y_i', \ldots, y_i^{(r-1)})^t$ satisfies $Y' = A_L Y$, where A_L is the companion matrix. Lemma 1.2.11 implies that the v_i and therefore the y_i are linearly dependent over C_k.

3) Apply Lemma 1.2.11 to $Y' = A_L Y$. $\square$

In general, the dimension of the solution space of a linear differential equation in a field is less than n. In the next section we will construct a field such that over this field the solution space has dimension n. It will be useful to have the following definition:

Definition 1.2.13 *Let (k, ∂) be a differential field, $A \in \mathrm{gl}_n(k)$ and R a differential ring containing k. A matrix $Z \in \mathrm{GL}_n(R)$ such that $Z' = AZ$ is called a* fundamental solution matrix *of $Y' = AY$.*

Note that if R is a field and Z is a fundamental solution matrix, then the columns of Z form a C_R-basis of the solution space of $Y' = AY$ in R and that this solution space has dimension n.

Let (M, ∂) be a differential module over k. We define the solution space $\mathrm{Soln}_k(M)$ of (M, ∂) to be the kernel $\ker_M \partial$ of ∂ on M. As we have noted above, if $\{e_i\}$ is a basis of M and $Y' = AY$ is the associated matrix differential equation in this basis, then $\sum_i v_i e_i \in \ker \partial$ if and only if $v' = Av$ where $v = (v_1, \ldots, v_n)^t$. If $K \supset k$ is a differential extension of k, then $K \otimes_k M$ can be given the structure of a K-differential module, where $\partial(a \otimes m) = a' \otimes m + a \otimes \partial m$. We then define $\mathrm{Soln}_K(M)$ to be the kernel $\ker(\partial, K \otimes_k M)$ of ∂ on $K \otimes_k M$.

1.3 Basic Galois Theory and Applications

Galois theory of polynomials The idea behind the Galois theory of polynomials is to associate to a polynomial a group (the group of symmetries of the roots that preserve all the algebraic relations among these roots) and deduce properties of the roots from properties of this group (for example, a solvable group implies that the roots can be expressed in terms of radicals). This idea can be formalized in the following way.

Let k be a field (of characteristic 0 as usual) and let $P(X) \in k[X]$ be a polynomial of degree n without repeated roots (i.e., $\mathrm{GCD}(P, P') = 1$).Let

$$S = k[X_1, \ldots, X_n, \frac{1}{\prod(X_i - X_j)}]$$

(the reason for the term $\frac{1}{\prod(X_i - X_j)}$ will be explained below) and let $I = (P(X_1), \ldots, P(X_n)) \triangleleft S$ be the ideal generated by the $P(X_i)$. The ring S/I is generated by n distinct roots of P but does not yet reflect the possible algebraic relations among these roots. We therefore consider any maximal ideal M in S containing I. This ideal can be thought of as a maximally consistent set of algebraic relations among the roots. We define

Definition 1.3.1 *1) The* splitting ring *of the polynomial P over k is the ring*

$$R = S/M = k[X_1, \ldots, X_n, \frac{1}{\prod(X_i - X_j)}]/M \ .$$

2) The Galois group *of P (or of R over k) is the group of automorphisms* $\mathrm{Aut}(R/k)$.

Note that since M is a maximal ideal, R is actually a field. Furthermore, since S contains $\frac{1}{\prod(X_i - X_j)}$, the images of the X_i in R are *distinct* roots of P. So, in fact, R coincides with the usual notion of a splitting field (a field generated over k by the distinct roots of P) and as such, is unique up to k-isomorphism. Therefore, R is independent of the choice of maximal ideal M containing I. We will follow a similar approach with linear differential equations.

Galois theory of linear differential equations Let (k, ∂) be a differential field and $Y' = AY, A \in \mathrm{gl}_n(k)$ a matrix differential equation over k. We now want the Galois group to be the group of symmetries of solutions preserving all algebraic *and differential* relations. We proceed in a way similar to the above.

Let

$$S = k[Y_{1,1}, \ldots, Y_{n,n}, \frac{1}{\det(Y_{i,j})}]$$

where $Y = (Y_{i,j})$ is an $n \times n$ matrix of indeterminates. We define a derivation on S by setting $Y' = AY$. The columns of Y form n independent solutions of the matrix linear differential equation $Y' = AY$ but we have not yet taken into account other possible algebraic and differential relations. To do this, let M be any maximal *differential* ideal and let $R = S/M$. We have now constructed a ring that satisfies the following definition

Definition 1.3.2 *Let (k, ∂) be a differential field and $Y' = AY, A \in \mathrm{gl}_n(k)$ a matrix differential equation over k. A* Picard-Vessiot ring *(PV-ring) for $Y' = AY$ is a differential ring R over k such that*

(i) *R is a simple differential ring (i.e., the only differential ideals are (0) and R).*

(ii) *There exists a fundamental matrix $Z \in \mathrm{GL}_n(R)$ for the equation $Y' = AY$.*

(iii) *R is generated as a ring by k, the entries of Z and $\frac{1}{\det Z}$.*

One can show that a PV-ring must be a domain and, if C_k is algebraically closed, then any PV-rings for the same equation are k-isomorphic as differential rings and $C_k = C_R$. Since a PV-ring is a domain we define a *PV-field* K to be the quotient field of a PV-ring. Assuming that C_k is algebraically closed this is the same as saying that K is generated over k by the entries of a fundamental solution matrix and that $C_K = C_k$. These facts are proven in Chapter 1.2 of [5]. *From now on, we will always assume that C_k is algebraically closed* (see [18, 19, 3, 20] for results when this is not the case) .

Examples 1.3.3 *1)* $k = \mathbb{C}(x)$ $x' = 1$ $\alpha \in \mathbb{C}$ $Y' = \frac{\alpha}{x}Y$ $S = k[Y, \frac{1}{Y}]$ $Y' = \frac{\alpha}{x}Y$

Case 1: $\alpha = \frac{n}{m}$, $GCD(n,m) = 1$. I claim that $\mathcal{R} = k[Y, \frac{1}{Y}]/[Y^m - x^n] = k(x^{\frac{m}{n}})$. To see this note that the ideal $(Y^m - x^n) \lhd k[Y, \frac{1}{Y}]$ is a maximal ideal and closed under differentiation since $(Y^m - x^n)' = \frac{n}{x}(Y^m - x^n)$.

Case 2: $\alpha \notin \mathbb{Q}$. In this case, I claim that (0) is a maximal differential ideal and so $R = k[Y, \frac{1}{Y}]$. To see this, first note that for any $f \in \mathbb{C}(x)$,

$$\frac{f'}{f} = \sum \frac{n_i}{(x - \alpha_i)}, n_i \in \mathbb{Z}, \alpha_i \in \mathbb{C}$$

and this can never equal $\frac{N\alpha}{x}$. Therefore $Y' = \frac{N\alpha}{x}Y$ has no nonzero solutions in $\mathbb{C}(x)$. Now assume I is a proper nonzero differential ideal in $K[Y, \frac{1}{Y}]$. One sees that I is of the form $I = (f)$ where $f = Y^N + a_{N-1}Y^{N-1} + \ldots + a_0$, $N > 0$. Since Y is invertible, we may assume that $a_0 \neq 0$. Since $f' \in (f)$, by comparing leading terms we have that $f' = \frac{N\alpha}{x}f$ and so $a_0' = \frac{N\alpha}{x}a_0$, a contradiction.

2) $k = \mathbb{C}(x)$ $x' = 1$ $Y' = Y$. Let $S = k[Y, \frac{1}{Y}]$. An argument similar to Case 2 above shows that (0) is the only proper differential ideal so the PV-ring is $k[Y, \frac{1}{Y}]$.

As expected the differential Galois group is defined as

Definition 1.3.4 *Let (k, ∂) be a differential field and R a PV-ring over k. The differential Galois group of R over k, $\mathrm{DGal}(R/k)$ is the group $\{\sigma : R \to R \mid \sigma \text{ is a differential } k\text{-isomorphism}\}$.*

If R is the PV-ring associated with $Y' = AY, A \in \mathrm{gl}_n(k)$, we sometimes denote the differential Galois group by $\mathrm{DGal}(Y' = AY)$. If K is a PV-field over k, then the k-differential automorphisms of K over k are can be identified with the differential k-automorphisms of R over k where R is a PV-ring whose quotient field is K.

Let R be a PV-ring for the equation $Y' = AY$ over k and $\sigma \in \mathrm{DGal}(R/k)$. Fix a fundamental solution matrix Z and let $\sigma \in \mathrm{DGal}(R/k)$. We then have that $\sigma(Z)$ is again a fundamental solution matrix of $Y' = AY$ and a calculation shows that $(Z^{-1}\sigma(Z))' = 0$. Therefore, $\sigma(Z) = Zm_\sigma$ for some $m_\sigma \in \mathrm{GL}_n(C_k)$. This gives an injective group homomorphism $\sigma \mapsto m_\sigma$ of $\mathrm{DGal}(R/k)$ into $\mathrm{GL}_n(C_\sigma)$. If we select a different fundamental solution matrix, the images of the resulting maps would be conjugate. The image has a further property.

Definition 1.3.5 *A subgroup G of $\mathrm{GL}_n(C_k)$ is said to be a* linear algebraic group *if it is the zero set in $\mathrm{GL}_n(C_k)$ of a system of polynomials over C_k in n^2 variables.*

With the identification above $\mathrm{DGal}(R/k) \subset \mathrm{GL}_n(C_k)$ we have

Proposition 1.3.6 $\mathrm{DGal}(R/k) \subset \mathrm{GL}_n(C_k)$ *is a linear algebraic group.*

Proof Let $R = k[Y, \frac{1}{\det Y}]/M$, $M = (f_1, \ldots, f_m)$. We may assume that the $f_i \in k[Y_{1,1}, \ldots, Y_{n,n}]$. We may identify $\mathrm{DGal}(R/k)$ with the subgroup of $\sigma \in \mathrm{GL}_n(C_k)$ such that $\sigma(M) \subset M$. Let W_N be the C_k-space of polynomials in $k[Y_{1,1}, \ldots, Y_{n,n}]$ of degree at most N. Note that $\mathrm{GL}_n(C_k)$ acts on $k[Y_{1,1}, \ldots, Y_{n,n}]$ by substitution and leaves W_N invariant. If N is the maximum of the degrees of the f_i then $V_N = W_N \cap M$ generates M and we may identify $\mathrm{DGal}(R/k)$ with the group of $\sigma \in \mathrm{GL}_n(C_k)$ such that $\sigma(V_N) \subset V_N$. Let $\{e_\alpha\}_{\mathcal{A}}$ be a C-basis of W_N with $\{e_\alpha\}_{\alpha \in \mathcal{B} \subset \mathcal{A}}$ a C-basis of V_N. For any $\beta \in \mathcal{A}$ there exist polynomials $p_{\alpha\beta}(x_{ij})$ such that for any $\sigma = (c_{ij}) \in \mathrm{GL}_n$, $\sigma(e_\beta) = \sum_\alpha p_{\alpha\beta}(c_{ij})e_\alpha$. We then have that $\sigma = (c_{ij}) \in \mathrm{DGal}(R/k)$ if and only if $p_{\alpha\beta}(c_{ij}) = 0$ for all $\beta \in \mathcal{B}$ and $\alpha \notin \mathcal{B}$. $\qquad\square$

Examples 1.3.7 *1)* $k = \mathbb{C}(x)$ $x' = 1$ $Y' = \frac{\alpha}{x}Y$ $DGal \subset \mathrm{GL}_1(\mathbb{C})$. *Form the above description of the differential Galois group as the group laving a certain ideal invariant, we see (using the information of Example 1.3.3):*

Case 1: $\alpha = \frac{n}{m}, GCD(n,m) = 1$. *We then have* $R = \mathbb{C}[Y, \frac{1}{Y}]/(Y^m - x^n) = \mathbb{C}(x)[x^{\frac{n}{m}}]$. *Therefore* $\mathrm{DGal} = \mathbb{Z}/m\mathbb{Z} = \{(c) \mid c^n - 1 = 0\} \subset \mathrm{GL}_1(\mathbb{C})$

Case 2: $\alpha \notin \mathbb{Q}$. *we have* $R = \mathbb{C}(x)[Y, \frac{1}{Y}]/(0)$. *Therefore* $\mathrm{DGal} = \mathrm{GL}_1(\mathbb{C})$.

2) $k = \mathbb{C}(x)$ $x' = 1$ $Y' = Y$. *As above, one sees that* $\mathrm{DGal} = \mathrm{GL}_1(\mathbb{C})$

In general, it is not easy to compute Galois groups (although there is a complete algorithm due to Hrushovski [21]). We will give more examples in Sections 1.4, 1.5, and 1.6.

One can define the differential Galois group of a differential module in the following way. Let M be a differential module. Once we have selected a basis, we may assume that $M \simeq M_A$, where M_A is the differential module associated with $Y' = AY$. We say that $Y' = AY$ is a matrix linear differential equation associated with M.

Definition 1.3.8 *Let M be a differential module over k. The* differential Galois group *of M is the differential Galois group of an associated matrix linear differential equation.*

To show that this definition makes sense we must show (see the discussion before Definition 1.2.5) that equivalent matrix differential equations have the same differential Galois groups. Let $Y' = A_1 Y$, $Y' = A_2 Y$ be equivalent equations and let $A_1 = B^{-1} A_2 B - B^{-1} B'$, $B \in \mathrm{GL}_n(k)$. If $R = k[Z_1, \frac{1}{Z_2}]$ is a PV-ring for $Y' = A_1 Y$, then a computation shows that $Z_2 = BZ_1$ is a fundamental solution matrix for $Y' = A_2 Y$ and so R will also be a PV-ring for $Y' = A_2 Y$. The equation $Z_2 = BZ_1$ furthermore shows that the action of the differential Galois group (as a matrix group) on Z_1 and Z_2 is the same.

We now state

Theorem 1.3.9 (Fundamental Theorem of Differential Galois Theory) *Let k be a differential field with algebraically closed constants C_k. Let K be a Picard-Vessiot field with differential Galois group G.*

1) There is a bijective correspondence between Zariski-closed subgroups $H \subset G$ and differential subfields F, $k \subset F \subset K$ given by

$$H \subset G \quad \mapsto \quad K^H = \{a \in K \mid \sigma(a) = a \text{ for all } \sigma \in H\}$$
$$k \subset F \subset K \quad \mapsto \quad \mathrm{DGal}(K/F) = \{\sigma \in G \mid \sigma(a) = a \text{ for all } a \in F\}$$

2) A differential subfield $k \subset F \subset K$ is a Picard-Vessiot extension of k if and only if $\mathrm{DGal}(K/F)$ is a normal subgroup of G, in which case $\mathrm{DGal}(F/k) \simeq \mathrm{DGal}(K/k)\mathrm{DGal}(K/F)$.

The Zariski closed sets form the collection of closed sets of a topology on GL_n and one can speak of closures, components, etc. in this topology. In this topology, each linear algebraic group G may be written (uniquely) as the finite disjoint union of connected closed subsets (see Appendix A of [22] or [23] for a further discussion of linear algebraic groups). The connected component containing the identity matrix is denoted by G^0 and is referred to as the identity component. It is a normal subgroup of G. As a consequence of the Fundamental Theorem, one can show the following:

Corollary 1.3.10 *Let k, K and G be as above.*

1) For $a \in K$, we have that $a \in k \Leftrightarrow \sigma(a) = a \ \forall \sigma \in DGal(K/k)$

2) For $H \subset DGal(K/k)$, the Zariski closure $\overline{H}$ is $DGal(K/k)$ if and only if $K^H = k$

3) K^{G^0} is precisely the algebraic closure $\tilde{k}$ of k in K.

I refer the reader to Chapter 1.3 of [5] for the proof of the Fundamental Theorem and its corollary.

Example 1.3.11 $y' = \frac{\alpha}{x}y \quad \alpha \notin \mathbb{Q}, \quad k = \mathbb{C}(x) \quad K = k(x^\alpha) \quad DGal = GL_1 = \mathbb{C}^*$

The Zariski-closed subsets of $\mathbb{C}^*$ are finite so the closed subgroups are finite and cyclic. We have the following correspondence

$$
\begin{array}{ccc}
\textit{Groups} & & \textit{Fields} \\
\{1\} & \Leftrightarrow & k(x^\alpha) \\
\cap & & \cup \\
\{1, e^{2\pi i/n}, \ldots, e^{2(n-1)\pi i/n}\} & \Leftrightarrow & k(x^{n\alpha}) \\
\cap & & \cup \\
\mathbb{C}^* & \Leftrightarrow & k
\end{array}
$$

The Fundamental Theorem also follows from a deeper fact giving geometric meaning to the Picard-Vessiot ring and relating this to the Galois group.

Galois theory and torsors Let k be a field (not necessarily a differential field) and $\bar{k}$ its algebraic closure. I wish to refine the notion of a Zariski closed set in order to keep track of the field over which these objects are defined.

Definition 1.3.12 *1) A Zariski closed subset $V \subset \bar{k}^n$ is called a* variety *defined over k or k-variety if it is the set of common zeroes of polynomials in $k[X_1, \ldots X_n]$.*

2) If V is a variety defined over k, the k-ideal of V, $I_k(V)$, is $\{f \in k[X_1, \ldots, X_n] \mid f(a) = 0 \text{ for all } a \in X\}$.

3) If V is a variety defined over k, the k-coordinate ring, $k[V]$, of V is the ring $k[X_1, \ldots, X_n]/I_k(V)$.

4) If $V \subset \bar{k}^n$ and $W \subset \bar{k}^n$, a k-morphism $f : V \to W$ is an n-tuple of polynomials $f = (f_1, \ldots, f_n)$ in $k[X_1, \ldots, X_n]$ such that $f(a) \in W$ for all $a \in V$.

5) We say two k-varieties V and W are k-isomorphic if there are k-morphisms $f : V \to W$ and $g : W \to V$ such that fg and gf are the identity maps.

Note that the coordinate ring is isomorphic to the ring of k-morphisms from V to k. Also note that a k-morphism $f : V \to W$ induces a k-algebra homomorphism $f^* : k[W] \to k[V]$ given by $f^*(g)(v) = g(f(v))$ for $v \in V$ and $g \in k[W]$.

Examples 1.3.13 *1) The affine spaces k^n are $\mathbb{Q}$-varieties and, as such, their coordinate rings are just the rings of polynomials $\mathbb{Q}[X_1, \ldots, X_n]$.*

2) The group of invertible matrices $\mathrm{GL}_n(\bar{k})$ is also a $\mathbb{Q}$-variety. In order to think of it as a closed set, we must embed each invertible matrix A in the $n+1 \times n+1$ matrix $\begin{pmatrix} A & 0 \\ 0 & \det(A) \end{pmatrix}$. The defining ideal in $\mathbb{Q}[X_{1,1}, \ldots, X_{n+1,n+1}]$ is generated by

$$X_{n+1,1}, \ldots, X_{n+1,n}, X_{1,n+1}, \ldots, X_{n,n+1}, (\det X) - X_{n+1,n+1}$$

where X is the $n \times n$ matrix with entries $X_{i,j}, 1 \leq i \leq n, 1 \leq j \leq n$. The coordinate ring is the ring $\mathbb{Q}[X_{1,1}, \ldots, X_{n,n}, \frac{1}{\det X}]$

Examples 1.3.14 *1) Let $k = \mathbb{Q}$ and $V = \{a \mid a^2 - 2 = 0\} \subset \bar{\mathbb{Q}}$. We have that $I_{\mathbb{Q}}(V) = (X^2 - 2) \subset \mathbb{Q}[X]$ and $k[V] = \mathbb{Q}[X]/(X^2 - 2) = \mathbb{Q}(\sqrt{2})$. Note that in this setting "$\sqrt{2}$" is a function on V, defined over $\mathbb{Q}$ with values in $\bar{\mathbb{Q}}$ and not a number!*

2) Let $k = \mathbb{Q}$ and $W = \{a \mid a^2 - 1 = 0\} = \{\pm 1\} \subset \bar{\mathbb{Q}}$. We have that $I_{\mathbb{Q}}(W) = (X^2 - 1)$ and $k[W] = \mathbb{Q}[X]/(X^2 - 1) = \mathbb{Q} \oplus \mathbb{Q}$.

3) Note that V and W are not $\mathbb{Q}$-isomorphic since such an isomorphism *would induce an isomorphism of their $\mathbb{Q}$-coordinate rings. If we consider V and W as $\bar{\mathbb{Q}}$-varieties, then the map $f(x) = (\sqrt{2}(x-1) + \sqrt{2}(x+1))/2$ gives a $\bar{\mathbb{Q}}$-isomorphism of W to V.*

We shall be interested in certain k-subvarieties of $\mathrm{GL}_n(\bar{k})$ that have a group acting on them. These are described in the following definition (a more general definition of torsor is given in Appendix A of [5] but the definition below suffices for our present needs).

Definition 1.3.15 *Let $G \subset \mathrm{GL}_n(\bar{k})$ be a linear algebraic group defined over k and $V \subset \mathrm{GL}_n(\bar{k})$ a k-variety. We say that V is a G-torsor (or G-principal homogeneous space) over k if for all $v, w \in V$ there exists a unique $g \in G$ such that $vg = w$. Two G-torsors $V, W \in \mathrm{GL}_n(\bar{k})$ are said to be k-isomorphic if there is a k-isomorphism between the two commuting with the action of G.*

Examples 1.3.16 *1) G itself is always a G-torsor.*

2) Let $k = \mathbb{Q}$ and $V = \{a \mid a^2 - 2 = 0\} \subset \bar{\mathbb{Q}}$. We have that $V \subset \bar{\mathbb{Q}}^* = \mathrm{GL}_1(\bar{\mathbb{Q}})$. Let $G = \{\pm 1\} \subset \mathrm{GL}_1(\bar{\mathbb{Q}})$. It is easy to see that V is a G-torsor. As noted in the previous examples, the torsors V and G are not isomorphic over $\mathbb{Q}$ but are isomorphic over $\bar{\mathbb{Q}}$.

We say a G-torsor over k is *trivial* if it is k-isomorphic to G itself. One has the following criterion.

Lemma 1.3.17 *A G-torsor V over k is trivial if and only if it contains a point with coordinates in k. In particular, any G-torsor over an algebraically closed field is trivial.*

Proof If V contains a point v with coordinates in k then the map $g \mapsto xg$ is defined over k and gives a k-isomorphism between G and V (with inverse $v \mapsto x^{-1}v$). If V is trivial and $f : G \to V$ is an isomorphism defined over k, then $f(id)$ has coordinates in k and so is a k-point of V.   □

For connected groups G, there are other fields for which all torsors are trivial.

Proposition 1.3.18 *Let G be a connected linear algebraic group defined over k and V a G-torsor over k. If k is an algebraic extension of $\mathbb{C}(x), \mathbb{C}((x))$, or $\mathbb{C}(\{x\})$, then V is a trivial G-torsor.*

Proof This result is due to Steinberg, see [24], Chapter III.2.3 □

Another way of saying that V is a G torsor is to say that V is defined over k and is a left coset of G in $\mathrm{GL}_n(\bar{k})$. Because of this one would expect that V and G share many geometric properties. The one we will be interested in concerns the dimension. We say that a k-variety V is *irreducible* if it is not the union of two proper k-varieties. This is equivalent to saying that $I_k(V)$ is a prime ideal or, equivalently, that $k[V]$ is an integral domain. For an irreducible k-variety V, one defines the *dimension of V*, $\dim_k(V)$ to be the transcendence degree of $k(V)$ over k, where $k(V)$ is the quotient field of $k[V]$. For example, if $k = \mathbb{C}$ and V is a k-variety that is nonsingular as a complex manifold, then the dimension in the above sense is just the dimension as a complex manifold (this happens for example when V is a subgroup of $\mathrm{GL}_n(\mathbb{C})$). In general, any k-variety V can be written as the irredundant union of irreducible k-varieties, called the irreducible components of V, and the dimension of V is defined to be the maximum of the dimensions of its irreducible components. One can show that if $k \subset K$, then $\dim_k(V) = \dim_K(V)$ and in particular, $\dim_k(V) = \dim_{\bar{k}} V$ (see [5], Appendix A for more information about k-varieties).

Proposition 1.3.19 *If V is a G-torsor, then $\dim_k(V) = \dim_k(G)$.*

Proof As noted above, any G-torsor over k is trivial when considered as a $\bar{k}$ torsor. We therefore have $\dim_k(V) = \dim_{\bar{k}}(V) = \dim_{\bar{k}}(G) = \dim_k(G)$. □

The reason for introducing the concept of a G-torsor in this paper is the following proposition. We use the fact that if V is a G-torsor over k and $g \in G$, then the isomorphism $\rho_g : v \mapsto vg$ induces an isomorphism $\rho_g^* : k[V] \to k[V]$.

Proposition 1.3.20 *Let R be a Picard-Vessiot extension with differential Galois group G. Then R is the coordinate ring of a G-torsor V defined over k. Furthermore, for $\sigma \in G$, the Galois action of σ on $R = k[V]$ is the same as ρ_σ^*.*

Corollary 1.3.21 *1) If K is a Picard-Vessiot field over k with Galois group G, then $\dim_{C_k} G = tr.deg.(K/k)$.*

2) If R is a Picard-Vessiot ring over k with Galois group G where k is either algebraically closed, $\mathbb{C}(x)$, $\mathbb{C}((x))$, or $\mathbb{C}(\{x\})$, then $R \simeq k[G]$.

We refer to Chapter 1.3 of [5] for a proof of the proposition and note that the Fundamental Theorem of Galois Theory is shown there to follow from this proposition.

Applications and ramifications. I shall discuss the following applications and ramifications of the differential Galois theory

- Monodromy
- Factorization of linear differential operators
- Transcendental numbers
- Why doesn't $\sec x$ satisfy a linear differential equation?
- Systems of linear partial differential equations

Another application concerning solving linear differential equations in terms of special functions and algorithms to do this will be discussed in Section 1.5. The Picard-Vessiot theory has also been used to identify obstructions to complete integrability of Hamiltonian systems (see [25] or [26]).

Monodromy. We will now consider differential equations over the complex numbers, so let $k = \mathbb{C}(x)$, $x' = 1$ and

$$\frac{dY}{dx} = A(x)Y, \ A(x) \in \mathrm{gl}_n(\mathbb{C}(x)) \tag{1.1}$$

Definition 1.3.22 *A point $x_0 \in \mathbb{C}$ is an* ordinary point *of equation (1.1) if all the entries of $A(x)$ are analytic at x_0, otherwise x_0 is a* singular point.

One can also consider the point at infinity, ∞. Let $t = \frac{1}{x}$. We then have that $\frac{dY}{dt} = -\frac{1}{t^2}A(\frac{1}{t})Y$. The point $x = \infty$ is an ordinary or singular point precisely when $t = 0$ is an ordinary or singular point of the transformed equation.

Example 1.3.23 $\frac{dy}{dx} = \frac{\alpha}{x}y \Rightarrow \frac{dy}{dt} = -\frac{\alpha}{t}y$ *so the singular points are* $\{0, \infty\}$.

Let $S = \mathbb{C}\mathbb{P}^1$ be the Riemann Sphere and let $X = S^2 - \{$singular points of (1.1)$\}$ and let $x_0 \in X$. From standard existence theorems, we know that in a small neighborhood $\mathcal{O}$ of x_0, there exists a fundamental matrix Z whose entries are analytic in $\mathcal{O}$. Let γ be a closed curve in X based at x_0. Analytic continuation along

γ yields a new fundamental solution matrix Z_γ at x_0 which must be related to the old fundamental matrix as

$$Z_\gamma \;=\; Z D_\gamma$$

where $D_\gamma \in \mathrm{GL}_n(\mathbb{C})$. One can show that D_γ just depends on the homotopy class of γ. This defines a homomorphism

$$\mathrm{Mon} : \pi_1(X, x_0) \;\longrightarrow\; \mathrm{GL}_n(\mathbb{C})$$
$$\gamma \;\longmapsto\; D_\gamma$$

Definition 1.3.24 *The homomorphism* Mon *is called the* monodromy map *and its image is called the* monodromy group.

Note that the monodromy group depends on the choice of Z so it is only determined up to conjugation. Since analytic continuation preserves analytic relations, we have that the monodromy group is contained in the differential Galois group.

Examples 1.3.25 *1)* $y' = \frac{\alpha}{x}y$ *(*$y = e^{\alpha \log x} = x^\alpha$*) Singular points =* $\{0, \infty\}$

$$\mathrm{Mon}(\pi_1(X, x_0)) = \{(e^{2\pi i \alpha})^n \mid n \in \mathbb{Z}\}$$

If $\alpha \in \mathbb{Q}$ *this image is finite and so equals the differential Galois group. If* $\alpha \notin \mathbb{Q}$, *then this image is infinite and so Zariski dense in* $\mathrm{GL}_1(\mathbb{C})$. *In either case the monodromy group is Zariski dense in the differential Galois group.*

2) $y' = y$ *(*$y = e^x$*),   Singular points =* $\{\infty\}$

$$\mathrm{Mon}(\pi_1(X, x_0)) = \{1\}$$

In this example the differential Galois group is $\mathrm{GL}_1(\mathbb{C})$ *and the monodromy group is not Zariski dense in this group.*

The second example shows that the monodromy group is not necessarily Zariski dense in the differential Galois group but the first example suggests that under certain conditions it may be. To state these conditions we need the following definitions.

Definition 1.3.26 *1) An open sector* $S = S(a, b, \rho)$ *at a point* p *is the set of complex numbers* $z \neq p$ *such that* $\arg(z - p) \in (a, b)$ *and* $|z - p| < \rho(\arg(z - p))$ *where* $\rho : (a, b) \to \mathbb{R}_{>0}$ *is a continuous function.*

2) A singular point p *of* $Y' = AY$ *is a regular singular point if for any open sector* $\mathcal{S}(a, b, \rho)$ *at* p *with* $|a - b| < 2\pi$, *there exists a fundamental solution matrix* $Z = (z_{ij})$, *analytic in* $\mathcal{S}$ *such that for some positive constant* c *and integer* N, $|z_{ij}| < c|x^N|$ *as* $x \to p$ *in* $\mathcal{S}$.

Note that in the first example above, the singular points are regular while in the second they are not. For scalar linear differential equations $L(y) = y^{(n)} + a_{n-1}y^{(n-1)} + \ldots + a_0 y = 0$, the criterion of Fuchs states that p is a regular singular point if and only if each a_i has a pole of order $\leq n - i$ at p (see Chapter 5.1 of [5] for a further discussion of regular singular points). For equations with only regular singular points we have

Proposition 1.3.27 *(Schlesinger) If* $Y' = AY$ *has only regular singular points, then* $\mathrm{Mon}(\pi_1(X, x_0))$ *is Zariski dense in* $DGal$.

Proof Let $K = $ PV extension field of k for $Y' = AY$. We have $Mon(\pi_1(X, x_0)) \subset DGal(K/k)$. By Corollary 1.3.10 it is enough to show that if $f \in K$ is left fixed by $Mon(\pi_1(X, x_0))$ then $f \in k = \mathbb{C}(x)$. If f is left invariant by analytic continuation on X, then f is a single valued meromorphic function on X. Let $Z = (z_{ij}), z_{ij} \in K$ be a fundamental solution matrix and let $f \in \mathbb{C}(x, z_{ij})$. Regular singular points imply that f has polynomial growth near the singular points of $Y' = AY$ and is at worst meromorphic at other points on the Riemann Sphere. Cauchy's Theorem implies that f is meromorphic on the Riemann Sphere and Liouville's Theorem implies that f must be a rational function. $\qquad\square$

In Section 1.4, we shall see what other analytic information one needs to determine the differential Galois group in general.

Factorization of linear differential operators. In Section 1.5, we will see that factorization properties of linear differential operators and differential modules play a crucial role in algorithms to calculate properties of Galois groups. Here I shall relate factorization of operators to the Galois theory.

Let k be a differential field, $k[\partial]$ the ring of differential operators over k and $L \in k[\partial]$. Let K be the PV-field over k corresponding to $L(y) = 0$, G its Galois group and $V = \mathrm{Soln}_K(L(y) = 0)$.

Proposition 1.3.28 *There is a correspondence among the following objects*

   (i) *Monic $L_0 \in k[\partial]$ such that $L = L_1 \cdot L_0$.*

   (ii) *Differential submodules of $k[\partial]/k[\partial]L$.*

   (iii) *G-invariant subspaces of V.*

Proof Let $W \subset V$ be G-invariant and $B = \{b_1, \ldots, b_t\}$ a C_k-basis of W. For $\sigma \in G$, let $[\sigma]_B$ be the matrix of $\sigma|_W$ with respect to B,

$$
\sigma\left(\begin{bmatrix} b_1 & \cdots & b_t \\ \vdots & \vdots & \vdots \\ b_1^{(t-1)} & \cdots & b_t^{(t-1)} \end{bmatrix}\right) = \begin{bmatrix} b_1 & \cdots & b_t \\ \vdots & \vdots & \vdots \\ b_1^{(t-1)} & \cdots & b_t^{(t-1)} \end{bmatrix} [\sigma]_B .
$$

Taking determinants we have $\sigma(wr(b_1, \ldots, b_t)) = wr(b_1, \ldots, b_t) \det([\sigma]_B)$. Define $L_W(Y) \stackrel{def}{=} \frac{wr(Y, b_1, \ldots, b_t)}{wr(b_1, \ldots, b_t)}$. Using the above, we have that the coefficients of L_W are left fixed by the differential Galois group and so lie in k. Therefore $L_W \in k[\partial]$. Furthermore, Corollary 1.2.12 implies that L_W vanishes on W. If we write $L = L_1 \cdot L_W + R$ where $\mathrm{ord}(R) < \mathrm{ord}(L_W)$, one sees that R also vanishes on W. Since the dimension of $W = \mathrm{ord}(L_W)$, Corollary 1.2.12 again implies that $R = 0$. Therefore W is the solution space of the right factor L_W of L.

Let $L_0 \in k[\partial]$ with $L = L_1 \cdot L_0$. Let $W = Soln_K(L_0(y) = 0)$. We have that L_0 maps V into $Soln_K(L_1(y) = 0)$. Furthermore we have the following implications

- $Im(L_0) \subset Soln_K(L_1(y) = 0) \Rightarrow \dim_{C_k}(Im(L_0)) \le \mathrm{ord}(L_1)$
- $Ker(L_0) = Soln_K(L_0(y) = 0) \Rightarrow \dim_{C_k}(Ker(L_0)) \le \mathrm{ord}(L_0)$
- $n = \dim_{C_k}(V) = \dim_{C_k}(Im(L_0)) + \dim_{C_k}(Ker(L_0)) \le$
  $$\mathrm{ord}(L_1) + \mathrm{ord}(L_0) = n$$

Therefore $\dim_{C_k} Soln_K(L_0(y) = 0) = \mathrm{ord}(L_0)$. This now implies that the association of G-invariant subspaces of V to right factors of L is a bijection.

I will now describe the correspondence between right factors of L and submodules of $k[\partial]/k[\partial]L$. One easily checks that if $L = L_1 L_0$, then $k[\partial]L \subset k[\partial]L_0$ so $M = k[\partial]L_0/k[\partial]L \subset k[\partial]/k[\partial]L$. Conversely let M be a differential submodule of $k[\partial]/k[\partial]L$. Let $\pi : k[\partial] \to k[\partial]/k[\partial]L$ and $\overline{M} = \pi^{-1}(M)$. $\overline{M} = k[\partial]L_0$ for some $L_0 \in k[\partial]$. Since $L \in \overline{M}, L = L_1 L_0$. $\qquad\square$

By induction on the order, one can show that any differential operator can be written as a product of irreducible operators. This need not be unique (for example $\partial^2 = \partial \cdot \partial = (\partial + \frac{1}{x})(\partial - \frac{1}{x})$) but a version of the Jordan-Hölder Theorem implies that if $L \, L_1 \cdot \ldots \cdot L_r = \tilde{L}_1 \cdot \ldots \cdot \tilde{L}_s$ then $r = s$ and, after possibly renumbering, $k[\partial]/k[\partial]L_i \simeq k[\partial]/k[\partial]\tilde{L}_i$ ([5], Chapter 2).

One can also characterize submodules of differential modules in terms of corresponding matrix linear differential equations. Let $Y' = AY$ correspond to the differential module M of dimension n. One then has that M has a proper submodule of dimension t if and only if $Y' = AY$ is equivalent to $Y' = BY, B = \begin{pmatrix} B_0 & B_1 \\ 0 & B_2 \end{pmatrix}$ $B_0 \in \mathrm{gl}_t(k)$.

Transcendental numbers. Results of Siegel and Shidlovski (and subsequent authors) reduce the problem of showing that the values of certain functions are algebraically independent to showing that these functions are themselves algebraically independent. The Galois theory of linear differential equations can be used to do this.

Definition 1.3.29 $f = \sum_{i=0}^{\infty} a_n \frac{z^n}{n!} \in \mathbb{Q}[[z]]$ *is an E-function if* $\exists \, c \in \mathbb{Q}$ *such that, for all* n, $|a_n| \leq c^n$ *and the least common denominator of* $a_0, \ldots, a_n \leq c^n$.

Examples 1.3.30 $e^x = \sum \frac{z^n}{n!}, \qquad J_0(z) = \sum \frac{z^{2n}}{(n!)^2}$

Theorem 1.3.31 *(Siegel-Shidlovski, c.f., [27]): If* $f_1, \ldots, f_s$ *are E-functions,* <u>algebraically independent</u> *over* $\mathbb{Q}(z)$ *and satisfy a linear differential equation*

$$L(y) = y^{(m)} + a_{m-1}y^{(n-1)} + \ldots + a_0 y, \quad a_i \in \mathbb{Q}(z)$$

then for $\alpha \in \mathbb{Q}, \alpha \neq 0$ *and* α *not a pole of the* a_i, *the numbers* $f_1(\alpha), \ldots, f_s(\alpha)$ *are algebraically independent over* $\mathbb{Q}$.

The differential Galois group of $L(y) = 0$ measures the algebraic relations among solutions. In particular, Corollary 1.3.21 implies that if $y_1, \ldots y_n$ are a $\bar{\mathbb{Q}}$-basis of the solution space of $L(y) = 0$ then the transcendence degree of $\bar{\mathbb{Q}}(x)(y_1, \ldots y_n, y_1', \ldots y_n', \ldots, y_1^{(n-1)}, \ldots, y_n^{(n-1)})$ over $\bar{\mathbb{Q}}(x)$ is equal to the dimension of the differential Galois group of $L(y) = 0$.

Example 1.3.32 *The differential Galois group of the Bessel equations* $y'' + \frac{1}{z}y' + (1 - \frac{\lambda^2}{z^2})y = 0$, $\lambda - \frac{1}{2} \notin \mathbb{Z}$ *over* $\bar{\mathbb{Q}}(x)$ *can be shown to be* SL_2 *([28]) and examining the recurrence defining a power series solution of this equation, one sees that the solutions are E-functions. Given a fundamental set of solutions* $\{J_\lambda, Y_\lambda\}$ *we have that the transcendence degree of* $\bar{\mathbb{Q}}(x)(J_\lambda, Y_\lambda, J'_\lambda, Y'_\lambda)$ *over* $\bar{\mathbb{Q}}(x)$ *is 3 (one can show that* $J_\lambda Y'_\lambda - J'_\lambda, Y_\lambda \in \bar{\mathbb{Q}}(x)$). *Therefore one can apply the Siegel-Shidlovski Theorem to show that certain values of* J_λ, Y_λ *and* J'_λ *are algebraically independent over* $\bar{\mathbb{Q}}$.

Why does $\cos x$ *satisfy a linear differential equation over* $\mathbb{C}(x)$, *while* $\sec x$ *does not?* There are many reasons for this. For example, $\sec x = \frac{1}{\cos x}$ has an infinite number of poles which is impossible for a solution of a linear differential equation over $\mathbb{C}(x)$. The following result, originally due to Harris and Sibuya [29], also yields this result and can be proven using differential Galois theory [30].

Proposition 1.3.33 *Let* k *be a differential field with algebraically closed constants and* K *a PV-field of* k. *Assume* $y \neq 0, 1/y \in K$ *satisfy linear differential equations over* k. *Then* $\frac{y'}{y}$ *is algebraic over* k.

Proof (Outline) We may assume that k is algebraically closed and that K is the quotient field of a PV-ring R. Corollary 1.3.21 implies that $R = k[G]$ where G is an algebraic group whose C_k points form the Galois group of R over k. Since both y and $\frac{1}{y}$ satisfy linear differential equations over k, their orbits under the action of the Galois group span a *finite dimensional* vector space over C_k. This in turn implies the orbit of that y and $\frac{1}{y}$ induced by the the action of $G(k)$ on $G(k)$ by right multiplication is finite dimensional. The general theory of linear algebraic groups [23] tells us that this implies that these elements must lie in $k[G]$. A result of Rosenlicht [31, 32] implies that $y = af$ where $a \in k$ and $f \in k[G]$ satisfies $f(gh) = f(g)f(h)$ for all $g, h \in G$. In particular, for $\sigma \in G(C_k) = \text{Aut}(K/k)$, we have $\sigma(f(x)) = f(x\sigma) = f(x)f(\sigma)$. If $L \in k[\partial]$ is of minimal positive order such that $L(f) = 0$, then $0 = \sigma(L(f)) = L(f \cdot f(\sigma))$ and a calculation shows that this implies that $f(\sigma) \in C_k$.

We therefore have that, for all $\sigma \in \text{Aut}(K/k)$ there exists a $c_\sigma \in C_k$ such that $\sigma(y) = c_\sigma y$. This implies that $\frac{y'}{y}$ is left fixed by $\text{Aut}(K/k)$ and so must be in k. $\qquad\square$

If $y = \cos x$ and $\frac{1}{y} = \sec x$ both satisfy linear differential equations over $\bar{\mathbb{Q}}(x)$, then the above result implies that $\tan x$ would be algebraic over $\bar{\mathbb{Q}}(x)$. One can show that this is not the case by noting that a periodic function that is algebraic over $\bar{\mathbb{Q}}(x)$ must be constant, a contradiction. Another proof can be given by starting with a putative minimal polynomial $P(Y) \in \mathbb{C}(x)[Y]$ and noting that $\frac{dP}{dx} + \frac{dP}{dY}Y' = \frac{dP}{dx} + \frac{dP}{dY}(Y^2 + 1)$ must be divisible by P. Comparing coefficients, one sees that the coefficients of powers of Y must be constant and so $\tan x$ would be constant.

Systems of linear partial differential equations. Let $(k, \Delta = \{\partial_1, \ldots, \partial_m\})$ be a partial differential field. One can define the ring of linear partial differential operators $k[\partial_1, \ldots, \partial_m]$ as the noncommutative polynomial ring in $\partial_1, \ldots, \partial_m$ where $\partial_i \cdot a = \partial_i(a) + a\partial_i$ for all $a \in k$. In this context, a differential module is a finite dimensional k-space that is a left $k[\partial_1, \ldots, \partial_m]$-module. We have

Proposition 1.3.34 *Assume k contains an element a such that $\partial_i(a) \neq 0$ for some i. There is a correspondence between*

- *differential modules,*
- *left ideals $I \subset k[\partial_1, \ldots, \partial_m]$ such that $\dim_k k[\partial_1, \ldots, \partial_m]/I < \infty$, and*
- *systems $\partial_i Y = A_i Y, A \in \mathrm{gl}_n(k), i = 1, \ldots m$ such that*

$$\partial_i(A_j) + A_i A_j = \partial_j(A_i) + A_j A_i$$

  i.e., integrable systems.

If

$$\partial_i Y = A_i Y, A \in \mathrm{gl}_n(k), i = 1, \ldots m$$

is an integrable system, the solution space over any differential field is finite dimensional over constants. Furthermore, one can define PV extensions, differential Galois groups, etc. for such systems and develop a Galois theory whose groups are linear algebraic groups. For further details see [3, 33] and Appendix D of [5].

1.4 Local Galois Theory

In the previous section, we showed that for equations with only regular singular points the analytically defined monodromy group determines the differential Galois group. In this section we shall give an idea of what analytic information determines the Galois group for general singular points. We shall use the following notation:

- $\mathbb{C}[[x]] = \{\sum_{i=0}^{\infty} a_i x^i \mid a_i \in \mathbb{C}\}$, $x' = 1$
- $\mathbb{C}((x)) =$ quotient field of $\mathbb{C}[[x]] = \mathbb{C}[[x]][x^{-1}]$
- $\mathbb{C}\{\{x\}\} = \{f \in \mathbb{C}[[x]] \mid f$ converges in a neighborhood $\mathcal{O}_f$ of $0\}$
- $\mathbb{C}(\{x\}) =$ quotient field of $\mathbb{C}\{\{x\}\} = \mathbb{C}\{\{x\}\}[x^{-1}]$

Note that

$$\mathbb{C}(x) \subset \mathbb{C}(\{x\}) \subset \mathbb{C}((x))$$

We shall consider equations of the form

$$Y' = AY \tag{1.2}$$

with $A \in \mathrm{gl}_n(\mathbb{C}(\{x\}))$ or $\mathrm{gl}_n(\mathbb{C}((x)))$ and discuss the form of solutions of such equations as well as the Galois groups relative to the base fields $\mathbb{C}(\{x\})$ and $\mathbb{C}((x))$. We start with the simplest case:

Ordinary points. We say that $x = 0$ is an *ordinary point* of equation (1.2) if $A \in \mathrm{gl}_n(\mathbb{C}[[x]])$. The standard existence theory (see [34] for a classical introduction or [35]) yields the following:

Proposition 1.4.1 *If $A \in \mathrm{gl}_n(\mathbb{C}[[x]])$ then there exists a unique $Y \in \mathrm{GL}_n(\mathbb{C}[[x]]$ such that $Y' = AY$ and $Y(0)$ is the identity matrix. If $A \in \mathrm{gl}_n(\mathbb{C}(\{x\}))$ then we furthermore have that $Y \in \mathrm{GL}_n(\mathbb{C}(\{x\}))$.*

This implies that for ordinary points with $A \in \mathrm{gl}_n(k)$ where $k = \mathbb{C}(\{x\})$ or $\mathbb{C}((x))$, the Galois group of the associated Picard-Vessiot extension over k is trivial. The next case to consider is

Regular singular points. From now on we shall consider equations of the form (1.2) with $A \in \mathbb{C}(\{x\})$ (although some of what we say can be adapted to the case when $A \in \mathbb{C}((x))$). We defined the notion of a regular singular point in Definition 1.3.26. The following proposition gives equivalent conditions (see Chapter 5.1.1 of [5] for a proof of the equivalences).

Proposition 1.4.2 *Assume $A \in \mathrm{gl}_n(\mathbb{C}(\{x\}))$. The following are equivalent:*

(i) $Y' = AY$ *is equivalent over $\mathbb{C}((x))$ to $Y' = \frac{A_0}{x}Y$, $A_0 \in \mathrm{gl}_n(\mathbb{C})$.*
(ii) $Y' = AY$ *is equivalent over $\mathbb{C}(\{x\})$ to $Y' = \frac{A_0}{x}Y$, $A_0 \in \mathrm{gl}_n(\mathbb{C})$.*

(iii) *For any small sector* $S(a, b, \rho)$, $|a-b| < 2\pi$, *at* 0, $\exists$ *a fundamental solution matrix* $Z = (z_{ij})$, *analytic in* S *such that* $|z_{ij}| < |x|^N$ *for some integer* N *as* $x \to 0$ *(regular growth of solutions).*

(iv) $Y' = AY$ *has a fundamental solution matrix of the form* $Y = x^{A_0} Z$, $A_0 \in \mathrm{gl}_n(\mathbb{C})$, $Z \in \mathrm{GL}_n(\mathbb{C}(\{x\}))$.

(v) $Y' = AY$ *is equivalent to* $y^{(n)} + a_{n-1}y^{(n-1)} + \ldots + a_0 y$, $a_i \in \mathbb{C}(\{x\})$, *where* $\mathrm{ord}_{x=0} a_i \geq i - n$.

(vi) $Y' = AY$ *is equivalent to* $y^{(n)} + a_{n-1}y^{(n-1)} + \ldots + a_0 y$, $a_i \in \mathbb{C}(\{x\})$, *having* n *linearly independent solutions of the form* $x^\alpha \sum_{i=0}^r c_i(x)(\log x)^i$, $\alpha \in \mathbb{C}$, $c_i(x) \in \mathbb{C}\{\{x\}\}$.

We now turn to the Galois groups. We first consider the Galois group of a regular singular equation over $\mathbb{C}(\{x\})$. Let $\mathcal{O}$ be a connected open neighborhood of 0 such that the entries of A are regular in $\mathcal{O}$ with at worst a pole at 0. Let $x_0 \neq 0$ be a point of $\mathcal{O}$ and γ a simple loop around 0 in $\mathcal{O}$ based at x_0. There exists a fundamental solution matrix Z at x_0 and the field $K = k(Z, \frac{1}{\det Z})$, $k = \mathbb{C}(\{x\})$ is a Picard-Vessiot extension of k. Analytic continuation around γ gives a new fundamental solution matrix $Z_\gamma = Z M_\gamma$ for some $M_\gamma \in \mathrm{GL}_n(\mathbb{C})$. The matrix M_γ depends on the choice of Z and so is determined only up to conjugation. Nonetheless, it is referred to as the *monodromy matrix at* 0. As in our discussion of monodromy in Section 1.5, we see that M_γ is in the differential Galois group $\mathrm{DGal}(K/k)$ and, arguing as in Proposition 1.3.27, one sees that M_γ generates a cyclic group that is Zariski dense in $\mathrm{DGal}(K/k)$. From the theory of linear algebraic groups, one can show that a linear algebraic group which has a Zariski dense cyclic subgroup must be of the form $(\mathbb{C}^*, \times)^r \times (\mathbb{C}, +)^s \times \mathbb{Z}/m\mathbb{Z}$, $r \geq 0$, $s \in \{0, 1\}$ and that any such group occurs as a differential Galois group of a regular singular equation over k. (Exercise 5.3 of [5]). This is not too surprising since property (vi) of the above proposition shows that K is contained in a field of the form $k(x^{\alpha_1}, \ldots, x^{\alpha_r}, \log x)$ for some $\alpha_i \in \mathbb{C}$.

Since $\mathbb{C}(\{x\}) \subset \mathbb{C}((x)) = \tilde{k}$, we can consider the the field $\tilde{K} = \tilde{k}(Z, \frac{1}{\det Z})$ where Z is as above. Again, using (vi) above, we have that $\tilde{K} \subset F = \tilde{k}(x^{\alpha_1}, \ldots, x^{\alpha_r}, \log x)$ for some $\alpha_i \in \mathbb{C}$. Since we are no longer dealing with analytic functions, we cannot appeal to analytic continuation but we can define a *formal* monodromy, that is a differential automorphism of F over $\tilde{k}$ that sends $x^\alpha \mapsto x^\alpha e^{2\pi i \alpha}$ and $\log x \mapsto \log x + 2\pi i$. One can show that the only elements of F that are left fixed by this automorphism lie in $\tilde{k}$. This automorphism restricts to an automorphism of $\tilde{K}$. Therefore the Fundamental Theorem implies that the group generated by the formal monodromy is Zariski dense in the differential Galois group $\mathrm{DGal}(\tilde{K}/\tilde{k})$. The restriction of the formal monodromy to K (as above) is just the usual monodromy, so one sees that the *convergent Galois group* $\mathrm{DGal}(K/k)$ equals the *formal Galois group* $\mathrm{DGal}(\tilde{K}/\tilde{k})$.

We note that the existence of n linearly independent solutions of the form described in (vi) goes back to Fuchs in the 19th Century and they can be constructed using the Frobenius method (see [36] for an historical account of the development of linear differential equations in the 19th Century).

Irregular singular points. We say $x = 0$ is an irregular singular point if it is not ordinary or regular. Fabry, in the 19th Century, constructed a fundamental set of formal solutions of a scalar linear differential equation at an irregular singular point. In terms of linear systems, every equation of the form $Y' = AY$, $A \in \mathrm{gl}_n(\mathbb{C}((x)))$ has a fundamental solution matrix of the form

$$\hat{Y} = \hat{\phi}(t) x^L e^{Q(1/t)}$$

where

- $t^{\nu} = x$, $\nu \in \{1, 2, 3, \ldots, n!\}$
- $\hat{\phi}(t) \in \mathrm{GL}_n(\mathbb{C}((t)))$
- $L \in \mathrm{gl}_n(\mathbb{C})$
- $Q = \mathrm{diag}(q_1, \ldots, q_n)$, $q_i \in \frac{1}{t}\mathbb{C}[\frac{1}{t}]$
- L and Q commute.

One refers to the integer ν as the *ramification* at 0 and the polynomial q_i as the *eigenvalues* at 0 of the equation. We note that even if $A \in \mathrm{gl}_n(\mathbb{C}(\{x\}))$, the matrix $\hat{\phi}(t)$ does not necessarily lie in $\mathrm{gl}_n(\mathbb{C}(\{t\}))$. There are various ways to make this representation unique but we will not deal with this here. We also note that by adjusting the term x^L, we can always assume that no ramification occurs in $\hat{\phi}(t)$, that is, $\hat{\phi} \in \mathrm{GL}_n(\mathbb{C}((x)))$ (but one may lose the last commutativity property above). As in the regular case, there are algorithms to determine a fundamental matrix of this form. We refer to Chapter 3 of [5] for a detailed discussion of the existence of these formal solutions and algorithms to calculate them and p. 98 of [5] for historical references and references to some recent work.

I will now give two examples (taken from [37]).

Example 1.4.3 *(Euler Eqn)* $x^2 y' + y = x$

This inhomogeneous equation has a solution $\hat{y} = \sum_{n=0}^{\infty} (-1)^n n! x^{n+1}$. By applying the operator $x \frac{d}{dx} - 1$ to both sides of this equation we get a second order homogeneous equation $x^3 y'' + (x^2 + x) y' - y = 0$ which, in matrix form becomes

$$\frac{dY}{dx} = \begin{pmatrix} 0 & 1 \\ \frac{1}{x^3} & -(\frac{1}{x} + \frac{1}{x^2}) \end{pmatrix} Y, \quad Y = \begin{pmatrix} y \\ y' \end{pmatrix}$$

This matrix equation has a fundamental solution matrix of the form $\hat{Y} = \hat{\phi}(x) e^Q$, where

$$Q = \begin{pmatrix} \frac{1}{x} & 0 \\ 0 & 0 \end{pmatrix} \quad \hat{\phi}(x) = \begin{pmatrix} 1 & \hat{f} \\ -\frac{1}{x^2} & \hat{f}' \end{pmatrix} \quad \hat{f} = \sum_{n=0}^{\infty} (-1)^n n! x^{n+1}$$

There is no ramification here (i.e., $\nu = 1$).

Example 1.4.4 *(Airy Equation)* $y'' = xy$

The singular point of this equation is at ∞. If we let $z = \frac{1}{x}$ the equation becomes $z^5 y'' + 2z^4 y' - y = 0$ or in matrix form

$$Y' = \begin{pmatrix} 0 & 1 \\ \frac{1}{z^5} & -\frac{2}{z} \end{pmatrix} Y$$

This equation has a fundamental matrix of the form $\hat{Y} = \hat{\phi}(z) z^J U e^{Q(t)}$ where

$$\hat{\phi}(z) = \ldots, \quad U = \begin{pmatrix} 1 & 1 \\ 1 & -1 \end{pmatrix}, \quad J = \begin{pmatrix} \frac{1}{4} & 0 \\ 0 & -\frac{3}{4} \end{pmatrix}$$

$$t = z^{\frac{1}{2}}, \quad Q = \begin{pmatrix} -\frac{2}{3t^3} & 0 \\ 0 & \frac{2}{3t^3} \end{pmatrix}$$

The ramification is $\nu = 2$. The precise form of $\hat{\phi}(z)$ is not important at the moment but we note that it lies in $\mathrm{GL}_n(\mathbb{C}((z)))$ (see [37] for the precise form).

We now turn to the Galois groups over $\mathbb{C}((x))$ and $\mathbb{C}(\{x\})$.

Galois Group over $\mathbb{C}((x))$: the Formal Galois Group at $x = 0$. We consider the differential equation $Y' = AY, A \in \mathrm{gl}_n(\mathbb{C}((x)))$. To simplify the presentation while capturing many of the essential ideas we shall assume that there is no ramification in $\hat{\phi}$, that is, the equation has a fundamental solution matrix $\hat{Y} = \hat{\phi}(x) x^L e^{Q(1/t)}$, with $\hat{\phi}(x) \in \mathrm{gl}_n(\mathbb{C}((x)))$. The map $Y_0 \mapsto \hat{\phi}(x) Y_0$ defines an equivalence over $\mathbb{C}((x))$ between the equations

$$Y_0' = A_0 Y_0 \text{ and } Y' = AY$$

where $A_0 = \hat{\phi}^{-1} A \hat{\phi} - \hat{\phi}^{-1} \hat{\phi}'$.

Definition 1.4.5 *The equation* $Y' = A_0 Y$ *is called a* Normal Form *of* $Y' = AY$. *It has* $Y_0 = x^L e^{Q(1/x)}$ *as a fundamental solution matrix.*

Example 1.4.6 *(Euler Equation) For this equation (Example 1.4.3), the normal form is* $\frac{dY_0}{dx} = A_0 Y_0$, $A_0 = \begin{pmatrix} -\frac{1}{x^2} & 0 \\ 0 & 0 \end{pmatrix}$ $Y_0 = e^{\begin{pmatrix} \frac{1}{x} & 0 \\ 0 & 0 \end{pmatrix}}$

Example 1.4.7 *(Airy Equation) Using the solution in Example 1.4.4, we get a normal form* $\frac{dY_0}{dx} = A_0 Y_0$ *with solution* $Y_0 = x^L U e^{Q(1/t)}$ *with* L, Q, t *as in Example 1.4.4.*

The map $Y_0 \mapsto \hat{\phi}(x) Y_0$ defines an isomorphism between the solutions spaces of $Y_0' = A_0 Y_0$ and $Y' = AY$ and a calculation shows that $F = \hat{\phi}(x)$ satisfies the differential equation

$$\frac{dF}{dx} = AF - FA_0.$$

Since the normal form is equivalent over $\mathbb{C}((x))$ to the original equation, its differential Galois group over $\mathbb{C}((x))$ is the same as that of the original equation.

Let K be the Picard-Vessiot extension of $\mathbb{C}((x))$ for $\frac{dY_0}{dx} = A_0 Y_0$ (and therefore also for $\frac{dY}{dx} = AY$). We have that

$$K \subset E = \mathbb{C}((x))(x^{\frac{1}{\nu}}, x^{\alpha_1}, \ldots, x^{\alpha_n}, \log x, e^{q_1(x^{1/\nu})}, \ldots, e^{q_n(x^{1/\nu})}).$$

We shall describe some differential automorphisms of E over $\mathbb{C}((x))$:

(1) Formal Monodromy γ: This is a differential automorphism of E given by the following conditions

- $\gamma(x^r) = e^{2\pi i r} x^r$
- $\gamma(\log x) = \log x + 2\pi i$
- $\gamma(e^{q(x^{1/\nu})}) = e^{q(\zeta x^{1/\nu})}$, $\zeta = e^{\frac{2\pi i}{\nu}}$

(2) Exponential Torus $\mathcal{T}$: This is the group of differential automorphisms $\mathcal{T} = \mathrm{DGal}(E/F)$ where $E = F(e^{q_1(x^{1/\nu})}, \ldots, e^{q_n(x^{1/\nu})})$ and $F = \mathbb{C}((x))(x^{\frac{1}{\nu}}, x^{\alpha_1}, \ldots, x^{\alpha_n}, \log x)$. Note that for $\sigma \in \mathcal{T}$:

- $\sigma(x^r) = x^r$
- $\sigma(\log x) = \log x$
- $\sigma(e^{q(x^{1/\nu})}) = c_\sigma e^{q(x^{1/\nu})}$

so we may identify $\mathcal{T}$ with a subgroup of $(\mathbb{C}^*)^n$. One can show (Proposition 3.25 of [5])

Proposition 1.4.8 *If $f \in E$ is left fixed by γ and $\mathcal{T}$, then $f \in \mathbb{C}((x))$.*

The formal monodromy and the elements of the exponential torus restrict to differential automorphisms of K over $\mathbb{C}((x))$ and are again called by these names. Therefore the above proposition and the Fundamental Theorem imply

Theorem 1.4.9 *The differential Galois group of $Y' = AY$ over $\mathbb{C}((x))$ is the Zariski closure of the group generated by the formal monodromy and the exponential torus.*

One can furthermore show (Proposition 3.25 of [5]) that $x = 0$ is a regular singular point if and only if the exponential torus is trivial.

Example 1.4.10 *(Euler Equation) We continue using the notation of Examples 1.4.3 and 1.4.6. One sees that $K = E = \mathbb{C}((x))(e^{\frac{1}{x}})$ in this case. Furthermore, the formal monodromy is trivial. Therefore* $\mathrm{DGal}(K/\mathbb{C}((x)))$ *is equal to the exponential torus* $\mathcal{T} = \{ \begin{pmatrix} c & 0 \\ 0 & 1 \end{pmatrix} \mid c \in \mathbb{C}^* \}$

Example 1.4.11 *(Airy Equation) We continue using the notation of Examples 1.4.4 and 1.4.7. In this case $F = \mathbb{C}((z))(z^{\frac{1}{4}})$ and $E = F(e^{\frac{2}{3}z^{3/2}})$. The formal monodromy* $\gamma = \begin{pmatrix} 0 & \sqrt{-1} \\ \sqrt{-1} & 0 \end{pmatrix}$ *and the exponential torus ·* $\mathcal{T} = \{ \begin{pmatrix} c & 0 \\ 0 & c^{-1} \end{pmatrix} \mid c \neq 0 \}$. *Therefore*

$$\mathrm{DGal}(K/\mathbb{C}((x))) = D_\infty = \{ \begin{pmatrix} c & 0 \\ 0 & c^{-1} \end{pmatrix} \mid c \neq 0 \} \cup \{ \begin{pmatrix} 0 & c \\ -c^{-1} & 0 \end{pmatrix} \mid c \neq 0 \}$$

Galois Group over $\mathbb{C}(\{x\})$: *the Convergent Galois Group at* $x = 0$. We now assume that we are considering a differential equation with $\frac{dY}{dx} = AY$ with $A \in \mathbb{C}(\{x\})$ and we let K be the Picard-Vessiot extension of $\mathbb{C}(\{x\})$ for this equation. Since $\mathbb{C}(\{x\}) \subset \mathbb{C}((x))$, we have that the formal Galois group $DGal(K/\mathbb{C}((x)))$ is a subgroup of the convergent Galois group $DGal(K/\mathbb{C}(\{x\}))$. If $x = 0$ is a regular singular point then these Galois groups are the same but in general the convergent group is larger than the formal group. I will describe analytic objects that measure the difference. The basic idea is the following. We have a formal solution $\hat{Y} = \hat{\phi}(x)x^L e^{Q(1/t)}$ for the differential equation. For each small enough sector $\mathcal{S}$ at $x = 0$ we can give the terms x^L and $e^{Q(1/t)}$ meaning in an obvious way. The matrix $\hat{\phi}(x)$ is a little more problematic but it has been shown that one can find a matrix $\Phi_\mathcal{S}$ of functions, analytic is $\mathcal{S}$, "asymptotic" to $\hat{\phi}(x)$ and such that $\Phi_\mathcal{S}(x)x^L e^{Q(1/t)}$ satisfies the differential equation. In overlapping sectors $\mathcal{S}_1$ and $\mathcal{S}_2$ the corresponding $\Phi_{\mathcal{S}_1}$ and $\Phi_{\mathcal{S}_2}$ may not agree but they will satisfy $\Phi_{\mathcal{S}_1} = \Phi_{\mathcal{S}_2} St_{12}$ for some $St_{12} \in GL_n(\mathbb{C})$. The $St_{i,j}$ ranging over a suitable choice of small sectors $\mathcal{S}_\ell$ together with the formal Galois group should generate a group that is Zariski dense in the convergent Galois group. There are two (related) problems with this idea. The first is what do we mean by asymptotic and the second is how do we select the sectors and the Φ_ℓ in some canonical way? We shall first consider these questions for fairly general formal series and then specialize to the situation where the series are entries of matrices $\hat{\phi}(x)$ that arise in the formal solutions of linear differential equations.

Asymptotics and Summability. In the 19th century, Poincaré proposed the following notion of an asymptotic expansion.

Definition 1.4.12 *Let f be analytic on a sector $\mathcal{S} = \mathcal{S}(a, b, \rho)$ at $x = 0$. The series $\sum_{n \geq n_0} c_n x^n$ is an* asymptotic expansion *of f if $\forall N \geq 0$ and every closed sector $W \subset \mathcal{S}$ there exists a constant $C(N, W)$ such that*

$$|f(x) - \sum_{n_0 \leq n \leq N-1} c_n x^n| \leq C(N, W)|x|^N \quad \forall x \in W.$$

A function f analytic on a sector can have at most one asymptotic expansion on that sector and we write $J(f)$ for this series if it exists. We denote by $\mathcal{A}(\mathcal{S}(a, b, \rho))$ the set of functions analytic on $\mathcal{S}(a, b, \rho)$ having asymptotic expansions on this sector. We define $\mathcal{A}(a, b)$ to be the direct limit of the $\mathcal{A}(\mathcal{S}(a, b, \rho))$ over all ρ (here we say $f_1 \in \mathcal{A}(\mathcal{S}(a, b, \rho_1))$ and

$f_2 \in \mathcal{A}(\mathcal{S}(a, b, \rho_2))$ are equivalent if they agree on the intersection of the sectors). If one thinks of $a, b \in S^1$, the unit circle, then one can see that the sets $\mathcal{A}(a, b)$ form a sheaf on S^1. We have the following facts (whose proofs can be found in Chapters 7.1 and 7.2 of [5] or in [38]).

- $\mathcal{A}(a, b)$ is a differential $\mathbb{C}$-algebra.
- $\mathcal{A}(S^1) = \mathbb{C}(\{x\})$.
- (Borel-Ritt) For all $(a, b) \neq S^1$, the map $J : \mathcal{A}(a, b) \to \mathbb{C}((x))$ is surjective.

Given a power series in $\mathbb{C}((x))$ that satisfies a linear differential equation, one would like to say that it is the asymptotic expansion of an analytic function that also satisfies the differential equation. Such a statement is contained in the next result, originally due to Hukuhara and Turrittin, with generalizations and further developments concerning nonlinear equations due to Ramis and Sibuya (see Chapter 7.2 of [5] for a proof and references).

Theorem 1.4.13 *Let $A \in \mathrm{gl}_n(\mathbb{C}(\{x\}))$, $w \in (\mathbb{C}(\{x\}))^n$ and $\hat{v} \in (\mathbb{C}((x)))^n$ such that*

$$\hat{v}' - A\hat{v} = w \ .$$

For any direction $\theta \in S^1$ there exist a, b, $a < \theta < b$ and $v \in (\mathcal{A}(a, b))^n$ such that $J(v) = \hat{v}$ and

$$v' - Av = w$$

In general, there may be many functions v satisfying the conclusion of the theorem. To guarantee uniqueness we will refine the notion of asymptotic expansion and require that this be valid on large enough sectors. The refined notion of asymptotics is given by

Definition 1.4.14 *(1) Let k be a positive real number, $\mathcal{S}$ an open sector. A function $f \in \mathcal{A}(\mathcal{S})$ with $J(f) = \sum_{n \geq n_0} c_n x^n$ is Gevrey of order k if for all closed $W \subset \mathcal{S}$, $\exists A > 0, c > 0$ such that $\forall N \geq 1$, $x \in W$, $|x| \leq c$ one has*

$$\left| f(z) - \sum_{n_0 \leq n \leq N-1} c_n x^n \right| \leq A^N (N!)^{\frac{1}{k}} |x|^N$$

(One usually uses $\Gamma(1 + \frac{N}{k})$ instead of $(N!)^{\frac{1}{k}}$ but this makes no difference.)

(2) We denote by $\mathcal{A}_{\frac{1}{k}}(\mathcal{S})$ the ring of all Gevrey functions of order k on $\mathcal{S}$ and let $\mathcal{A}_{\frac{1}{k}}^{0}(\mathcal{S}) = \{f \in \mathcal{A}_{\frac{1}{k}}(\mathcal{S}) \mid J(f) = 0\}$

The following is a useful criterion to determine if a function is in $\mathcal{A}_{\frac{1}{k}}^{0}(\mathcal{S})$.

Lemma 1.4.15 $f \in \mathcal{A}_{\frac{1}{k}}^{0}(\mathcal{S})$ *if and only if* $\forall$ *closed* $W \subset \mathcal{S}, \exists\, A, B > 0$ *such that*

$$|f(x)| \leq A \exp\left(-B|x|^{-k}\right) \quad \forall x \in W$$

Example 1.4.16 $f(x) = e^{-\frac{1}{x^k}} \in \mathcal{A}_{\frac{1}{k}}^{0}(\mathcal{S})$, *for* $\mathcal{S} = \mathcal{S}(-\frac{\pi}{2k}, \frac{\pi}{2k})$ *but* not on a larger sector. This is a hint of what is to come.

Corresponding to the notion of a Gevrey function is the notion of a Gevrey series.

Definition 1.4.17 $\sum_{n \geq n_0} c_n x^n$ *is a Gevrey series of order k if $\exists A > 0$ such that $\forall n > 0\ |c_n| \leq A^n (n!)^{\frac{1}{k}}$.*

(2) $\mathbb{C}((x))_{\frac{1}{k}} = $ Gevrey series of order k.

Note that if $k < \ell$ then $\mathbb{C}((x))_{\frac{1}{\ell}} \subset \mathbb{C}((x))_{\frac{1}{k}}$. Using these definitions one can prove the following key facts:

- (Improved Borel-Ritt) If $|b - a| < \frac{\pi}{k}$, the map $J : \mathcal{A}_{\frac{1}{k}}(a, b) \to \mathbb{C}((x))_{\frac{1}{k}}$ is surjective but not injective.
- (Watson's Lemma) If $|b - a| > \frac{\pi}{k}$, the map $J : \mathcal{A}_{\frac{1}{k}}(a, b) \to \mathbb{C}((x))_{\frac{1}{k}}$ is injective but not surjective.

Watson's lemma motivates the next definition.

Definition 1.4.18 *(1) $\hat{y} \in \mathbb{C}((x))$ is k-summable in the direction $d \in S^1$ if there exists an $f \in \mathcal{A}_{\frac{1}{k}}(d - \frac{\alpha}{2}, d + \frac{\alpha}{2})$ with $J(f) = \hat{y}$, $\alpha > \frac{\pi}{k}$.*

(2) $\hat{y} \in \mathbb{C}((x))$ is k-summable if it is k-summable in all but a finite number of directions

A k-summable function has unique analytic "lifts" in many large sectors (certainly enough to cover a deleted neighborhood of the origin). I would like to say that the entries of the matrix $\hat{\phi}(x)$ appearing in the formal solution of a linear differential equation are k-summable for some k but regrettably this is not exactly true and the situation is more complicated (each is a sum of k_i-summable functions for several possible k_i). Before I describe what does happen, it is useful to have criteria for deciding if a series $\hat{y} \in \mathbb{C}((x))$ is k-summable. Such criteria can be given in terms of Borel and Laplace transforms.

Definition 1.4.19 *The* Formal Borel Transform $\mathcal{B}_k$ *of order* k:

$$\mathcal{B}_k : \mathbb{C}[[x]] \to \mathbb{C}[[\zeta]]$$

$$\mathcal{B}_k\left(\sum c_n x^n\right) = \sum \frac{c_n}{\Gamma(1 + \frac{n}{k})} \zeta^n$$

Note: If $\hat{y} = \sum c_n x^n \in \mathbb{C}((x))_{\frac{1}{k}} \Rightarrow \mathcal{B}(\hat{y})$ has a nonzero radius of convergence.

Example 1.4.20 $\mathcal{B}_1\left(\sum(-1)^n n! x^{n+1}\right) = \sum(-1)^n \frac{\zeta^{n+1}}{n+1} = \log(1 + \zeta)$

Definition 1.4.21 *The* Laplace Transform $\mathcal{L}_{k,d}$ *of order* k *in the direction* d: *Let* f *satisfy* $|f(\zeta)| \le A \exp(B|\zeta|^k)$ *along the ray* r_d *from* 0 *to* ∞ *in direction* d. *Then*

$$\mathcal{L}_{k,d} f(x) = \int_{r_d} f(\zeta) \exp(-(\tfrac{\zeta}{x})^k) d((\tfrac{\zeta}{x})^k)$$

Note: $\mathcal{L}_{k,d} \circ \mathcal{B}_k(x^n) = x^n$ and $\mathcal{L}_{k,d} \circ \mathcal{B}_k(f) = f$ for $f \in \mathbb{C}\{\{x\}\}$.

Example 1.4.22 *For any ray* r_d *not equal to the negative real axis, the function* $f(\zeta) = \log(1 + \zeta)$ *has an analytic continuation along* r_d. *Furthermore, for such a ray, we have*

$$\mathcal{L}_{1,d}(\log(1 + \zeta))(x) = \int_d \log(1 + \zeta) e^{-\frac{\zeta}{x}} d(\tfrac{\zeta}{x}) = \int_d \frac{1}{1 + \zeta} e^{-\frac{\zeta}{x}} d\zeta$$

Note that there are several slightly different definitions of the Borel and Laplace transforms scattered in the literature but for our purposes these differences do not affect the final results. I am following the definitions in [38] and [5]. The following gives useful criteria for a Gevrey series of order k to be k-summable (Chapter 3.1 of [38] or Chapter 7.6 of [5]).

Proposition 1.4.23 $\hat{y} \in \mathbb{C}[[x]]_{\frac{1}{k}}$ *is* k-summable in direction d

$$\Updownarrow$$

$\mathcal{B}_k(\hat{y})$ *has an analytic continuation* h *in a full sector* $\{\zeta \mid 0 < |\zeta| < \infty, \; |\arg \zeta - d| < \epsilon\}$ *and has exponential growth of order* $\le k$ *at* ∞ *in this sector* $(|h(\zeta)| \le A \exp(B|\zeta|^k))$. *In this case,* $f = \mathcal{L}_{k,d}(h)$ *is its* k-sum.

Example 1.4.24 *Consider $\hat{y} = \sum(-1)^n n! x^{n+1}$, a power series of Gevrey order 1 that is also a solution of the Euler equation. We have seen that $f(\zeta) = \mathcal{B}_1(\hat{y}) = \sum(-1)^n \frac{\zeta^{n+1}}{n+1} = \log(1+\zeta)$ for $|\zeta| < 1$. As we have noted in Example 1.4.22, this function can be continued outside the unit disk in any direction other than along the negative real axis. On any ray except this one, it has exponential growth of order at most 1 and its Laplace transform is*

$$y_d(x) = \mathcal{L}_{1,d}(\log(1+\zeta))(x) = \int_d \log(1+\zeta) e^{-\frac{\zeta}{x}} d(\frac{\zeta}{x}) = \int_d \frac{1}{1+\zeta} e^{-\frac{\zeta}{x}} d\zeta.$$

We note that this function again satisfies the Euler equation. To see this note that $y(x) = Ce^{\frac{1}{x}}$ is a solution of $x^2 y' + y = 0$ and so variation of constants gives us that $f(x) = \int_0^x e^{\frac{1}{x} - \frac{1}{t}} \frac{dt}{t}$ is a solution of $x^2 y' + y = x$. For convenience let $d = 0$ and define a new variable ζ by $\frac{\zeta}{x} = \frac{1}{t} - \frac{1}{x}$ which gives $f(x) = \int_{d=0} \frac{1}{1+\zeta} e^{-\frac{\zeta}{x}} d\zeta$. Therefore, for any ray except the negative real axis, we are able to canonically associate a function, analytic in a large sector around this ray, with the divergent series $\sum(-1)^n n! x^{n+1}$ and so this series is 1-summable. Furthermore these functions will again satisfy the Euler equation. This is again a hint for what is to come.

Linear Differential Equations and Summability. It has been known since the early 20th Century that a formal series solution of an analytic differential equation $F(x, y, y', \ldots, y^{(n)}) = 0$ must be Gevrey of order k for some k ([39]). Regrettably, even for linear equations, these formal solutions need not be k summable (see [37]). Nonetheless, if $\hat{y}$ is a vector of formal series that satisfy a linear differential equation it can be written as a sum $\hat{y} = \hat{y}_1 + \ldots + \hat{y}_t$ where each $\hat{y}_i$ is k_i summable for some k_i. This result is the culmination of the work of several authors including Ramis, Malgrange, Martinet, Sibuya and Écalle (see [37], [40] and Chapter 7.8 of [5]). The k_i that occur as well as the directions along which the $\hat{y}_i$ are not summable can be seen from the normal form of the differential equation. We need the following definitions. Let $Y' = AY$, $A \in \mathrm{gl}_n(\mathbb{C}(\{x\}))$ and let $\hat{Y} = \hat{\phi}(x) x^L e^{Q(1/t)}$ be a formal solution with $t^\nu = x$ and $Q = \mathrm{diag}(q_1, \ldots, q_n), q_i \in \frac{1}{t}\mathbb{C}[\frac{1}{t}]$.

Definition 1.4.25 *(1) The rational number $k_i = \frac{1}{\nu}$(the degree of q_i in $\frac{1}{t}$) is called a* slope *of $Y' = AY$.*

(2) A Stokes direction *d for a $q_i = cx^{-k_i} + \ldots$ is a direction such that $\mathrm{Re}(cx^{-k_i}) = 0$, i.e., where e^{q_i} changes behavior.*

(3) Let d_1, d_2 be consecutive Stokes directions. We say that (d_1, d_2) is a negative Stokes pair if $Re(cx^{-k_i}) < 0$ for $\arg(x) \in (d_1, d_2)$.

(4) A singular direction *is the bisector of a negative Stokes pair.*

Example 1.4.26 *For the Euler equation we have that $\nu = 1$ and $Q(x) = diag(\frac{1}{x}, 0)$. Therefore the slopes are $\{0, 1\}$. The Stokes directions are $\{\frac{\pi}{2}, \frac{3\pi}{2}\}$. The pair $(\frac{\pi}{2}, \frac{3\pi}{2})$ is a negative Stokes pair and so π is the only singular direction.*

With these definitions we can state

Theorem 1.4.27 *(Multisummation Theorem) Let $1/2 < k_1 < \ldots < k_r$ be the slopes of $Y' = AY$ and let $\hat{Y} \in (\mathbb{C}((x)))^n$ satisfy this differential equation. For all directions d that are not singular directions, there exist $\hat{Y}_i \in \mathbb{C}((x))$ such that $\hat{Y} = \hat{Y}_1 + \ldots + \hat{Y}_r$, each $\hat{Y}_i$ is k_i-summable in direction d and, for $y_i =$ the k_i-sum of $\hat{Y}_i$, $y_d = y_1 + \ldots + y_r$ satisfies $y' = Ay$.*

The condition that $1/2 < k_1$ is a needed technical assumption but, after a suitable change of coordinates, one can always reduce to this case. We shall refer to the element y_d as above as the *multisum of $\hat{Y}$ in the direction d*.

Stokes Matrices and the Convergent Differential Galois Group. We are now in a position to describe the analytic elements which, together with the formal Galois group, determine the analytic Galois group. Once again, let us assume that we have a differential equation $Y' = AY$ with $A \in gl_n(\mathbb{C}(\{x\}))$ and assume this has a formal solution $\hat{Y} = \hat{\phi}(x)x^L e^{Q(1/t)}$ with $\hat{\phi}(x) \in GL_n(\mathbb{C}((x)))$. The matrix $F = \hat{\phi}(x)$ is a formal solution of the $n^2 \times n^2$ system

$$\frac{dF}{dx} = AF - FA_0 \tag{1.3}$$

where $Y' = A_0 Y$ is the normal form of $Y' = AY$. We will want to associate functions, analytic in sectors, to $\hat{\phi}(x)$ and therefore will apply the Multisummation Theorem to equation 1.3 (and not to $Y' = AY$). One can show that the eigenvalues of equation (1.3) are of the form $q_i - q_j$ where the q_i, q_j are eigenvalues of $Y' = AY$.

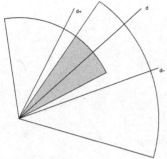

Let d be a singular direction of $Y' = AY - YA_0$. Select d^+ and d^- nearby so that the associated sectors overlap and contain d. Let ϕ^+, ϕ^- be the associated multisums. We must have $\phi^+ x^L e^{Q(1/t)} = \phi^- x^L e^{Q(1/t)} \cdot St_d$ for some $St_d \in \mathrm{GL}_n(\mathbb{C})$ on the overlap of the regions (the shaded area in the figure). The matrix St_d is independent of the choice of d^+ and d^-.

Definition 1.4.28 *The matrix St_d is called a* Stokes matrix in the direction d.

Example 1.4.29 *(Euler equation)*
- *Eigenvalues of $Y' = AY - YA_0$:* $\{\frac{1}{x}, 0\}$
- *Singular direction of $Y' = AY - YA_0$:* $d = \pi$
- $\hat{\phi}(x) = \begin{pmatrix} 1 & \hat{f} \\ -\frac{1}{x^2} & \hat{f}' \end{pmatrix}$ $\hat{f} = \sum_{n=0}^{\infty}(-1)^n x^{n+1}$

For $d+$ and d^- close to the negative real axis, we have

$$f^+ = \int_{d+} \frac{1}{1+\zeta} e^{-\frac{\zeta}{x}} d\zeta \text{ and } f^- = \int_{d-} \frac{1}{1+\zeta} e^{-\frac{\zeta}{x}} d\zeta.$$

Since $f^+ - f^- = 2\pi i \, Res_{\zeta=-1}\left(\frac{e^{-\frac{\zeta}{x}}}{1+\zeta}\right) = 2\pi i$, we have

$$\begin{pmatrix} 1 & f^+ \\ -\frac{1}{x^2} & (f^+)' \end{pmatrix} = \begin{pmatrix} 1 & f^- \\ \frac{1}{x^2} & (f^-)' \end{pmatrix}\begin{pmatrix} 1 & 2\pi i \\ 0 & 1 \end{pmatrix}$$

so $St_\pi = \begin{pmatrix} 1 & 2\pi i \\ 0 & 1 \end{pmatrix}$

The following result is due to Ramis [41, 42, 43, 44]. Another proof appears in the work of Loday-Richaud [45]. An exposition of this result and its proof appears in Chapter 8 of [5].

Theorem 1.4.30 *The convergent differential Galois group* $\mathrm{DGal}(K/\mathbb{C}(\{x\}))$ *is the Zariski closure of the group generated by*
- (i) *the formal differential Galois group* $\mathrm{DGal}(K/\mathbb{C}((x)))$, *and*
- (ii) *The collection of Stokes matrices $\{St_d\}$ where d runs over the set of singular directions of the polynomials $\{q_i - q_j\}$, q_i, q_j eigenvalues of $Y' = AY$.*

Examples 1.4.31 *(1) (Euler equation)*

- *The formal differential Galois group:*

$$\left\{ \begin{pmatrix} c & 0 \\ 0 & 1 \end{pmatrix} \mid c \neq 0 \right\}$$

- *Stokes matrix:* $\begin{pmatrix} 1 & 2\pi i \\ 0 & 1 \end{pmatrix}$

Therefore the convergent differential Galois group is

$$\left\{ \begin{pmatrix} c & d \\ 0 & 1 \end{pmatrix} \mid c \neq 0, d \in \mathbb{C} \right\}.$$

(2) (Airy equation)

- *The formal differential Galois group:*

$$D_\infty = \left\{ \begin{pmatrix} c & 0 \\ 0 & c^{-1} \end{pmatrix}, \begin{pmatrix} 0 & c \\ -c^{-1} & 0 \end{pmatrix} \mid c \neq 0 \right\}$$

- *The Stokes matrices:*

$$St_0 = \begin{pmatrix} 1 & * \\ 0 & 1 \end{pmatrix}, St_{\frac{2\pi}{3}} = \begin{pmatrix} 1 & 0 \\ * & 1 \end{pmatrix}, St_{\frac{4\pi}{3}} = \begin{pmatrix} 1 & * \\ 0 & 1 \end{pmatrix}$$

(see Example 8.15 of [22] for the calculation of the Stokes matrices.)

Therefore the convergent differential Galois group is $\mathrm{SL}_2(\mathbb{C})$

We highly recommend the book [40] of Ramis, the papers of Loday-Richaud [46, 37] and the papers of Varadarajan [47, 48] for further introductions to this analytic aspect of linear differential equations. One can find references to the original papers in these works as well as in [5].

1.5 Algorithms

In this section we shall consider the question: Can we compute differential Galois groups and their properties? We shall not give a complete algorithm to determine the Galois groups although one exists - see [21]. This general algorithm has not been implemented at present and in its present form seems not very efficient. We will rather give a flavor of the various techniques that are presently implemented and used.

Categories motivate algorithms. We will show how results in a categorical setting concerning representations of groups lead to algorithms to determine Galois groups. We begin with an example from the Galois theory of polynomials.

Example 1.5.1 *Let*

$$P(Y) = Y^3 + bY + c = 0 \in \mathbb{Q}[Y].$$

The usual Galois group G of this equation is a subgroup of the full permutation group $\mathcal{S}_3$ on 3 elements. We will describe how to compute G. If $P(Y) = (Y+\alpha)(Y^2+\beta Y+\gamma), \alpha, \beta, \gamma \in \mathbb{Q}$, then G is the Galois group of $Y^2 + \beta Y + \gamma$, which we assume we know how to compute. Therefore we will assume that $P(Y)$ is irreducible over $\mathbb{Q}$. This implies that G acts transitively on the roots $\{\rho_1, \rho_2, \rho_3\}$ of P. From group theory we know that the only transitive subgroups of $\mathcal{S}_3$ are $\mathcal{A}_3$, the alternating group on 3 elements, and $\mathcal{S}_3$. To decide which we have, consider the following associated polynomial

$$Z^2 + 4b^3 + 27c^2 = (Z + \delta)(Z - \delta) \quad \delta = (r_1 - r_2)(r_2 - r_3)(r_3 - r_1).$$

One then has that $G = \mathcal{A}_3$ if $Z^2 + 4b^3 + 27c^2$ factors over $\mathbb{Q}$ and $G = \mathcal{S}_3$ if $Z^2 + 4b^3 + 27c^2$ irreducible over $\mathbb{Q}$.

In the above example we determined the Galois group by considering factorization properties of the polynomial and associated polynomials. Why does this work?

To answer this question, let us consider a square-free polynomial $f(Y) \in \mathbb{Q}[Y]$ with roots $r_1, \ldots r_n \in \bar{\mathbb{Q}}$, the algebraic closure of Q. Let G be the Galois group of f over $\mathbb{Q}$. The group G acts as a group of permutations on the set $S = \{r_1, \ldots, r_n\}$. We have that G acts transitively on S if and only if the polynomial f is irreducible. More generally, G leaves a subset $\{r_{i_1}, \ldots, r_{i_t}\}$ invariant if and only if the polynomial $g(Y) = \prod_{j=1}^{t}(Y - r_{i_j}) \in \mathbb{Q}[Y]$. Therefore the orbits of G in the set S correspond to irreducible factors of f. In fact, for any $h \in \mathbb{Q}[Y]$ having all its roots in the splitting field of f, G-orbits in the set of roots correspond to irreducible factors of h. In the above example, we were able to distinguish between groups because they have different orbits in certain "well-selected" sets and we were able to see this phenomenon via factorization properties of associated polynomials. The key is that, in general, *one can distinguish between finite groups by looking at the sets on which they act.*

This strange looking statement can be formalized in the following way (see Appendix B1 of [5] for details). For any finite set, let Perm(S) be the group of permutations of S. One constructs a category Perm$_G$ of G-sets whose objects are pairs (S, ρ), where $\rho : G \to$ Perm(S) is the map defining an action of G on S. The morphisms of this category are just maps between G-sets that commute with the action of G. One has the forgetful functor $\omega :$ Perm$_G \to$ Sets which takes (S, ρ) to S. One then has the formalization of the above statement as: one can recover the group from the pair (Perm$_G, \omega$). A consequence of this is that given two groups $H \subset G$ there is a set S on which G (and therefore H) acts so that the orbit structures with respect to these two groups are different.

These facts lead to the following strategy to compute Galois groups of a polynomial $f \in \mathbb{Q}[X]$. For each pair of groups $H \subset G \subset \mathcal{S}_n$ we construct a polynomial $f_{H,G} \in \mathbb{Q}[X]$ whose roots can be expressed in terms of the roots of f and such that H and G can be distinguished by their orbit structures on the roots of $f_{H,G}$ (by the above facts, we know that there is a set that distinguishes H from G and one can show that this set can be constructed from the roots of f). The factorization properties of $f_{H,G}$ will then distinguish H from G. This is what was done in Example 1.5.1.

To generalize the above ideas to calculating differential Galois groups of linear differential equations, we must first see what determines a linear algebraic group. In fact, just as finite groups are determined by their *permutation* representations, linear algebraic groups are determined by their *linear* representations. To make this more precise, let G be a linear group algebraic group defined over an algebraically closed field C. I will denote by Rep$_G$ the category of linear representations of G. The objects of this category are pairs (V, ρ) where V is a finite dimensional vector space over C and $\rho : G \to$ GL$_n(V)$ is a representation of G. The morphisms $m : (V_1, \rho_1) \to (V_2, \rho_2)$ are linear maps $m : V_1 \to V_2$ such that $m \circ \rho_1 = \rho_2 \circ m$. One can define subobjects, quotients, duals, direct sums and tensor products in obvious ways (see Appendix B2 and B3 of [5] for details). One again has a forgetful functor $\omega :$ Rep$_G \to$ Vect$_C$ (where Vect$_C$ is the category of vector spaces over C) given by $\omega(V, \rho) = V$. In analogy to the finite group case, we have Tannaka's Theorem, that says that we can recover G from the pair (Rep$_G, \omega$).

Let T be a category with a notion of quotients, duals, direct sums, tensor products and a functor $\tilde{\omega} : T \to \text{Vect}_C$. An example of such a category is the category of differential modules $\mathcal{D}$ over a differential field k with algebraically closed constants and $\tilde{\omega}(M) = \ker(\partial, K \otimes M)$ (see the discussion following Definition 1.2.13) where K is a differential field large enough to contain solutions of the differential equations corresponding to the modules in $\mathcal{D}$. One can give axioms that guarantee that $\tilde{\omega}(T)$ "=" $\omega(\text{Rep}_G)$ for some linear algebraic group G (see [49], [50] or Appendix B3 of [5]). Such a pair $(T, \tilde{\omega})$ is called a *neutral tannakian category*. Applying this to a subcategory of $\mathcal{D}$, one has:

Theorem 1.5.2 *Let (k, ∂) be a differential field with algebraically closed constants, let $Y' = AY$ be a linear differential equation over (k, ∂), let M_A be the associated differential module and K the associated Picard-Vessiot extension. Let $T = \{\{M_A\}\}$ be the smallest subcategory of $\mathcal{D}$ containing M_A and closed under the operations of taking submodules, quotients, duals, direct sums, and tensor products. For $N \in \{\{M_A\}\}$, let $\tilde{\omega}(N) = \ker(\partial, N \otimes K)$.*

Then $(T, \tilde{\omega})$ is a neutral tannakian category and the image of $\tilde{\omega}$ is the category of vector spaces on which $\text{DGal}(K/k)$ acts.

This theorem and Tannaka's Theorem have the following implications:

- One can construct all representations of $\text{DGal}(K/k)$ by:
  - using the constructions of linear algebra (submodules, quotients, duals, direct sums, tensor products) on M_A, and
  - taking the solution spaces of the corresponding differential equations.

- $\text{DGal}(K/k)$ is determined by knowing which subspaces of these solution spaces are $\text{DGal}(K/k)$-invariant.

At first glance, it seems that one would need to consider an infinite number of differential modules but in many cases, it is enough to consider only a finite number of modules constructed from M_A. Regrettably, there are groups (for example, any group of the form $C^* \oplus \ldots \oplus C^*$) for which one needs an infinite number of representations to distinguish it from other groups. Nonetheless, there are many groups where a finite number of representations suffice. Furthermore, Proposition 1.3.28 states that the DGal-invariant subspaces of the solution space of a differential module N correspond to factors of the associated scalar equation

$L_N(Y) = 0$. Therefore, in many cases, one can give criteria for a differential equation to have a given group as Galois group in terms of factorization properties of certain associated differential operators. We shall give examples for second order equations. We begin with a definition.

Let k be a differential field with algebraically closed constants C, let $L = \partial^2 - s \in k[\partial]$, and let K be the Picard-Vessiot extension corresponding to $L(Y) = 0$. Let $\{y_1, y_2\}$ be a basis for the solution space V of $L(Y) = 0$ in K and let $V_m = C - \text{span of } \{y_1^m, y_1^{m-1}y_2, \ldots, y_2^m\}$. One can see that V_m is independent of the selected basis of V and that it is invariant under the action of $\text{DGal}(K/k)$. As in the proof of Proposition 1.3.28 one can show that this implies that V_m is precisely the solution space of a scalar linear differential equation over k.

Definition 1.5.3 *The mth symmetric power $L^{\circledS m}$ of L is the monic operator whose solution space is V_m.*

One has that the order of $L^{\circledS m}$ is $m + 1$. To see this it is enough to show that the $y_1^{m-i}y_2^i$, $0 \le i \le m$ are linearly independent over C. Since any homogeneous polynomial in two variables factors over C, if we had $0 = \sum c_i y_1^{m-i}y_2^i = \prod(a_i y_1 + b_i y_2)$, then y_1 and y_2 would be linearly dependent over C. One can calculate $L^{\circledS m}$ by starting with y^m and formally differentiating it m times. One uses the relation $y'' - sy = 0$ to replace $y^{(j)}$, $j > 1$ with combinations of y and y'. One then has $m+2$ expressions $y^m, (y^m)', \ldots, (y^m)^{(m+1)}$ in the $m+1$ "indeterminates" $y^{m-i}(y')^i$, $0 \le i \le m$. These must be linearly dependent and one can find a dependence $(y^m)^{(m+1)} + b_m(y^m)^{(m)} + \ldots + b_0(y^m) = 0$. One can show that $L^{\circledS m} = \partial^{m+1} + b_m \partial^m + \ldots + b_0$ (see Chapter 2.3 of [5]).

Example 1.5.4 $m = 2$:

$$\begin{array}{rcl}
y^2 & = & y^2 \\
(y^2)' & = & 2yy' \\
(y^2)'' & = & 2(y')^2 + 2sy^2 \\
(y^2)''' & = & 8syy' + 2s'y^2
\end{array} \quad \Rightarrow \quad L^{\circledS 2}(y) = y''' - 4sy' - 2s'y$$

The vector spaces V_m correspond to representations of $\text{DGal}(K/k)$ on the mth symmetric power of V and are well understood for many groups [51] (the mth symmetric power of a vector space is the vector space with a *basis* formed by the monomials of degree m in a basis of V). Using representation theory, one can show:

Proposition 1.5.5 *Let* $L(y) = y'' - sy$ *and let* $G = \mathrm{DGal}(K/k)$. *Assume* L *and* $L^{\circledS 2}$ *are irreducible.*

- *If* $L^{\circledS 3}$ *is reducible then* $G/ \pm I \simeq A_4$.
- *If* $L^{\circledS 3}$ *is reducible and* $L^{\circledS 4}$ *is irreducible then* $G/ \pm I \simeq S_4$.
- *If* $L^{\circledS 4}$ *is reducible, and* $L^{\circledS 6}$ *is irreducible, then* $\Rightarrow G/ \pm I \simeq A_5$.
- *If* $L^{\circledS 6}$ *is irreducible then* $G \simeq \mathrm{SL}_2(C)$.

In order to implement these criteria, one needs good algorithms to factor linear operators. Such algorithms exist and I refer to Chapter 4.2 of [5] for descriptions of some algorithms that factor linear operators as well as further references to the extensive literature. The calculation of symmetric powers and factorization of operators has been implemented in MAPLE in the DEtools package.

One can develop similar criteria for higher order operators. Given a differential operator L of order n with $\{y_1, \ldots, y_n\}$ being a basis of the solution space V in some Picard-Vessiot extension, one can consider the C-space V_m of homogeneous polynomials of degree m in the y_i. This space has dimension at most $\begin{pmatrix} m + n - 1 \\ n - 1 \end{pmatrix}$ but may have dimension that is strictly smaller and so is not isomorphic to the mth symmetric power of V (although it is isomorphic to a quotient). This complicates matters but this situation can be dealt with by casting things directly in terms of differential modules, forming the symmetric powers of differential modules and developing algorithms to find submodules of differential modules. This is explained in Chapter 4.2 of [5]. Examples of criteria for third order equations are given in Chapter 4.3.5 of [5] where further references are given as well.

Solving $L(y) = 0$ in terms of exponentials, integrals and algebraics. I shall give a formal definition that captures the meaning of this term.

Definition 1.5.6 *Let* k *be a differential field,* $L \in k[\partial]$, *and* K *be the associated Picard-Vessiot extension.*

1) A liouvillian tower *over* k *is a tower of differential fields* $k = K_0 \subset \ldots \subset K_m$ *with* $K \subset K_m$ *and, for each* $i, 0 \le i < m$, $K_{i+1} = K_i(t_i)$, *with*

- t_i *algebraic over* K_i, *or*
- $t_i' \in K_i$, *i.e.,* $t_i = \int u_i, u_i \in K_i$, *or*
- $t_i'/t_i \in K_i$, *i.e.,* $t_i = e^{\int u_i}, u_i \in K_i$

2) An element of a liouvillian tower is said to be liouvillian over k.

3) $L(y) = 0$ is solvable in terms of liouvillian functions if the associated PV-extension K lies in a liouvillian tower over k.

Picard and Vessiot stated the following Galois theoretic criteria and this was given a formal modern proof by Kolchin [52, 3].

Theorem 1.5.7 *Let k, L, and K be as above. $L(y) = 0$ is solvable in terms of liouvillian functions if and only if the identity component (in the Zariski topology) of $\mathrm{DGal}(K/k)$ is solvable.*

Recall that a group G is solvable if there exists a tower of subgroups $G = G_0 \supset G_1 \supset \ldots \supset G_m = \{e\}$ such that G_{i+1} is normal in G_i and G_{i+1}/G_i is abelian. This above theorem depends on the

Theorem 1.5.8 *(Lie-Kolchin Theorem) Let C be an algebraically closed field. A Zariski connected solvable group $G \subset \mathrm{GL}_n(C)$ is conjugate to a group of triangular matrices. In particular, G leaves a one dimensional subspace invariant.*

I shall show how the Lie-Kolchin Theorem can be strengthened resulting in a strengthened version of Theorem 1.5.7 that leads to an algorithm to decide if a given linear differential equation (over $C(x)$) can be solved in terms of liouvillian functions. I begin with

Proposition 1.5.9 *Let k be a differential field, $L \in k[\partial]$, K the corresponding Picard-Vessiot extensions and G the differential Galois group of K over k. The following are equivalent:*

(i) *$L(y) = 0$ has a liouvillian solution $\neq 0$.*

(ii) *G has a subgroup H, $|G : H| = m < \infty$ such that H leaves a one dimensional subspace invariant.*

(iii) *$L(y) = 0$ has a solution $z \neq 0$ such that z'/z is algebraic over k of degree $\leq m$.*

Proof (Outline; see Chapter 4.3 of [5]) (iii) $\Rightarrow$(i): Clear.

(i) $\Rightarrow$(ii): One can reduce this to the case where all solutions are liouvillian. In this case, the Lie-Kolchin Theorem implies that the identity component G^0 of G leaves a one dimensional subspace invariant.

(ii) $\Rightarrow$(iii): Let $V = Soln(L)$ and $v \in V$ span an H-invariant line. We then have that $\forall \sigma \in H, \exists c_\sigma \in C_k$ such that $\sigma(v) = c_\sigma v$. This implies that $\forall \sigma \in H, \sigma(\frac{v'}{v}) = \frac{(cv)'}{cv} = \frac{v'}{v}$. Therefore, the Fundamental Theorem implies that $\frac{v'}{v} \in E =$ fixed field of H. One can show that $[E : k] = |G : H| = m$ so $\frac{v'}{v}$ is algebraic over k of degree at most m. $\square$

In order to use this result, one needs a bound on the integer m that can appear in (ii) and (iii) above. This is supplied by the following group theoretic result (see Chapter 4.3.1 of [5] for references).

Lemma 1.5.10 *There exists a function $I(n)$ such that if $G \subset \mathrm{GL}_n(\mathbb{C})$, $H \subset G$ satisfies*

(i) $|G : H| < \infty$ and

(ii) H leaves a one dimension subspace invariant

then there exists a subgroup $\tilde{H} \subset G$ such that

(i) $|G : \tilde{H}| < I(n)$ and

(ii) $\tilde{H}$ leaves a one dimensional subspace invariant.

In general we know that $I(n) \leq n^{2n^2+2}$. For small values of n we have exact values, (*e.g.*, I(2) = 12, I(3) = 360). This clearly allows us to deduce the following corollary to Proposition 1.5.9.

Corollary 1.5.11 *Let k, L, K be as in Proposition 1.5.9 with the order of L equal to n. The following are equivalent*

(i) $L(y) = 0$ has a liouvillian solution $\neq 0$.

(ii) G has a subgroup $H, |G : H| \leq I(n)$ such that H leaves a one dimensional subspace invariant.

(iii) $L(y) = 0$ has a solution $z \neq 0$ such that z'/z is algebraic $/k$ of deg $\leq I(n)$.

This last result leads to several algorithms to decide if a differential equation $L(y) = 0$, $L \in C(x)[\partial]$, C a finitely generated subfield of $\mathbb{C}$, has a liouvillian solution. The one presented in [53] (containing ideas going back to [54]) searches for a putative minimal polynomial $P(u) = a_m u^m + a_{m-1} u^{m-1} + \ldots + a_0$, $a_i \in \mathbb{C}[x]$ of an element $u = z'/z$ where z is a solution of $L(y) = 0$ and $m \leq I(n)$. This algorithm shows how the degrees of the a_i in x can be bounded in terms of information calculated at each singular point of L. Once one has degree bounds, the actual coefficients a_i can be shown to satisfy a system of polynomial equations and one can (in theory) use various techniques (*e.g.*, Gröbner bases) to solve these. Many improvements and new ideas have been given since then (see Chapter 4 of [5]). We shall present criteria that form the basis of one method, describe what one needs to do to use this in general and give details for finding liouvillian solutions of second order differential equations.

Proposition 1.5.12 *(1) Let G be a subgroup of GL_n. There exists a subgroup $H \subset G, |G : H| \leq I(n)$ with H leaving a one dimensional space invariant if and only if G permutes a set of at most $I(n)$ one dimensional subspaces.*

(2) Let k, L, K, G be as in Proposition 1.5.9 with the order of L equal to n. Then $L(y) = 0$ has a liouvillian solution if and only if $G \subset \mathrm{GL}_n$ permutes a set of at most $I(n)$ one dimensional subspaces.

Proof (1) Let ℓ be a one dimensional space left invariant by H. The orbit of ℓ under the action of G has dimension at most $|G : H|$. Now, let ℓ be a one dimensional space whose orbit under G is at most $I(n)$ and let H be the stabilizer of ℓ. We then have $|G : H| \leq I(n)$. (2) is an immediate consequence of (1) and the previous proposition. $\qquad\square$

To apply this result, we need the following definition:

Definition 1.5.13 *Let V be a vector space of dimension n and $t \geq 1$ be an integer. The tth symmetric power $\mathrm{Sym}^t(V)$ of V is the quotient of the $t-fold$ tensor product $V^{\otimes t}$ by the subspace generated by elements $v_1 \otimes \ldots \otimes v_t - v_{\pi(1)} \otimes \ldots \otimes v_{\pi(t)}$, for all $v_i \in V$ and π a permutation.*

We denote by $v_1 v_2 \cdots v_t$ the image of $v_1 \otimes \ldots \otimes v_t$ in $\mathrm{Sym}^t(V)$. If $e_1, \ldots, e_n$ is a basis of V and $1 \leq t \leq n$, then $\{e_{i_1} e_{i_2} \cdots e_{i_t} \mid i_1 < i_2 < \ldots < i_t\}$ is a basis of $\mathrm{Sym}^t(V)$. Furthermore, if $G \subset \mathrm{GL}_n(V)$, then G acts on $\mathrm{Sym}^t(V)$ as well via the formula $\sigma(v_1 \cdots v_t) = \sigma(v_1) \cdots \sigma(v_t), \sigma \in G$.

Definition 1.5.14 *(1) An element $w \in \mathrm{Sym}^t(V)$ is decomposable if $w = w_1 \cdots w_t$ for some $w_i \in V$.*

(2) A one dimensional subspace $\ell \subset \mathrm{Sym}^t(V)$ is decomposable if ℓ is spanned by a decomposable vector.

We note that if we fix a basis $\{e_1, \ldots, e_n\}$ of V then an element

$$\sum_{i_1 < \ldots < i_t} c_{i_1 \ldots i_t} e_{i_1} \cdots e_{i_t}$$

is decomposable if and only if the $c_{i_1 \ldots i_t}$ satisfy a system of equations called the *Brill equations* ([55], p. 120,140). Using these definitions, it is not hard to show

Proposition 1.5.15 *(1) A group $G \subset \mathrm{GL}_n(V)$ permutes a set of t one dimensional subspaces if and only if G leaves a decomposable one dimensional subspace of $\mathrm{Sym}^t(V)$ invariant.*

(2) Let k, L, K, G be as in Proposition 1.5.9 with the order of L equal to n. Then $L(y) = 0$ has a liouvillian solution if and only if for some $t \leq I(n)$ G leaves invariant a one dimensional subspace of $\mathrm{Sym}^t(\mathrm{Soln}(L))$.

Note that $\mathrm{Sym}^t(V)$ is constructed using tensor products and quotients. When we apply this construction to differential modules, we get

Definition 1.5.16 *Let k be a differential field and M a differential module over k. The tth symmetric power $\mathrm{Sym}^t(M)$ of M is the quotient of the t-fold tensor product $M^{\otimes t}$ by the k-subspace generated by elements $v_1 \otimes \ldots \otimes v_t - v_{\pi(1)} \otimes \ldots \otimes v_{\pi(t)}$, for all $v_i \in V$ and π a permutation (note that this subspace is a differential submodule as well).*

At the end of Section 1.2, I defined the solution space of a k-differential module M in a differential field $K \supset k$ to be $\mathrm{Soln}_K(M) = \ker(\partial, K \otimes_k M)$. From our discussion of tannakian categories, we have that $\mathrm{Soln}_K(\mathrm{Sym}^t(M)) = \mathrm{Sym}^t(\mathrm{Soln}_K(M))$. This latter fact suggests the following algorithm to decide if a linear differential equation has a liouvillian solution (we continue to work with scalar equations although everything generalizes easily to systems). Let M_L be the differential module associated to the equation $L(y) = 0$. For each $t \leq I(n)$

- Calculate $N_t = \mathrm{Sym}^t(M)$ and find all one dimensional submodules.
- Decide if any of these is decomposable as a subspace of N_t.

The tannakian formalism implies that a one dimensional submodule of N_t is decomposable if and only if its solution space is a decomposable G-invariant subspace of $\mathrm{Sym}^t(\mathrm{Soln}_K(M_L))$.

Much work has been done on developing algorithms for these two steps. The problem of finding one dimensional submodules of a differential module was essentially solved in the 19th and early 20th Centuries (albeit in a different language). More recently work of Barkatou, Bronstein, van Hoeij, Li, Weil, Wu, Zhang and others has produced good algorithms to solve this problem (see [56], [57], [58] and Chapter 4.3.2 of [5] for references). As noted above, the second problem can be solved (in principle)using the Brill equations (see [59] and Chapter 4.3.2 of [5] for a fuller discussion). Finally, one can modify the above to produce the liouvillian solutions as well.

Before I explicate these ideas further in the context of second order equations, I will say a few words concerning finding invariant one dimensional subspaces of solution spaces of linear scalar equations.

Proposition 1.5.17 *Let k be a differential field with algebraically closed constants C. $L \in k[\partial]$, K the Picard-Vessiot extension of k for $L(Y) = 0$ and G the differential Galois group of K over k. An element z in the solution space V of $L(Y) = 0$ spans a one dimensional G-invariant subspace if and only if $u = z'/z$ is left fixed by G. In this case, $\partial - u$ is a right divisor of L. Conversely if $\partial - u, u \in k$, is a right divisor of L, then there exists a solution $z \in K$ such that $z'/z = u$, in which case, z spans a G-invariant space.*

Proof This follows from Proposition 1.3.28 and its proof applied to first order factors of L.   □

Therefore, to find one dimensional G-invariant subspaces of the solution space of $L(Y) = 0$, we need to be able to find elements $u \in k$ such that $z = e^{\int u}$ satisfies $L(z) = 0$. This is discussed in detail in Chapter 4.1 of [5] and I will give a taste of the idea behind a method to do this for $L \in \mathbb{C}(x)$ assuming L has only regular singular points.

Let $u \in \mathbb{C}(x)$ satisfy $L(e^{\int u}) = 0$ and let $x = \alpha$ be a pole of u. The assumption that x_0 is at worst a regular singular point implies that $e^{\int u}$ is dominated by a power of $(x - \alpha)$ near α. We must therefore have that u has a pole of order at most 1 at $x = x_0$ and so $z = e^{\int u} = (x - \alpha)^a h(x)$ where $a \in \mathbb{C}$ and h is analytic near $\alpha \in \mathbb{C}$. To determine a, we write

$$L(Y) = Y^{(n)} + (\frac{b_{n-1}}{(x - \alpha)} + \text{h.o.t.})Y^{(n-1)} + \ldots + (\frac{b_0}{(x - \alpha)^{n-1}} + \text{h.o.t.})Y.$$

Substituting $Y = (x - \alpha)^a(c_0 + c_1(x - \alpha) + \text{h.o.t.})$ and setting the coefficient of $(x - \alpha)^{a-n}$ equal to zero, we have

$$c_0(a(a-1)\ldots(a-(n-1)) + b_{n-1}(a(a-1)\ldots(a-(n-2)) + \ldots + b_0)) = 0$$

or

$$a(a-1)\ldots(a-(n-1)) + b_{n-1}(a(a-1)\ldots(a-(n-2)) + \ldots + b_0 = 0$$

This latter equation is called the *indicial equation* at $x = \alpha$ and its roots are called the *exponents at* α. If α is an ordinary point, then $b_{n-1} = \ldots = b_0 = 0$ so $a \in \{0, \ldots, n-1\}$. One can also define exponents at ∞. We therefore have that

$$y = \prod_{\alpha_i = \text{ finite sing. pt.}} (x - \alpha_i)^{a_i} P(x)$$

where the a_i are exponents at α_i and $-\sum a_i - \deg P$ is an exponent at ∞. The a_i and the degree of P are therefore determined up to a finite set of choices. Note that P is a solution of $\tilde{L}(Y) = \prod(x - \alpha_i)^{-a_i} L(\prod(x - \alpha_i)^{a_i} Y)$. Finding polynomial solutions of $\tilde{L}(Y) = 0$ of a fixed degree can be done by substituting a polynomial of that degree with undetermined coefficients and equating powers of x to reduce this to a problem in linear algebra.

Liouvillian Solutions of Second Order Equations. The method outlined above can be simplified for second order equations $L(Y) = Y'' - sY$ because of several facts that we summarize below (see Chapter 4.3.4 of [5]). The resulting algorithm is essentially the algorithm presented by Kovacic in [60] but put in the context of the general algorithm mentioned above. Kovacic's algorithm predated and motivated much of the work on liouvillian solutions of general linear differential equation presented above.

- It can be shown that the fact that no Y' term appears implies that the differential Galois group must be a subgroup of SL_2 (Exercise 1.35.5, p. 27 [5]).
- An examination of the algebraic subgroups of SL_2 implies that SL_2 has no subgroup of finite index leaving a one dimensional subspace invariant and any proper algebraic subgroup of SL_2 has a subgroup of index $1, 2, 4, 6$, or 12 that leaves a one dimensional subspace invariant (*c.f.*, [60]).
- As shown in the paragraph following Definition 1.5.3, the dimension of the solution space of $L^{\circledS t}$ is the same as the dimension of $\mathrm{Sym}^t(\mathrm{Soln}(L))$ and so these two spaces are the same.
- Any element $z = \sum_{i=0}^{t} y_1^{t-i} y_2^{i} \in \mathrm{Sym}^t(\mathrm{Soln}(L))$ can be written as a product $\prod_{i=0}^{t}(c_i y_1 + d_i y_2)$ and so all elements of $\mathrm{Sym}^t(\mathrm{Soln}(L))$ are decomposable.

Combining these facts with the previous results we have

$$L(y) = 0 \text{ has a nonzero liouvillian solution}$$

$$\Updownarrow$$

The differential Galois group G permutes $1, 2, 4, 6$ or 12 one dimensional subspaces in $Sol(L)$

$$\Updownarrow$$

For $t = 1, 2, 4, 6,$ or 12, G leaves invariant a line in $Sol(L^{\circledS t})$

$$\Updownarrow$$

For $t = 1, 2, 4, 6,$ or 12, $L^{\circledS t}$ has a solution z such that $u = z'/z \in k$

So, to check if $Y'' - sY = 0$ has a nonzero liouvillian solution (and therefore, by variation of parameters, only liouvillian solutions), one needs to check this last condition.

Example 1.5.18

$$L(Y) = Y'' + \frac{3(x^2 - x + 1)}{16(x-1)^2 x^2} Y$$

This is an equation with only regular singular points. One can show that $L(Y) = 0$ has no nonzero solutions z with $z'/z \in \mathbb{C}(x)$. The second symmetric power is

$$L^{\circledS 2}(Y) = Y'' + 3/4 \frac{(x^2 - x + 1)}{x^2 (x-1)^2} Y' - 3/8 \frac{(2x^3 - 3x^2 + 5x - 2)}{x^3 (x-1)^3} Y.$$

The singular points of this equation are at $0, 1$ and ∞. The exponents there are

At 0 : $\{1, \frac{1}{2}, \frac{3}{2}\}$
At 1 : $\{1, \frac{1}{2}, \frac{3}{2}\}$
At ∞ : $\{-1, -\frac{1}{2}, -\frac{3}{2}\}$

For $Y = x^{a_0}(x-1)^{a_1} P(x)$, try $a_0 = \frac{1}{2}, a_1 = 1$, $\deg P = 0$. One sees that $Y = x^{\frac{1}{2}}(x-1)$ is a solution of $L^{\circledS 2}(y) = 0$. Therefore $L(Y) = 0$ is solvable in terms of liouvillian functions.

For second order equations one can also easily see how to modify the above to find liouvillian solutions when they exist. Assume that one has found a nonzero element $u \in k$ such that $L^{\circledS m}(Y) = 0$ has a solution z with $z'/z = u$. We know that

$$z = y_1 \cdot \ldots \cdot y_m, \quad y_1, \ldots, y_m \in Soln(L) .$$

Since $\sigma(z) \in C \cdot z$ for all $\sigma \in G$, we have that for any $\sigma \in Gal(L)$, there exists a permutation π and constants c_i such that $\sigma(y_i) = c_i y_{\pi(i)}$.

Therefore, for $v_i = \frac{y_i'}{y_i}$ and $\sigma \in G$ we have $\sigma(v_i) = v_{\pi(i)}$, *i.e.*, G permutes the v_i. Let

$$P(Y) = \prod_{i=1}^{m}(Y - v_i) = Y^m + a_{m-1}Y^{m-1} + \frac{a_{m-2}}{2!}Y^{m-2} + \ldots + \frac{a_0}{m!}.$$

The above reasoning implies that the $a_i \in k$ and

$$a_{m-1} = -(v_1 + \ldots + v_m) = -(\frac{y_1'}{y_1} + \ldots + \frac{y_m'}{y_m})$$

$$= -\frac{(y_1 \cdot \ldots \cdot y_m)'}{y_1 \cdot \ldots \cdot y_m} = -\frac{z'}{z} = -u$$

The remaining a_i can be calculated from $a_{m-1} = -u$ using the following fact from [61], [60] and [62] (see also Chapter 4.3.4 of [5]).

Lemma 1.5.19 *Using the notation above, we have*

$$a_{i-1} = -a_i' - a_{m-i}a_i - (m - i)(i + 1)sa_{i+1}, \quad i = m - 1, \ldots, 0$$

where $a_{-1} = 0$.

In particular we can find a nonzero polynomial satisfied by y'/y for some nonzero solution of $L(y) = 0$. This gives a liouvillian solution and variation of parameters yields another.

Example 1.5.20 *We continue with the example*

$$L(Y) = Y'' + \frac{3(x^2 - x + 1)}{16(x - 1)^2 x^2}Y$$

We have that $y = x^{\frac{1}{2}}(x - 1)$ is a soln of $L^{\otimes 2}(Y) = 0$. Let

$$a_1 = -\frac{y'}{y} = -(\frac{1}{2x} + \frac{1}{x - 1}) = -\frac{3x - 1}{2x(x - 1)},$$

then if v is a root of

$$P(Y) = Y^2 - \frac{3x - 1}{2x(x - 1)}Y + \frac{9x^2 - 7x + 1}{16x^2(x - 1)^2}$$

we have $y = e^{\int v}$ satisfies $L(y) = 0$. This yields

$$\sqrt[4]{x - x^2}\sqrt{1 + \sqrt{x}} \quad and \sqrt[4]{x - x^2}\sqrt{1 - \sqrt{x}}$$

as solutions. DGal$= D_4 =$ symmetries of a square.

Another approach to second order linear differential equations was discovered by Felix Klein. It has been put in modern terms by Dwork and Baldassari and made more effective recently by van Hoeij and Weil (see [63] for references).

Solving in Terms of Lower Order Equations. Another way of defining the notion of "solving in terms of liouvillian functions" is to say that a linear differential equation can be solved in terms of solutions of first order equations $Y' + aY = b$ and algebraic functions. With this in mind it is natural to make the following definition.

Definition 1.5.21 *Let k be a differential field, $L \in k[\partial]$ and K the associated Picard-Vessiot extension. The equation $L(Y) = 0$ is solvable in terms of lower order linear equations if there exists a tower $k = K_0 \subset K_1 \subset \ldots \subset K_m$ with $K \subset K_m$ and for each $i, 0 \leq i < m$, $K_{i+1} = K_i < t_i >$ (K_{i+1} is generated as a differential field by t_i and K_i), with*

- *t_i algebraic over K_i, or*
- *$L_i(t_i) = b_i$ for some $L_i \in K_i[\partial]$ $\mathrm{ord}(L_i) < \mathrm{ord}(L)$, $b_i \in K_i$*

One has the following characterization of this property in terms of the differential Galois group.

Theorem 1.5.22 *([64]) Let k, L, K be as above with $\mathrm{ord}(L) = n$. The equation $L(Y) = 0$ is <u>not</u> solvable in terms of lower order linear equations if and only if the lie algebra of $\mathrm{DGal}(K/k)$ is simple and has no faithful representations of dimension less than n.*

If $\mathrm{ord}(L) = 3$, then an algorithm to determine if this equation is solvable in terms of lower order linear equations is given in [65]. Refinements of this algorithm and extensions to higher order L are given in [66], [67], [68], [69], [70].

1.6 Inverse Problems

In this section we consider the following problem:

Given a differential field k, characterize those linear differential algebraic groups that appear as differential Galois groups of Picard-Vessiot extensions over k.

We begin with considering the inverse problem over fields of constants.

Differential Galois groups over $\mathbb{C}$. Let $L \in \mathbb{C}[\partial]$ be a linear differential operator with constant coefficients. All solutions of $L(Y) = 0$, $L \in \mathbb{C}[\partial]$ are of the form $\sum_i P_i(x)e^{\alpha_i x}$ where $P_i(x) \in \mathbb{C}[x]$, $\alpha_i \in \mathbb{C}$, $x' = 1$. This implies that the associated Picard-Vessiot K is a subfield of a field $E = \mathbb{C}(x, e^{\alpha_1 x}, \dots, e^{\alpha_t x})$. One can show that $\mathrm{DGal}(E/k) = (\mathbb{C}, +) \times ((\mathbb{C}^*, \cdot) \times \dots \times (\mathbb{C}^*, \cdot))$. The group $\mathrm{DGal}(K/\mathbb{C})$ is a quotient of this latter group and one can show that it therefore must be of the form $(\mathbb{C}, +)^a \times (\mathbb{C}^*, \cdot)^b, a = 0, 1, \ b \in \mathbb{N}$.

One can furthermore characterize in purely group theoretic terms the groups that appear as differential Galois groups over $\mathbb{C}$. Note that

$$(\mathbb{C}, +) \simeq \left\{ \begin{pmatrix} 1 & a \\ 0 & 1 \end{pmatrix} \mid a \in \mathbb{C} \right\}$$

and that all these elements are unipotent matrices (*i.e.*, $(A - I)^m = 0$ for some $m \neq 0$). This motivates the following definition.

Definition 1.6.1 *If G is a linear differential algebraic group, the* unipotent radical R_u *of G is the largest normal subgroup of G all of whose elements are unipotent.*

Note that for a group of the form $G = \mathbb{C} \times (\mathbb{C}^* \times \dots \times \mathbb{C}^*)$ we have that $R_u(G) = \mathbb{C}$. Using facts about linear algebraic groups (see [71] or [23]) one can characterize those linear algebraic groups that appear as differential Galois groups over an algebraically closed field of constants C as

Proposition 1.6.2 *A linear algebraic group G is a differential Galois group of a Picard-Vessiot extension of $(\mathbb{C}, \partial), \partial c = \forall c \in \mathbb{C}$ if and only if G is connected, abelian and $R_u(G) \leq 1$.*

All of the above holds equally well for any algebraically closed field C.

Differential Galois groups over $\mathbb{C}(x), x' = 1$. The inverse problem for this field was first solved by C. and M. Tretkoff, who showed

Theorem 1.6.3 *[72] Any linear algebraic group is a differential Galois group of a Picard-Vessiot extension of $\mathbb{C}(x)$.*

Their proof (which I will outline below) depends on the solution of Hilbert's 21st Problem. This problem has a weak and strong form. Let $\mathcal{S} = \{\alpha_1, \dots, \alpha_m, \infty\} \subset S^2$, $\alpha_0 \in S^2 - \mathcal{S}$ and $\rho : \pi_1(S^2 - \mathcal{S}, \alpha_0) \rightarrow$

$GL_n(\mathbb{C})$ be a homomorphism.

<u>Weak Form</u> Does there exist $A \in gl_n(\mathbb{C}(x))$ such that

- $Y' = AY$ has only regular singular points, and
- The monodromy representation of $Y' = AY$ is ρ.

<u>Strong Form</u> Do there exist $A_i \in GL_n(\mathbb{C})$ such that the monodromy representation of

$$Y' = (\frac{A_1}{x - \alpha_1} + \ldots + \frac{A_m}{x - \alpha_m})Y$$

is ρ.

We know that a differential equation $Y' = AY$ that has a regular singular point is locally equivalent to one with a simple pole. The strong form of the problem insists that we find an equation that is *globally* equivalent to one with only simple poles. A solution of the strong form of the problem of course yields a solution of the weak form. Many special cases of the strong form were solved before a counterexample to the general case was found by Bolibruch (see [73], [74] or Chapters 5 and 6 of [5] for a history and exposition of results and [75] for more recent work and generalizations).

  (i) Let $\gamma_1, \ldots, \gamma_m$ be generators of $\pi_1(S^2 - \mathcal{S}, \alpha_0)$, each enclosing just one α_i. If some $\rho(\gamma_i)$ is diagonalizable, then the answer is yes. (Plemelj)
  (ii) If all $\rho(\gamma_i)$ are sufficiently close to I, the answer is yes. (Lappo-Danilevsky)
  (iii) If $n = 2$, the answer is yes. (Dekkers)
  (iv) If ρ is irreducible, the answer is yes. (Kostov, Bolibruch)
  (v) Counterexample for $n = 3, |\mathcal{S}| = 4$ and complete characterization for $n = 3, 4$ (Bolibruch, Gladyshev)

In a sense, the positive answer to the weak form follows from Plemelj's result above - one just needs to add an additional singular point α_{m+1} and let γ_{m+1} be the identity matrix. A modern approach to give a positive answer to the weak form was given by Röhrl and later Deligne (see [15]). We now turn to

Proof of Theorem 1.6.3 (outline, see [72] of Chapter 5.2 or [5] for details) Let G be a linear algebraic group. One can show that there exist $g_1, \ldots, g_m \in G$ that generate a Zariski dense subgroup H. Select points $\mathcal{S} = \{\alpha_1, \ldots \alpha_m\} \subset S^2$ and define $\rho : \pi_1(S^2 - \mathcal{S}, \alpha_0) \rightarrow H$ via $\rho(\gamma_i) = g_i$. Using the solution of the weak form of Hilbert's 21st, there exist $A \in \mathrm{gl}_n(\mathbb{C}(x))$ such that $Y' = AY$ has regular singular points and monodromy group H.Schlesinger's Theorem, Theorem 1.3.27, implies that the differential Galois group of this equation is the Zariski closure of H. $\square$

The above result leads to the following two questions:

- What is the minimum number of singular points (not necessarily regular singular points) that a linear differential system must have to realize a given group as its Galois group?
- Can we realize all linear algebraic groups as differential Galois groups over $C(x), x' = 1$ where C is an arbitrary algebraically closed field?

To answer the first question, we first will consider

Differential Galois groups over $\mathbb{C}(\{x\})$. To characterize which groups occur as differential Galois groups over this field, we need the following definitions.

Definition 1.6.4 *Let C be an algebraically closed field.*

(1) A torus *is a linear algebraic group isomorphic to $(C^*, \cdot)^r$ for some r.*

(2) If G is a linear algebraic group then we define $L(G)$ to be the group generated by all tori in G. This is a normal, Zariski closed subgroup of G (see [76] and Chapter 11.3 of [5]).

Examples 1.6.5 *(1) If G is reductive and connected (e.g., tori, $\mathrm{GL}_n, \mathrm{SL}_n$), then $L(G) = G$ (A group G is reductive if $R_u(G) = \{I\}$).*

(2) If $G = (C, +)^r, r \geq 2$, then $L(G) = \{I\}$.

Ramis [76] showed the following (see also Chapter 11.4 of [5]).

Theorem 1.6.6 *A linear algebraic group G is a differential Galois group of a Picard-Vessiot extension of $\mathbb{C}(\{x\})$ if and only if $G/L(G)$ is the Zariski closure of a cyclic group.*

Any linear algebraic group G with G/G^0 cyclic and G reductive satisfies these criteria while $G = (C, +)^r, r \geq 2$ does not. Ramis showed that his characterization of local Galois groups in terms of formal monodromy, exponential torus and Stokes matrices yields a group of this type and conversely any such group can be realized as a local Galois group. Ramis was furthermore able to use analytic patching techniques to get a global version of this result.

Theorem 1.6.7 *A linear algebraic group G is a differential Galois group of a Picard-Vessiot extension of $\mathbb{C}(x)$ with at most $r-1$ regular singular points and one (possibly) irregular singular point if and only if $G/L(G)$ is the Zariski closure of a group generated by $r-1$ elements.*

Examples 1.6.8 *(1) We once again have that any linear algebraic group is a differential Galois group over $\mathbb{C}(x)$ since such a group satisfies the above criteria for some r.*

(2) SL_n can be realized as a differential Galois group of a linear differential equation over $\mathbb{C}(x)$ with only one singular point.

(3) One needs r singular points to realize the group $(\mathbb{C}, +)^{r-1}$ as a differential Galois group over $\mathbb{C}(x)$.

Differential Galois groups over $C(x)$, C an algebraically closed field. The proofs of the above results depend on analytic techniques over $\mathbb{C}$ that do not necessarily apply to general algebraically closed fields C of characteristic zero. The paper [77] examines the question of the existence of "transfer principles" and shows that for certain groups G, the existence of an equation over $\mathbb{C}(x)$ having differential Galois group G over $\mathbb{C}(x)$ implies the existence of an equation over $C(x)$ having differential Galois group G over $C(x)$ but these results do not apply to all groups. Algebraic proofs for various groups over $C(x)$ (in fact over any differential field finitely generated over its algebraically closed field of constants) have been given over the years by Bialynicki-Birula (nilpotent groups), Kovacic (solvable groups), Mitschi/Singer(connected groups) (see [78] for references). Finally, Hartmann [79] showed that any linear algebraic group can be realized over $C(x)$ (see also [80]). An exciting recent development is the announcement of a new proof by Harbater and Hartmann [81, 82] of this fact based on a generalization of algebraic patching techniques. I will give two examples of a constructive technique (from [78]) that allows one to produce explicit equations for connected linear algebraic groups. This technique is based on (see Proposition 1.31 of [5])

Proposition 1.6.9 *Let $Y' = AY$ be a differential equation over a differential field k with algebraically closed constants C. Let G, H be linear* *algebraic subgroups of* $\mathrm{GL}_n(C)$ *with lie algebras* $\mathcal{G}, \mathcal{H}$. *Assume* G *is connected.*

1) Let $Y' = AY$ *is a differential equation with* $A \in \mathcal{G} \otimes_C k$ *and assume that* G *is connected. Then the differential Galois group of this equation is conjugate to a subgroup of* G.

(2) Assume $k = C(x)$, *that* $A \in \mathcal{G} \otimes_C C(x)$ *and that the differential Galois group of* $Y' = AY$ *is* $H \subset G$, *where* H *is assumed to be connected. Then there exists* $B \in G(C(x))$ *such that* $\tilde{A} = B^{-1}AB - B^{-1}B' \in \mathcal{H} \otimes_C C(x)$, *that is* $Y' = AY$ *is equivalent to an equation* $Y' = \tilde{A}Y$ *with* $\tilde{A} \in \mathcal{G} \otimes_C C(x)$ *and the equivalence is given by an element of* $G(C(x))$.

These results allow us to formulate the following strategy to construct differential equations with differential Galois group over $C(x)$ a given connected group G. Select $A \in \mathcal{G} \otimes_C C(x)$ such that

(i) the differential Galois group of $Y' = AY$ over $C(x)$ is connected, and

(ii) for any proper connected linear algebraic subgroup H of G, there is no $B \in G(C(x))$ such that $B^{-1}AB - B^{-1}B' \in \mathcal{H} \otimes_C C(x)$.

We will assume for convenience that $C \subset \mathbb{C}$. To insure that the differential Galois group is connected, we will select an element $A \in \mathcal{G} \otimes_C C[x]$, that is a matrix with *polynomial entries*. This implies that there is a fundamental solution matrix whose entries are entire functions. Since these functions and their derivatives generate the associated Picard-Vessiot extension K, any element of K will be a function with at worst poles on the complex plane. If $E, C(x) \subset E \subset K$ is the fixed field of G^0, then the elements of E are algebraic functions that can only be ramified at the singular points of $Y' = AY$, that is, only at ∞. But such an algebraic function must be rational so $E = C(x)$ and therefore $G = G^0$.

The next step is to select an $A \in \mathcal{G} \otimes_C C[x]$ satisfying condition (ii) above. This is the heart of [78] and we shall only give two examples.

Example 1.6.10 $G = C^* \times C^* =$

$$\{ \begin{pmatrix} a & 0 \\ 0 & b \end{pmatrix} \mid ab \neq 0 \}$$

If $\mathcal{G}$ *is the lie algebra of* G *then* $\mathcal{G} \otimes_C C[x] =$

$$\{ \begin{pmatrix} f_1 & 0 \\ 0 & f_2 \end{pmatrix} \mid f_1, f_2 \in C[x] \}$$

The proper closed subgroups of G *are of the form*

$$G_{m,n} = \left\{ \begin{pmatrix} c & 0 \\ 0 & d \end{pmatrix} \mid c^n d^m = 1 \right\},$$

m, n *integers, not both zero. The connected subgroups are those* $G_{m,n}$ *with* m, n *relatively prime. If* $\mathcal{G}_{m,n}$ *is the lie algebra of* $G_{m,n}$ *then* $\mathcal{G}_{m,n} \otimes_C C(x) =$

$$\left\{ \begin{pmatrix} g_1 & 0 \\ 0 & g_2 \end{pmatrix} \mid mg_1 + ng_2 = 0 \right\}$$

We need to find $A = \begin{pmatrix} f_1 & 0 \\ 0 & f_2 \end{pmatrix}$, $f_1, f_2 \in C[x]$ *such that for any* $B = \begin{pmatrix} u & 0 \\ 0 & v \end{pmatrix}$, $u, v \in C(x)$, $BAB^{-1} - BB^{-1} \notin \mathcal{G}_{m,n} \otimes_C C(x)$, *that is*

$$m(f_1 - \frac{u'}{u}) + n(f_2 - \frac{v'}{v}) = 0 \Rightarrow m = n = 0.$$

It is sufficient to find $f_1, f_2 \in C[x]$ *such that*

$$mf_1 + nf_2 = \frac{h'}{h} \text{ for some } h \in C(x) \Rightarrow m = n = 0.$$

Selecting $f_1 = 1$ *and* $f_2 = \sqrt{2}$ *will suffice so*

$$Y' = \begin{pmatrix} 1 & 0 \\ 0 & \sqrt{2} \end{pmatrix} Y$$

has Galois group G.

Example 1.6.11 $G = \text{SL}_2$. *Its lie algebra is* $\text{sl}_2 = \text{trace } 0$ *matrices. Let*

$$A = A_0 + xA_1 = \begin{pmatrix} 0 & 1 \\ 1 & 0 \end{pmatrix} + x \begin{pmatrix} 1 & 0 \\ 0 & -1 \end{pmatrix}$$

It can be shown that any proper subalgebra of sl_2 *is solvable and therefore leaves a one dimensional subspace invariant. A calculation shows that no* $U'U^{-1} + UAU^{-1}$ *leaves a line invariant (see [78]) for details) and so this equation must have Galois group* SL_2.

This latter example can be generalized to any semisimple linear algebraic group G. Such a group can be realized as the differential Galois group of an equation of the form $Y' = (A_0 + xA_1)Y$ where A_0 is a sum of generators of the root spaces and A_1 is a sufficiently general element of the Cartan subalgebra. In general, we have

Theorem 1.6.12 *Let G be a connected linear algebraic group defined over an algebraic closed field C. One can construct $A_i \in \mathrm{gl}_n(C)$ and $A_\infty \in C[x]$ such that for distinct $\alpha_i \in C$*

$$Y' = (\frac{A_1}{x - \alpha_1} + \ldots + \frac{A_d}{x - \alpha_d} + A_\infty)Y$$

has Galois group G.

Furthermore, the number d in the above result coincides with the number of generators of a Zariski dense subgroup of $G/L(G)$ and the degree of the polynomials in A_∞ can be bounded in terms of the groups as well. For non-connected groups [83] gives a construction to realize solvable-by-finite groups and [84] gives a construction to realize certain semisimple-by-finite groups. Finally, many linear algebraic groups can be realized as differential Galois groups of members of classical families of differential equations [85, 86, 87, 88, 89, 90, 91, 92].

To end this section, I will mention a general inverse problem. We know that any Picard-Vessiot ring is the coordinate ring of a torsor for the differential Galois group. One can ask if every coordinate ring of a torsor for a linear algebraic group can be given the structure of a Picard-Vessiot ring. To be more precise, let k be a differential field withe derivation ∂ and algebraically closed constants C. Let G be a linear algebraic group defined over C and V a G-torsor defined over k.

Does there exist a derivation on $R = k[V]$ extending ∂ such that R is a Picard-Vessiot ring with differential Galois group G and such that the Galois action of G on R corresponds to the action induced by G on the torsor V?

When $k = C(x)$, this has can be answered affirmatively due to the work of [93]. For a general k, finitely generated over C, I do not know the answer. For certain groups and fields, Juan and Ledet have given positive answers, see [94, 95, 96, 97].

Finally, the inverse problem over $\mathbb{R}(x)$ has been considered by [98].

1.7 Families of Linear Differential Equations

In this section we will consider a family of parameterized linear differential equations and discuss how the differential Galois group depends on the parameter. I will start by describing the situation for polynomials

and the usual Galois groups. Let C be a field and $\overline{C}$ its algebraic closure.
Let $G \subset \mathcal{S}_m =$ fixed group of permutations and $\mathcal{P}(n, m, G) =$

$$\{P = \sum_{i=0}^{n} \sum_{j=0}^{m} a_{i,j} x^i y^j \mid a_{i,j} \in C, Gal(P/C(x)) = G\} \subset C^{(n+1)(m+1)}$$

To describe the structure of $\mathcal{P}(n, m, G)$, we need the following definition

Definition 1.7.1 *A set $S \subset \overline{C}^N$ is C-constructible if it is the finite union of sets*

$$\{a \in \overline{C}^N \mid f_1(a) = \ldots = f_t(a) = 0, g(a) \neq 0\}$$

where $f_i, g \in C[X_1, \ldots, X_N]$.

For example, any Zariski closed set defined over C is C-constructible and a subset $S \subset \overline{C}$ is $\overline{C}$-constructible if and only if it is finite or cofinite. In particular, $\mathbb{Q}$ is not $\mathbb{C}$-constructible in $\mathbb{C}$. We have the following results (see [99] and the references given there)

Theorem 1.7.2 $\mathcal{P}(m, n, G)$ *is $\mathbb{Q}$-constructible.*

From this we can furthermore deduce

Corollary 1.7.3 *Any finite group G is a Galois group over $\overline{\mathbb{Q}}(x)$.*

Proof (Outline) For $G \subset \mathcal{S}_m$, one can construct an m-sheeted normal covering of the Riemann Sphere with this as the group of deck transformations, [100]. The Riemann Existence Theorem implies that this Riemann Surface is an algebraic curve and that the Galois group of its function field over $\mathbb{C}(x)$ is G. Therefore for some n, $\mathcal{P}(n, m, G)$ has a point in $\mathbb{C}^{(n+1)(m+1)}$. Hilbert's Nullstellensatz implies it will have a point in $\overline{\mathbb{Q}}^{(n+1)(m+1)}$ and therefore that G is a Galois group over $\overline{\mathbb{Q}}(x)$. $\square$

Differential Galois Groups of Families of Linear Differential Equations. I would like to apply the same strategy to linear differential equations. We begin by fixing a linear algebraic group $G \subset \mathrm{GL}_m$ and set $\mathcal{L}(m, n, G) =$

$$\{L = \sum_{i=0}^{n} \sum_{j=0}^{m} a_{i,j} x^i \partial^j \mid a_{i,j} \in \mathbb{C}, Gal(L/\mathbb{C}(x)) = G\} \subset \mathbb{C}^{(n+1)(m+1)}$$

Regrettably, this set is not necessarily a $\mathbb{Q}$ or even a $\mathbb{C}$-constructible set.

Example 1.7.4 *We have seen that the Galois group of $\frac{dy}{dx} - \frac{\alpha}{x}y$ is* $GL_1(\mathbb{C})$ *if and only if $\alpha \notin \mathbb{Q}$. If $\mathcal{L}(1,1,GL_1)$ were $\mathbb{C}$-constructible then*

$$\mathcal{L}(1,1,GL_1) \cap \{x\partial - \alpha \mid \alpha \in \mathbb{C}\} = \mathbb{C}\backslash\mathbb{Q}$$

would be constructible. Therefore, even the set $\mathcal{L}(1,1,GL_1)$ is not $\mathbb{C}$-constructible.

Example 1.7.5 *Consider the family*

$$\frac{d^2y}{dx^2} - (\alpha_1 + \alpha_2)\frac{dy}{dx} + \alpha_1\alpha_2 y = 0 \qquad G = \mathbb{C}^* \oplus \mathbb{C}^*$$

This equation has solution $\{e^{\alpha_1 x}, e^{\alpha_2 x}\}$. Assume $\alpha_2 \neq 0$. If $\alpha_1/\alpha_2 \notin \mathbb{Q}$ then $DGal = \mathbb{C}^* \oplus \mathbb{C}^*$ while if $\alpha_1/\alpha_2 \in \mathbb{Q}$ then $DGal = \mathbb{C}^*$. Therefore, $\mathcal{L}(2,0,\mathbb{C}^* \oplus \mathbb{C}^*)$ is not $\mathbb{C}$-constructible.

Recall that for $L \in \mathbb{C}(x)[\partial]$ and x_0 a singular point $L(y) = 0$ has a fundamental set of solutions:

$$y_i = (x - \alpha)^{\rho_i} e^{P_i(1/t)}(\sum_{j=0}^{s_i} b_{ij}(\log(x - x_0)^j))$$

where $\rho_i \in \mathbb{C}$, $t^\ell = (x - x_0)$ for some $\ell \leq n!$, P_i are polynomials without constant term, $b_{ij} \in \mathbb{C}[[x - x_0]]$.

Definition 1.7.6 *The set $\mathcal{D}_{x_0} = \{\rho_1, \ldots, \rho_n, P_1, \ldots, P_n\}$ is called the* local data of L at x_0.

In Example 1.7.4, 0 and ∞ are the singular points and at both of these $\{\alpha\}$ is the local data. In Example 1.7.5, ∞ is the only singular point and the local data is $\{P_1 = \frac{\alpha_1}{x}, P_2 = \frac{\alpha_2}{x}\}$. In both cases, the parameters allow us to vary the local data. One would hope that by fixing the local data, parameterized families of linear differential equations with fixed Galois group G would be constructible. This turns out to be true for many *but not all* groups G.

Definition 1.7.7 *Let $\mathcal{D} = \{\rho_1, \ldots, \rho_r, P_1, \ldots, P_s\}$ be a finite set with $\rho_i \in \mathbb{C}$ and P_i polynomials without constant terms. We define*

$$\mathcal{L}(m,n,\mathcal{D},G) = \{L \in \mathcal{L}(m,n,G) \mid \mathcal{D}_a \subset \mathcal{D} \text{ for all sing pts } a \in S^2\}$$

Note that we do not fix the position of the singular points; we only fix the local data at putative singular points. Also note that $\mathcal{L}(m, n, \mathcal{D}) = \{L = \sum_{i=0}^{n} \sum_{j=0}^{m} a_{i,j} x^i \partial^j \mid a_{i,j} \in \mathbb{C}, \mathcal{D}_{x_0} \subset \mathcal{D}, \text{ for all sing pts } x_0\}$ is a constructible set. To describe the linear algebraic groups for which $\mathcal{L}(m, n, \mathcal{D}, G)$ is a constructible set, we need the following definitions.

Definition 1.7.8 *Let G a linear algebraic group defined over C, an algebraically closed field and let G^0 be the identity component.*

1) A character χ is a polynomial homomorphism $\chi : G \to C^*$.

2) $KerX(G^0)$ is defined to be the intersection of kernels of all characters $\chi : G^0 \to C^*$.

Examples 1.7.9 *1) If G is finite, then $KerX(G^0)$ is trivial.*

2) If G^0 is semisimple or unipotent, then $KerX(G^0) = G^0$.

One can show [77] that $KerX(G^0)$ is the smallest normal subgroup such that $G^0/KerX(G^0)$ is a torus. In addition it is the subgroup generated by all unipotent elements of G^0. Furthermore, $KerX(G^0)$ is normal in G and we have

$$1 \to G^0/KerX(G^0) \to G/KerX(G^0) \to G/G^0 \to 1$$

Finally, $G^0/KerX(G^0)$ is abelian, so we have that G/G^0 acts on $G^0/KerX(G^0)$.

Theorem 1.7.10 *[77] Assume that the action of G/G^0 on $G^0/KerX(G^0)$ is trivial. Then $\mathcal{L}(m, n, \mathcal{D}, G)$ is constructible.*

Examples 1.7.11 *The groups G that satisfy the hypothesis of this theorem include all finite groups, all connected groups and all groups that are either semisimple or unipotent. An example of a group that does not satisfy the hypothesis is the group $\mathbb{C}^* \rtimes \{1, -1\}$ where -1 sends c to c^{-1}.*

For groups satisfying the hypothesis of this theorem we can prove a result similar to Corollary 1.7.3.

Corollary 1.7.12 *Let G be as in Theorem 1.7.10, defined over $\overline{\mathbb{Q}}$. Then G can be realized as a differential Galois group over $\overline{\mathbb{Q}}(x)$.*

Proof (Outline) Let G be as in Theorem 1.7.10, defined over $\overline{\mathbb{Q}}$. We know that G can be realized as the differential Galois group of a linear differential equation over $\mathbb{C}(x)$, *e.g.*, Theorem 1.6.3. A small modification of Theorem 1.6.3 allows one to assume that the local data of such an equation is defined over $\overline{\mathbb{Q}}$ (see [77]). Therefore for some $n, m, \mathcal{D}$, $\mathcal{L}(m, n, \mathcal{D}, G)$ has a point in $\mathbb{C}^{(n+1)(m+1)}$. The Hilbert Nullstellensatz now implies that $\mathcal{L}(m, n, \mathcal{D}, G)$ has a point in $\overline{\mathbb{Q}}^{(n+1)(m+1)}$. Therefore $G \subset \mathrm{GL}_m$ is a differential Galois group over $\overline{\mathbb{Q}}(x)$. $\square$

The above proof ultimately depends on analytic considerations. We again note that Harbater and Hartman [81, 82] have given a direct algebraic proof showing all linear algebraic groups defined over $\overline{\mathbb{Q}}$ can be realized as a differential Galois group over $\overline{\mathbb{Q}}(x)$..

The hypotheses of Theorem 1.7.10 are needed. In fact one can construct (see [77], p. 384-385)) a two-parameter family of second order linear differential equations $L_{a,t}(Y) = 0$ such that

- $L_{a,t}(Y) = 0$ has 4 regular singular points $0, 1, \infty, a$ and exponents *independent of the parameters* a, t,
- for all values of the parameters, the differential Galois group is a subgroup of $\mathbb{C}^* \rtimes \{1, -1\}$ where -1 sends c to c^{-1}.
- there is an elliptic curve $\mathcal{C} : y^2 = f(x)$, f cubic with $(f, f') = 1$ such that if $(a, y(a))$ is a point of finite order on $\mathcal{C}$ then there exists a unique t such that the differential Galois group of $L_{a,t}(Y) = 0$ is finite and conversely, the differential Galois group being finite implies that $(a, y(a))$ is a point of finite order on $\mathcal{C}$.

One already sees in the Lamé equation

$$L_{n,B,e}(y) = f(x)y'' + \frac{1}{2}f'(x)y' - (n(n+1)x + B)y = 0$$

where $f(x) = 4x(x-1)(x-e)$, that the finiteness of the differential Galois group depends on relations between the fourth singular point e and the other parameters. For example, Brioschi showed that if $n + \frac{1}{2} \in \mathbb{Z}$, then there exists a $P \in \mathbb{Q}[u, v]$ such that $L_{n,B,e}(y) = 0$ has finite differential Galois group if and only if $P(e, B) = 0$. Algebraic solutions of the Lamé equation are a continuing topic of interest, see [101, 102, 103, 104, 105, 106, 107].

Finally, the results of [77] have been recast and generalized in [108, 109]. In particular the conditions of Theorem 1.7.10 are shown to be necessary and sufficient and existence and properties of moduli spaces are examined (see also Chapter 12 of [5]).

Parameterized Picard-Vessiot Theory. In the previous paragraphs, I examined how differential Galois groups depend algebraically on parameters that may appear in families of differential equations. Now I will examine the differential dependence and give an introduction to [33] where a Galois theory is developed that measures this. To make things concrete, let us consider parameterized differential equations of the form

$$\frac{\partial Y}{\partial x} = A(x, t_1, \ldots, t_m)Y \quad A \in \mathrm{gl}_n(\mathbb{C}(x, t_1, \ldots, t_m)). \quad (1.4)$$

If $x = x_0, t_1 = \tau_1, \ldots, t_m = \tau_m$ is a point where the entries of A are analytic, standard existence theory yields a solution $Y = (y_{i,j}(x, t_1, \ldots, t_m))$, with $y_{i,j}$ analytic near $x = x_0, t_1 = \tau_1, \ldots, t_m = \tau_m$. Loosely speaking, we want to define a Galois group that is the group of transformations that preserve all algebraic relations among $x, t_1, \ldots, t_m$, the $y_{i,j}$ and the derivatives of the $y_{i,j}$ and their derivatives with respect to *all the variables* $x, t_1, \ldots, t_m$.

To put this in an algebraic setting, we let $k = \mathbb{C}(x, t_1, \ldots, t_m)$ be a (partial) differential field with derivations $\Delta = \{\partial_0, \partial_1, \ldots, \partial_m\}, \partial_0 = \frac{\partial}{\partial x}, \partial_i = \frac{\partial}{\partial t_i}$ $i = 1, \ldots m$. We let

$$K = k(y_{1,1}, \ldots, y_{n,n}, \ldots, \partial_1^{n_1} \partial_2^{n_2} \cdots \partial_m^{n_m} y_{i,j}, \ldots) = k\langle y_{1,1}, \ldots, y_{n,n}\rangle_\Delta.$$

Note that the fact that the $y_{i,j}$ appear as entries of a solution of (1.4) implies that this field is stable under the derivation ∂_0. We shall define the parameterized Picard-Vessiot group (PPV-group) $\mathrm{DGal}_\Delta(K/k)$ to be the group of k-automorphisms σ of K that commute with all ∂_i, $i = 0, \ldots, m$.

Example 1.7.13 *Let* $n = 1, m = 1, k = \mathbb{C}(x, t), \Delta = \Delta = \{\partial_x, \partial_t\}$ *and*

$$\frac{\partial y}{\partial x} = \frac{t}{x}y.$$

This equation has $y = x^t = e^{t \log x}$ *as a solution. Differentiating with respect to* t *and* x *shows that* $K = \mathbb{C}(x, t, x^t, \log x)$. *Let* $\sigma \in DGal_\Delta(K/k)$.

Claim 1: $\sigma(x^t) = ax^t$, $a \in K$, $\partial_x(a) = 0$ *(and so* $a \in \mathbb{C}(t)$*)*. *To see this note that* $\partial_x \sigma(x^t) = \frac{t}{x}\sigma(x^t)$ *so* $\partial_x(\sigma(x^t)/x^t) = 0$.

Claim 2: $\sigma(\log x) = \log x + c$, $c \in \mathbb{C}$. *To see this, note that* $\partial_x \sigma(\log x) = \frac{1}{x}$ *so* $\partial_x(\sigma(\log x) - \log x) = 0$. *Since* $\partial_t(\log x) = 0$ *we also have that* $\partial_t(\sigma(\log x) - \log x) = 0$.

Claim 3: $\sigma(x^t) = ax^t, \partial_x a = 0, \partial_t a = ca, c \in \mathbb{C}$. *To see this note* *that $\sigma(\partial_t x^t) = \sigma(\log x \; x^t) = (\log x + c)ax^t = a\log x \; x^t + cax^t$ and $\partial_t \sigma(x^t) = \partial_t(ax^t) = (\partial_t a)x^t + a\log x \; x^t$.*

Therefore,

$$\mathrm{DGal}_\Delta(K/k) = \{a \in k^* \mid \partial_x a = 0, \; \partial_t(\frac{\partial_t a}{a}) = 0\}$$
$$= \{a \in k^* \mid \partial_x a = 0, \; \partial_t^2(a)a - (\partial_t(a))^2 = 0\}$$

Remarks: (1) Let $k_0 = \ker \partial_x = \mathbb{C}(t)$. This is a ∂_t-field. The PPV-group $\mathrm{DGal}_\Delta(K/k)$ can be identified with a certain subgroup of $\mathrm{GL}_1(k_0)$, that is, with a group of matrices whose entries satisfy a differential equations. This is an example of a linear differential algebraic group (to be defined more fully below).

(2) In fact, using partial fraction decompositions, one can see that

$$\{a \in k^* \mid \partial_x a = 0, \; \partial_t(\frac{\partial_t a}{a}) = 0\} = \mathbb{C}^*$$

This implies that for any $\sigma \in \mathrm{DGal}_\Delta(K/k)$, $\sigma(\log x) = \log x$. If we want a Fundamental Theorem for this new Galois theory, we will want to have enough automorphisms so that any element $u \in K \backslash k$ is moved by some automorphism. This is not the case here.

(3) If we specialize t to be an element $\tau \in \mathbb{C}$, we get a linear differential equation whose monodromy group is generated by the map $y \mapsto e^{2\pi\tau}y$. One can think of the map $y \mapsto e^{2\pi t}y$ as the "parameterized monodromy map". I would like to say that this map lies in the PPV-group and that this map generates a group that is, in some sense, dense in the PPV-group. We cannot do this yet.

Remarks (2) and (3) indicate that the PPV group does not have enough elements. When we considered the Picard-Vessiot theory, we insisted that the constants be algebraically closed. This guaranteed that (among other things) the differential Galois group (defined by polynomial equations) had enough elements. Since our PPV-groups will be defined by differential equations, it seems natural to require that the field of ∂_0-constants is "differentially closed". More formally,

Definition 1.7.14 *Let k_0 be a differential field with derivations $\Delta_0 = \{\partial_1, \ldots, \partial_m\}$. We say that k_0 is differentially closed if any system of polynomial differential equations*

$$f_1(y_1, \ldots, y_n) = \ldots = f_r(y_1, \ldots, y_n) = 0, g(y_1, \ldots, y_n) \neq 0$$

with coefficients in k_0 that has a solution in some Δ_0-differential extension field has a solution in k_0.

Differentially closed fields play the same role in differential algebra as algebraically closed fields play in algebraic geometry and share many (but not all) of the same general properties and, using Zorn's Lemma, they are not hard to construct. They have been studied by Kolchin (under the name of constrained closed differential fields [110]) and by many logicians [111]. We need two more definitions before we state the main facts concerning parameterized Picard-Vessiot theory.

Definition 1.7.15 *Let k_0 be a differential field with derivations $\Delta_0 = \{\partial_1, \ldots, \partial_m\}$.*

(1) A set $X \subset k_0^n$ is Kolchin closed if it is the zero set of a system of differential polynomials over k_0.

(2) A linear differential algebraic group is a subgroup $G \subset \mathrm{GL}_n(k_0)$ that is Kolchin closed in $\mathrm{GL}_n(k_0) \subset k_0^{n^2}$.

Kolchin closed sets form the closed sets of a topology called the Kolchin topology. I will therefore use topological language (open, dense, etc) to describe their properties. Examples and further description of linear differential algebraic groups are given after the following definition and result.

Definition 1.7.16 *Let k be a differential field with derivations $\Delta = \{\partial_0, \partial_1, \ldots, \partial_m\}$ and let*

$$\partial_0 Y = AY, \quad A \in \mathrm{gl}_n(k) \tag{1.5}$$

A Parameterized Picard-Vessiot extension (PPV-extension) of k for (1.5) is a Δ-ring $R \supset k$ such that R is a simple Δ-ring (i.e., its only Δ-ideals are (0) and R) and R is generated as a Δ-ring by the entries of some $Z \in \mathrm{GL}_n(R)$ and $\frac{1}{\det Z}$ where $\partial_0 Z = AZ$. A PPV-field is defined to be the quotient field of a PPV-ring.

As in the usual Picard-Vessiot theory, such rings exist and are domains. Under the assumption that $k_0 = \ker \partial_0$ is a differentially closed $\Delta_0 = \{\partial_1, \ldots, \partial_m\}$-field, PPV-rings can be shown to be unique (up to appropriate isomorphism).

Theorem 1.7.17 *Let k be a differential field with derivations $\Delta = \{\partial_0, \partial_1, \ldots, \partial_m\}$ and assume that $k_0 = \ker \partial_0$ is a differentially closed*

$\Delta_0 = \{\partial_1, \ldots, \partial_m\}$-*field. Let K be the PPV-field for an equation*

$$\partial_0 Y = AY, \quad A \in \mathrm{gl}_n(k).$$

(1) The group $\mathrm{DGal}_\Delta(K/k)$ (called the PPV-group) of Δ-differential k-automorphisms of K has the structure of a linear differential algebraic group.

(2) There is a Galois correspondence between Δ-subfields E of K containing k and Kolchin Δ_0 closed subgroups H of $\mathrm{DGal}_\Delta(K/k)$ with normal subgroups H corresponding to field K that are again PPV-extensions of k.

In the context of this theorem we can recast the above example. Let $k_0, \mathbb{C}(t) \subset k_0$ be a differentially closed ∂_t-field and $k = k_0(x)$ be a $\Delta = \{\partial_x, \partial_t\}$ field where $\partial_x(k_0) = 0, \partial_x(x) = 1, \partial_t(x) = 0$. One can show that the field $K = k < x^t >_\Delta = k(x^t, \log)$ is a PPV-field for the equation $\partial_x Y = \frac{t}{x} Y$ with PPV group as described above. Because k_0 is differentially closed, this PPV group has enough elements so that the Galois correspondence holds. Furthermore the element $e^{2\pi i t} \in k_0$ is an element of this group and, in fact, generates a Kolchin dense subgroup.

Remarks: (1) In general, let $\partial_x Y = AY$ is a parameterized equation with $A \in \mathbb{C}(x, t_1, \ldots, t_m)$. For each value of $\bar{t}$ of $t = (t_1, \ldots, t_m)$ at which the entries of A are defined and generators $\{\gamma_i\}$ of the Riemann Sphere punctured at the poles of $A(x, \bar{t}_1, \ldots \bar{t}_m)$, we can define monodromy matrices $\{M_i(\bar{t})\}$. For all values of t in a sufficiently small open set, the entries of M_i are analytic functions of t and one can show that these matrices belong to the PPV-group. If the equation $\partial_x Y = AY$ has only regular singular points, then one can further show that these matrices generate a Kolchin dense subgroup of the PPV group.

(2) One can consider parameterized families of linear partial differential equations as well. Let k be a $\Delta = \{\partial_0, \ldots, \partial_s, \partial_{s+1}, \ldots, \partial_m\}$ and let

$$\partial_i Y = A_i Y, \quad A_i \in \mathrm{gl}_n(k), \quad i = 0, \ldots, s$$

be a system of differential equations with $\partial_i(A_j) + A_i A_j = \partial_j(A_i) + A_j A_i$. One can develop a parameterized differential Galois theory for these as well.

(3) One can develop a theory of parameterized liouvillian functions and show that one can solve parameterized linear differential equations in terms of these if and only if the PPV group is solvable-by-finite.

To clarify and apply the PPV theory, we will describe some facts about linear differential algebraic groups.

Linear differential algebraic groups. These groups were introduced by Cassidy [112]. A general theory of differential algebraic groups was initiated by Kolchin [113] and developed by Buium, Hrushovski, Pillay, Sit and others (see [33] for references). We give here some examples and results. Let k_0 be a differentially closed $\Delta_0 = \{\partial_1, \ldots, \partial_m\}$ differential field.

Examples 1.7.18 *(1) All linear algebraic groups are linear differential algebraic groups.*

(2) Let $C = \cap_{i=1}^m \ker \partial_i$ and let $G(k_0)$ be a linear algebraic group defined over k_0. Then $G(C)$ is a linear differential algebraic group (just add $\{\partial_1 y_{i,j} = \ldots = \partial_m y_{i,j} = 0\}_{i,j=1}^n$ to the defining equations!).

(3) Differential subgroups of

$$G_a(k_0) = (k_0, +) = \{\begin{pmatrix} 1 & z \\ 0 & 1 \end{pmatrix} \mid z \in k\}.$$

The linear differential subgroups of this group are all of the form

$$G_a^{\{L_1, \ldots, L_s\}} = \{z \in k_0 \mid L_1(z) = \ldots = L_s(z) = 0\}$$

where the L_i are linear differential operators in $k_0[\partial_1, \ldots, \partial_m]$. When $m = 1$ we may always take $s = 1$.

For example, if $m = 1$,

$$G_a^\partial = \{z \in k_0 \mid \partial z = 0\} = G_a(C)$$

(4) Differential subgroups of $G_m(k_0) = (k_0^*, \times) = \mathrm{GL}_1(k_0)$. Assume $m = 1, \Delta_0 = \{\partial\}$ (for the general case, see [112]). The linear differential subgroups are:

- *finite and cyclic, or*
- *$G_m^L = \{z \in k_0^* \mid L(\frac{\partial z}{z}) = 0\}$ where the L is a linear differential operator in $k_0[\partial]$.*

For example the PPV group of Example 1.7.13, $G_m^\partial = \{z \in k_0 \mid \partial(\frac{\partial z}{z}) = 0\}$ is of this form. The above characterization follows from the exactness of

$$(1) \longrightarrow G_m(C) \longrightarrow G_m(k_0) \overset{z \mapsto \frac{\partial z}{z}}{\longrightarrow} G_a(k_0) \longrightarrow (0)$$

and the characterization of differential subgroups of G_a. Note that in the usual theory of linear algebraic groups there is no surjective homomorphism from G_m to G_a.

(5) A result of Cassidy [112] states: if $m = 1$, $\Delta_0 = \{\partial\}$ and H is a Zariski dense proper differential subgroup of $\mathrm{SL}_n(k_0)$, then there is an element $g \in \mathrm{SL}_n(k_0)$ such that $gHg^{-1} = \mathrm{SL}_n(C)$, where $C = \ker \partial$. This has been generalized by Cassidy and, later, Buium to the following. Let H be a Zariski dense proper differential subgroup of G, a simple linear algebraic group, defined over $\mathbb{Q}$. Then there exists $\mathbb{D} = \{D_1, \ldots D_s\} \subset k\Delta_0 = k - span$ of Δ_0 such that

 (i) $[D_i, D_j] = 0$ for $1 \leq i, j \leq s$ and

 (ii) There exists a $g \in G(k)$ such that $gHg^{-1} = G(C_{\mathbb{D}})$, where $C_{\mathbb{D}} = \cap_{i=1}^s \ker D_i$.

Inverse Problem. In analogy to the Picard-Vessiot theory, the question I will discuss is: Given a differential field k, which linear differential algebraic groups appear as PPV-groups over k? Even in simple cases the complete answer is not known and there are some surprises.

Let $\Delta = \{\partial_x, \partial_t\}$, $\Delta_0 = \{\partial_x\}$, k_0 a ∂_t-differentially closed ∂_t-field and $k = k_0(x)$ with $\partial_0(k_0) = 0$, $\partial_x(x) = 1$, an $\partial_t(x) = 0$. Let $C \subset k_0$ be the Δ-constants of k. I will show that all differential subgroups of $G_a = (k_0, +)$ appear as PPV-groups over this field but that the full group G_a does not appear as a PPV-group [33].

One can show that the task of describing the differential subgroups of G_a that appear as PPV-groups is equivalent to describing which linear differential algebraic groups appear as PPV-groups for equations of the form

$$\frac{dy}{dx} = a(t, x), \ a(t, x) \in k_0(x)$$

Using partial fraction decomposition, one sees that a solution of such an equation is of the form

$$y = R(t, x) + \sum_{i=1}^{s} a_i(t) \log(x - b_i(t)), R(t, x) \in k_0(x), a_i(t), b_i(t) \in k_0$$

and the associated PPV-extension is of the form $K = k(y)$. If H is the PPV-group of K over k and $\sigma \in H$, then one shows that

$$\sigma(y) = R(t,x) + \sum_{i=1}^{s} a_i(t) \log(x - b_i(t)) + \sum_{i=1}^{s} c_i a_i(t)$$

for some $c_i \in C$. Let $L \in k_0[\frac{d}{dt}]$ have solution space spanned by the $a_i(t)$ (note that $L \neq 0$). For any $\sigma \in H$, σ can be identified with an element of

$$H = \{z \in G_a(k_0) \mid L(z) = 0.\}$$

Conversely, any element of H can be shown to induce a differential automorphism of K, so H is the PPV-group. Therefore H must be a proper subgroup of G_a and not all of G_a. One can furthermore show that all proper differential algebraic subgroups of G_a can appear in this way.

A distinguishing feature of proper subgroups of G_a is that they contain a finitely generated Kolchin dense subgroup whereas G_a does not have such a subgroup. A possible conjecture is that a linear differential algebraic group H is a PPV-group over k if and only if H has a finitely generated Kolchin dense subgroup. If H has such a finitely generated subgroup, a parameterized solution of Hilbert's 22nd problem shows that one can realize this group as a PPV-group. If one could establish a parameterized version of Ramis's theorem that the formal monodromy, exponential torus and Stokes generate a dense subgroup of the differential Galois group (something of independent interest), the other implication of this conjecture would be true.

I will now construct a field over which one can realize G_a as a PPV-group. Let k be as above and let $F = k(\log x, x^{t-1} e^{-x})$. The *incomplete Gamma function*

$$\gamma(t,x) = \int_0^x s^{t-1} e^{-s} ds$$

satisfies

$$\frac{d\gamma}{dx} = x^{t-1} e^{-x}$$

over F. The PPV-group over k is $G_a(k_0)$ because $\gamma, \frac{d\gamma}{dt}, \frac{d^2\gamma}{dt^2}, \ldots$ are algebraically independent over k [114] and so there are no relations to preserve. Over $k = k_0(x)$, $\gamma(t,x)$ satisfies

$$\frac{d^2\gamma}{dx^2} - \frac{t-1-x}{x} \frac{d\gamma}{dx} = 0$$

and the PPV-group is

$$H \; = \; \{\begin{pmatrix} 1 & a \\ 0 & b \end{pmatrix} \mid a \in k_0, b \in k_0^*, \partial_t(\frac{\partial_t b}{b}) = 0\}$$

$$= \; G_a(k_0) \rtimes G_m^{\partial_t}, \; G_m^{\partial_t} = \{b \in k_0^* \mid \partial_t(\frac{\partial_t b}{b}) = 0$$

Isomonodromic families. The parameterized Picard-Vessiot theory can be used to characterize isomonodromic families of linear differential equations with regular singular points, that is, families of such equations where the monodromy is independent of the parameters.

Definition 1.7.19 *Let k be a $\Delta = \{\partial_0, \ldots, \partial_m\}$-differential field and let $A \in \mathrm{gl}_n(k)$. We say that $\partial_0 Y = AY$ is completely integrable if there exist $A_i \in \mathrm{gl}_n(k), i = 0, 1, \ldots, m$ with $A_0 = A$ such that*

$$\partial_j A_i - \partial_i A_j = A_j A_i - A_i A_j \text{ for all } i, j = 0, \ldots m$$

The nomenclature is motivated by the fact that the latter conditions on the A_i are the usual integrability conditions on the system of differential equations $\partial_i Y = AY, i =, \ldots, m$.

Proposition 1.7.20 *[33] Let k be a $\Delta = \{\partial_0, \ldots, \partial_m\}$-differential field and let $A \in \mathrm{gl}_n(k)$. Assume that $k_0 = \ker \partial_0$ is $\Delta_0 = \{\partial_1, \ldots, \partial_m\}$-differentially closed field and let K be the PPV-field of k for $\partial_0 Y = AY$ with $A \in \mathrm{gl}_n(k)$. Let $C = \cap_{i=0}^m \ker \partial_i$.*

There exists a linear algebraic group G, defined over C, such that $\mathrm{DGal}_\Delta(K/k)$ is conjugate to $G(C)$ over k_0 if and only if $\partial_0 Y = AY$ is completely integrable. If this is the case, then K is a Picard-Vessiot extension corresponding to this integrable system.

Let $k = \mathbb{C}(x, t_1, \ldots, t_n)$ and $A \in \mathrm{gl}_n(k)$. For $t = (t_1, \ldots, t_m)$ in some sufficiently small open set $\mathcal{O} \subset \mathbb{C}^n$, one can select generators $\{\gamma_i(t_i, \ldots, t_m)\}$ of the fundamental group of the Riemann Sphere minus the (parameterized) singular points such that the monodromy matrices $\{M_i(t_1, \ldots, t_m)\}$ depend analytically on the parameters. One says that the differential equation $\partial_x Y = AY$ is *isomonodromic* if for some sufficiently small open set $\mathcal{O}$, the M_i are independent of t. In [33] we deduce the following corollary from the above result.

Corollary 1.7.21 *Let $k, A, \mathcal{O}$ be as above and assume that $\partial_x Y = AY$ has only regular singular points for $t \in \mathcal{O}$. Then $\partial_x Y = AY$ is isomon-*

odromic if and only if the PPV-group is conjugate to a linear algebraic group $G(\mathbb{C}) \subset \mathrm{GL}_n(\mathbb{C})$.

Second order parameterized equations. A classification of the linear differential subgroups of SL_2 allows one to classify parameterized systems of second order linear differential equations. To simplify the discussion we will consider equations of the form

$$\frac{\partial Y}{\partial x} = A(x, t)Y$$

that depend on only one parameter and satisfy $A \in \mathrm{sl}_n(\mathbb{C}(x, t))$.

Let H be a linear differential algebraic subgroup of $\mathrm{SL}_2(k_0)$ (where k_0 is a ∂_t-differentially closed field containing $\mathbb{C}(t)$ and $\ker \partial_t = \mathbb{C}$) and $\bar{H}$ its Zariski closure. If $\bar{H} \neq \mathrm{SL}_2$, then one can show that $\bar{H}$ and therefore H is solvable-by-finite [60, 112]. Furthermore, Cassidy's result (Example 1.7.18(3)) states that if $\bar{H} = \mathrm{SL}_2$ but $H \neq \mathrm{SL}_2$, then H is conjugate to $\mathrm{SL}_2(\mathbb{C})$. From this discussion and the discussion in the previous paragraphs one can conclude the following.

Proposition 1.7.22 *Let k and $\frac{\partial Y}{\partial x} = AY$ be as above. Assume that for some small open set of values of t that this equation has only regular singular points. Then either*

- *the PPV group is SL_2, or*
- *the equation is solvable in terms of parameterized exponentials, integrals and algebraics, or*
- *the equation is isomonodromic.*

The notion of 'solvable in terms of parameterized exponentials, integrals and algebraics' is analogous to the similar notion in the usual Picard-Vessiot theory and is given a formal definition in [33], where it is also shown that this notion is equivalent to the PPV-group being solvable-by-finite.

1.8 Final Comments

I have touched on only a few of the aspects of the Galois theory of linear differential equations. I will indicate here some of the other aspects and give pointers to some of the current literature.

Kolchin himself generalized this theory to a theory of *strongly normal* extension fields [3]. The Galois groups in this theory are arbitrary algebraic groups. Recently Kovacic [115, 116] has recast this theory in terms of groups schemes and differential schemes. Umemura [117, 118, 119, 120, 121, 122] has developed a Galois theory of differential equations for general nonlinear equations. Instead of Galois groups, Umemura uses Lie algebras to measure the symmetries of differential fields. Malgrange [123, 124] has proposed a Galois theory of differential equations where the role of the Galois group is replaced by certain groupoids. This theory has been expanded and applied by Casale [125, 126, 127]. Pillay [128, 129, 130, 131, 132] develops a Galois theory where the Galois groups can be arbitrary *differential* algebraic groups and, together with Bertrand, has used these techniques to generalize Ax's work on Schanuel's conjecture (see [133, 134]). Landesman [135] has generalized Kolchin's theory of strongly normal extensions to a theory of strongly normal parameterized extensions. The theory of Cassidy and myself presented in Section 1.7 can be viewed as a special case of this theory and many of the results of [33] can be derived from the theories of Umemura and Malgrange as well.

A Galois Theory of linear *difference* equations has also been developed, initally by Franke ([136] and several subsequent papers), Bialynicki-Birula [137] and more fully by van der Put and myself [22]. André [138] has presented a Galois theory that treats both the difference and differential case in a way that allows one to see the differential case as a limit of the difference case. Chatzidakis, Hrushovski and Kamensky have used model theoretic tools to develop Galois theory of difference equations (see [139], [140] and [141] for a discussion of connections with the other theories). Algorithmic issues have been considered by Abramov, Barkatou, Bronstein, Hendriks, van Hoeij, Kauers, Petkovšek, Paule, Schneider, Wilf, M. Wu and Zeilberger and there is an extensive literature on the subject for which I will refer to Math-SciNet and the ArXiV. Recently work by André, Di Vizio, Etingof, Hardouin, van der Put, Ramis, Reversat, Sauloy, and Zhang, has been done on understanding the Galois theory of q-difference equation (see [142, 143, 144, 145, 146, 147, 148, 149, 150, 151, 152]). Fields with both derivations and automorphisms have been studied from a model theoretic point of view by Bustamente [153] and work of Abramov, Bronstein, Li, Petkovšek, Wu, Zheng and myself [154, 155, 156, 157, 158, 58, 159] considers algorithmic questions for mixed differential-difference systems.

The Picard-Vessiot theory of linear differential equations has been used by Hardouin [160] to study the differential properties of solutions of difference equations. A Galois theory, with linear differential algebraic groups as Galois groups, designed to measure the differential relations among solutions of difference equations, has been recently developed and announced in [161].

Finally, I have not touched on the results and enormous literature concerning the hypergeometric equations and their generalizations as well as arithmetic properties of differential equations. For a taste of the current research, I refer the reader to [162] and [163] for the former and [164] for the latter.

Bibliography

[1] Beukers, F. (1992). Differential Galois Theory, in *From Number Theory to Physics (Les Houches, 1989)*, ed. M. Waldschmidt *et al*, pages 413–439 (Springer-Verlag, Berlin).

[2] Kaplansky, I. (1976). *An Introduction to Differential Algebra*, second edition (Hermann, Paris).

[3] Kolchin, E.R. (1976). *Differential Algebra and Algebraic Groups* (Academic Press, New York).

[4] Magid, A. (1994). *Lectures on Differential Galois Theory*, Second Edition, University Lecture Series (American Mathematical Society, Providence, R.I.).

[5] van der Put, M. and Singer, M.F. (2003). *Galois Theory of Linear Differential Equations*, volume 328 of *Grundlehren der mathematischen Wissenshaften* (Springer, Heidelberg).

[6] Matzat, B.H. and van der Put, M. (2003). Constructive differential Galois theory, in *Galois groups and fundamental groups*, volume 41 of *Math. Sci. Res. Inst. Publ.*, pages 425–467 (Cambridge Univ. Press, Cambridge).

[7] Matzat, B.H. and van der Put, M. (2003). Iterative differential equations and the Abhyankar conjecture, *J. Reine Angew. Math.* **557**, 1–52.

[8] van der Put, M. (1995). Differential Equations in Characteristic p, *Compositio Mathematica* **97**, 227–251.

[9] van der Put, M. (1996). Reduction modulo p of differential equations, *Indagationes Mathematicae* **7**, 367–387.

[10] Seidenberg, A. (1958). Abstract differential algebra and the analytic case, *Proc. Amer. Math. Soc.* **9**, 159–164.

[11] Seidenberg, A. (1969). Abstract differential algebra and the analytic case. II, *Proc. Amer. Math. Soc.* **23**, 689–691.

[12] Churchill, R.C. and Kovacic, J.J. (2002). Cyclic vectors, in *Differential algebra and related topics (Newark, NJ, 2000)*, pages 191–218 (World Sci. Publ., River Edge, NJ).

[13] Cope, F. (1934). Formal solutions of irregular linear differential equations, I, *American Journal of Mathematics* **56**, 411–437.

[14] Cope, F. (1936). Formal solutions of irregular linear differential equations, II, *American Journal of Mathematics* **58**, 130–140.

[15] Deligne, P. (1970). *Equations Différentielles à Points Singuliers Réguliers*, volume 163 of *Lecture Notes in Mathematics* (Springer-Verlag, Heidelberg).

[16] Jacobson, N. (1937). Pseudo-Linear Transformations, *Annals of Mathematics* **38**, 484–507.

[17] Katz, N. (1987). A simple algorithm for cyclic vectors, *American Journal of Mathematics* **109**, 65–70.

[18] Dyckerhoff, T. (2005). Picard-Vessiot Extensions over Number Fields, Fakultät für Mathematik und Informatik der Universität Heidelberg Diplomarbeit.

[19] Hendriks, P.A. and van der Put, M. (1995). Galois Action on Solutions of a Differential Equation, *Journal of Symbolic Computation* **19**(6), 559 – 576.

[20] Ulmer, F. (1994). Irreducible Linear Differential Equations of Prime Order, *Journal of Symbolic Computation* **18**(4), 385–401.

[21] Hrushovski, E. (2002). Computing the Galois group of a linear differential equation, in *Differential Galois Theory*, volume 58 of *Banach Center Publications*, pages 97–138 (Institute of Mathematics, Polish Academy of Sciences, Warszawa).

[22] van der Put, M. and Singer, M.F. (1997). *Galois Theory of Difference Equations*, volume 1666 of *Lecture Notes in Mathematics* (Springer-Verlag, Heidelberg).

[23] Springer, T. (1998). *Linear Algebraic Groups, Second Edition*, volume 9 of *Progress in Mathematics* (Birkhäuser, Boston).

[24] Serre, J.P. (1964). *Cohomologie Galoisienne*, Number 5 in Lecture Notes in Mathematics (Springer-Verlag, New York).

[25] Audin, M. (2001). *Les Systèmes Hamiltoniens et leur Intégrabilité*, volume 8 of *Cours Spécialisés* (Société Mathématique de France, Paris).

[26] Morales-Ruiz, J. (1999). *Differential Galois Theory and Non-Integrability of Hamiltonian Systems* (Birkhäuser, Basel).

[27] Lang, S. (1966). *Introduction to Tanscendental Numbers* (Addison Wesley, New York).

[28] Kolchin, E.R. (1968). Algebraic Groups and Algebraic Dependence, *American Journal of Mathematics* **90**, 1151–1164.

[29] Harris, W. and Sibuya, Y. (1985). The reciprocals of solutions of linear ordinary differential equations, *Adv. in Math.* **58**, 119–132.

[30] Singer, M.F. (1986). Algebraic Relations Among Solutions of Linear Differential Equations, *Transactions of the American Mathematical Society* **295**, 753–763.

[31] Magid, A. (1976). Finite generation of class groups of rings of invariants, *Proc. Amer. Math. Soc.* **60**, 45–48.

[32] Rosenlicht, M. (1961). Toroidal Algebraic Groups, *Proc. Amer. Math. Soc.* **12**, 984–988.

[33] Cassidy, P.J. and Singer, M.F. (2006). Galois Theory of Parameterized Differential Equations and Linear Differential Algebraic Groups, in *Differential Equations and Quantum Groups*, ed. D. Bertrand, B. Enriquez, C. Mitschi, C. Sabbah, and R. Schaefke, volume 9 of *IRMA Lectures in Mathematics and Theoretical Physics*, pages 113–157 (EMS Publishing House, Zürich).

[34] Poole, E. (1960). *Introduction to the Theory of Linear Differential Equations* (Dover Publications, New York).

[35] Hartman, P. (1964). *Ordinary differential equations* (John Wiley & Sons Inc., New York).

[36] Gray, J.J. (2000). *Linear Differential Equations and Group Theory from Riemann to Poincaré*, second edition (Birkhäuser, Boston, Basel, Stuttgart).

[37] Loday-Richaud, M. (1995). Solutions formelles des systèmes différentiels linéaires méromorphes et sommation, *Expositiones Mathematicae* **13**, 116–162.

[38] Balser, W. (1994). *From Divergent Power Series to Analytic Functions*, volume 1582 of *Lecture Notes in Mathematics* (Springer-Verlag, Heidelberg).

[39] Maillet, E. (1903). Sur les séries divergentes et les équations différentielles, *Ann. Éc. Norm. Sup.* pages 487–518.

[40] Ramis, J.P. (1993). *Séries Divergentes et Théories Asymptotiques*, volume 121 of *Panoramas et Synthèses* (Société Mathématique de France, Paris).

[41] Ramis, J.P. (1985). Filtration Gevrey sur le groupe de Picard-Vessiot d'une équation différentielle irrégulière, Informes de Matematica, Preprint IMPA, Series A-045/85.

[42] Ramis, J.P. (1985). Phénomène de Stokes et resommation, *Comptes Rendus de l'Académie des Sciences, Paris* **301**, 99–102.

[43] Ramis, J.P. (1985). Phénomène de Stokes et filtration Gevrey sur le groupe de Picard-Vessiot, *Comptes Rendus de l'Académie des Sciences, Paris* **301**, 165–167.

[44] Martinet, J. and Ramis, J.P. (1989). Théorie de Galois différentielle et resommation, in *Computer Algebra and Differential Equations*, ed. E. Tournier, pages 115–214 (Academic Press, New York).

[45] Loday-Richaud, M. (1994). Stokes Phenomenon, Multisummability and Differential Galois Groups, *Annales de l'Institut Fourier* **44**(3), 849–906.

[46] Loday-Richaud, M. (1990). Introduction à la multisommabilité, *Gazette des Mathématiciens, SMF* **44**, 41–63.

[47] Varadarajan, V.S. (1991). Meromorphic Differential Equations, *Expositiones Mathematicae* **9**(2), 97–188.

[48] Varadarajan, V.S. (1996). Linear Meromorphic Differential Equations: A Modern Point of View, *Bulletin (New Series) of the American Mathematical Society* **33**(1), 1 – 42.

[49] Deligne, P. and Milne, J. (1982). Tannakian Categories, in *Hodge Cycles, Motives and Shimura Varieties*, ed. P. Deligne *et al.*, pages 101–228 (Springer, Heidelberg), Lecture Notes in Mathematics, Vol. 900.

[50] Deligne, P. (1990). Catégories Tannakiennes, in *Grothendieck Festschrift, Vol. 2*, ed. P. Cartier *et al.*, pages 111–195 (Birkhäuser, Boston, Basel, Stuttgart), Progress in Mathematics, Vol. 87.

[51] Fulton, W. and Harris, J. (1991). *Representation theory. A first course*, volume 129 of *Graduate Texts in Mathematics* (Springer, New York).

[52] Kolchin, E.R. (1948). Algebraic Matric Groups and the Picard-Vessiot Theory of homogeneous Linear Ordinary Differential Equations, *Annals of Mathematics* **49**, 1–42.

[53] Singer, M.F. (1981). Liouvillian solutions of nth order homogeneous linear differential equations, *American Journal of Mathematics* **103**, 661–681.

[54] Boulanger, A. (1898). Contribution à l'étude des équations linéaires homogènes intégrables algébriquement, *Journal de l'École Polytechnique, Paris* **4**, 1 – 122.

[55] Gelfand, I., Kapranov, M., and Zelevinsky, A.V. (1994). *Discriminants, Resultants and Multidimensional Determinants* (Birkhäuser, Boston, Basel, Stuttgart).

[56] van Hoeij, M. and Weil, J.A. (1996). An Algorithm for Computing Invariants of Differential Galois Groups, *Journal of Pure and Applied Algebra* 117/118, 353–379.

[57] van Hoeij, M., Ragot, J.F., Ulmer, F., and Weil, J. (1999). Liouvillian solutions of linear differential equations of order three and higher, *Journal of Symbolic Computation* 28(4/5), 589–610.

[58] Li, Z., Singer, M.F., Wu, M., and Zheng, D. (2006). A Recursive Method for Determining the One-Dimensional Submodules of Laurent-Ore Modules, in *Proceedings of the 2006 International Symposium on Symbolic and Algebraic Computation (ISSAC 2006)*, ed. J.G. Dumas, pages 200–208 (ACM Press, New York).

[59] Singer, M.F. and Ulmer, F. (1997). Linear Differential Equations and Products of Linear Forms, *Journal of Pure and Applied Algebra* 117, 549–564.

[60] Kovacic, J. (1986). An Algorithm for Solving Second Order Linear Homogeneous Differential Equations, *Journal of Symbolic Computation* 2, 3–43.

[61] Bronstein, M., Mulders, T., and Weil, J.A. (1997). On Symmetric Powers of Differential Operators, in *Proceedings of the 1997 International Symposium on Symbolic and Algebraic Computation (ISSAC'97)*, ed. W. Küchlin, pages 156–163 (ACM Press, New York).

[62] Ulmer, F. and Weil, J. (1996). A Note on Kovacic's Algorithm, *Journal of Symbolic Computation* 22(2), 179 – 200.

[63] van Hoeij, M. and Weil, J.A. (2005). Solving second order linear differential equations with Klein's theorem, in *ISSAC'05*, pages 340–347 (electronic) (ACM, New York).

[64] Singer, M. (1988). Algebraic Relations Among Solutions of Linear Differential Equations: Fano's Theorem, *American Journal of Mathematics* 110, 115–144.

[65] Singer, M.F. (1985). Solving Homogeneous Linear Differential Equations in Terms of Second Order Linear Differential Equations, *American Journal of Mathematics* 107, 663–696.

[66] van Hoeij, M. (2002). Decomposing a 4th order linear differential equation as a symmetric product, in *Differential Galois Theory*, volume 58 of *Banach Center Publications*, pages 89–96 (Institute of Mathematics, Polish Academy of Sciences, Warszawa).

[67] van Hoeij, M. (2007). Solving Third Order Linear Differential Equations in Terms of Second Order Equations, in *Proceedings of ISSAC2007*, ed. C. Brown, pages 355–360 (ACM Press, New York).

[68] Nguyen, K. (2007). On d-solvability for linear differential equations, Preprint, University of Groningen.

[69] Nguyen, K. and van der Put, M. (2007). Solving linear differential equations, Preprint, University of Groningen.

[70] Person, A. (2002). *Solving Homogeneous Linear Differential Equations of Order 4 in Terms of Equations of Lower Order*, North Carolina State University Ph.D. thesis, http://www.lib.ncsu.edu/theses/available/etd-08062002-104315/.

[71] Humphreys, J. (1975). *Linear Algebraic Groups*, Graduate Texts in Mathematics (Springer-Verlag, New York).

[72] Tretkoff, C. and Tretkoff, M. (1979). Solution of the Inverse Problem in Differential Galois Theory in the Classical Case, *American Journal of Mathematics* **101**, 1327–1332.

[73] Beauville, A. (1993). Monodromie des systèmes différentiels à pôles simples sur la sphère de Riemann (d'après A. Bolibruch), *Astérisque* **216**, 103–119, Séminaire Bourbaki, No. 765.

[74] Anosov, D.V. and Bolibruch, A.A. (1994). *The Riemann-Hilbert Problem* (Vieweg, Braunschweig, Wiesbaden).

[75] Bolibruch, A.A., Malek, S., and Mitschi, C. (2006). On the generalized Riemann-Hilbert problem with irregular singularities, *Expo. Math.* **24**(3), 235–272.

[76] Ramis, J.P. (1996). About the Inverse Problem in Differential Galois Theory: Solutions of the Local Inverse Problem and of the Differential Abhyankar Conjecture (Université Paul Sabatier, Toulouse).

[77] Singer, M.F. (1993). Moduli of Linear Differential Equations on the Riemann Sphere with Fixed Galois Group, *Pacific Journal of Mathematics* **106**(2), 343–395.

[78] Mitschi, C. and Singer, M.F. (1996). Connected linear groups as differential Galois groups, *Journal of Algebra* **184**, 333–361.

[79] Hartmann, J. (2005). On the inverse problem in differential Galois theory, *J. Reine Angew. Math.* **586**, 21–44.

[80] Oberlies, T. (2003). *Embedding problems in differential Galois theory*, University of Heidelberg Ph.D. thesis, (available at www.ub.uni-heidelberg.de/archiv/4550).

[81] Harbater, D. (2007). Patching over Fields (joint with Julia Hartmann), to appear Oberwohlfach Reports.

[82] Hartmann, J. (2007). Differential Galois groups and Patching (joint with David Harbater), to appear in Oberwohlfach Reports.

[83] Mitschi, C. and Singer, M.F. (2002). Solvable-by-finite groups as differential Galois groups, *Ann. Fac. Sci. Toulouse* **XI**(3), 403 – 423.

[84] Cook, W.J., Mitschi, C., and Singer, M.F. (2005). On the constructive inverse problem in differential Galois theory, *Comm. Algebra* **33**(10), 3639–3665.

[85] Beukers, F., Brownawell, D., and Heckman, G. (1988). Siegel Normality, *Annals of Mathematics* **127**, 279 – 308.

[86] Beukers, F. and Heckman, G. (1989). Monodromy for the hypergeometric function $_nF_{n-1}$, *Inventiones Mathematicae* **95**, 325–354.

[87] Katz, N. (1987). On the calculation of some differential Galois groups, *Inventiones Mathematicae* **87**, 13–61.

[88] Katz, N. (1990). *Exponential Sums and Differential Equations*, volume 124 of *Annals of Mathematics Studies* (Princeton University Press, Princeton).

[89] Katz, N. (1996). *Rigid Local Systems*, volume 139 of *Annals of Mathematics Studies* (Princeton University Press, Princeton).

[90] Duval, A. and Mitschi, C. (1989). Groupe de Galois des Équations Hypergéométriques Confluentes Généralisées, *C. R. Acad. Sci. Paris* **309**(1), 217–220.

[91] Mitschi, C. (1989). Matrices de Stokes et Groupe de Galois des Équations Hypergéométriques Confluentes Généralisées, *Pacific Journal of Mathematics* **138**(1), 25–56.

[92] Mitschi, C. (1996). Differential Galois groups of Confluent Generalized Hy-

pergeometric Equations: An Approach using Stokes Multipliers, *Pacific Journal of Mathematics* **176**(2), 365– 405.

[93] Hartmann, J. (2002). *On the inverse problem in differential Galois theory,* University of Heidelberg Thesis, (available at www.ub.uni-heidelberg.de/archiv/3085).

[94] Juan, L. (2007). Generic Picard-Vessiot extensions for connected-by-finite groups, *J. of Algebra* **312**, 194–2006.

[95] Juan, L. and Ledet, A. (2007). On generic differential SO_n-extensions (Texas Tech University, Lubbock), to appear in Proc. AMS.

[96] Juan, L. and Ledet, A. (2007). Equivariant vector fields on non-trivial SO_n-torsors and differential Galois theory (Texas Tech University, Lubbock), to appear in J. of Algebra.

[97] Juan, L. and Ledet, A. (2007). On Picard-Vessiot extensions with group PGL_3 (Texas Tech University, Lubbock).

[98] Dyckerhoff, T. (2007). The inverse problem of differential Galois theory over the field $\mathbb{R}(Z)$, to appear Oberwohlfach Reports.

[99] van den Dries, L. and Ribenboim, P. (1979). Application de la théorie des modèles aux groupes de Galois de corps de fonctions, *C. R. Acad. Sci. Paris Sér. A-B* **288**(17), A789–A792.

[100] Tretkoff, M. (1971). Algebraic extensions of the field of rational functions, *Comm. Pure Appl. Math.* **24**, 491–497.

[101] Baldassarri, F. (1981). On algebraic solutions of Lamé's differential equation, *J. Differential Equations* **41**(1), 44–58.

[102] Baldassarri, F. (1987). Algebraic solutions of the Lamé equation and torsion of elliptic curves, in *Proceedings of the Geometry Conference (Milan and Gargnano, 1987)*, volume 57, pages 203–213 (1989)).

[103] Dwork, B. (1990). Differential operators with nilpotent p-curvature, *Amer. J. Math.* **112**(5), 749–786.

[104] Liţcanu, R. (2002). Counting Lamé differential operators, *Rend. Sem. Mat. Univ. Padova* **107**, 191–208.

[105] Maier, R.S. (2004). Algebraic solutions of the Lamé equation, revisited, *J. Differential Equations* **198**(1), 16–34.

[106] Beukers, F. and van der Waall, A. (2004). Lamé equations with algebraic solutions, *J. Differential Equations* **197**(1), 1–25.

[107] Loray, F., van der Put, M., and Ulmer, F. (2005). The Lamé family of connections on the projective line (IRMAR, Rennes).

[108] Berkenbosch, M. (2002). Moduli spaces for linear differential equations, in *Differential equations and the Stokes phenomenon*, pages 15–33 (World Sci. Publishing, River Edge, NJ).

[109] Berkenbosch, M. (2004). *Algorithms and Moduli Spaces for Differential Equations*, University of Gröningen Ph.D. Thesis.

[110] Kolchin, E.R. (1974). Constrained extensions of differential fields, *Advances in Math.* **12**, 141–170.

[111] Marker, D. (2000). Model theory of differential fields, in *Model theory, algebra, and geometry*, volume 39 of *Math. Sci. Res. Inst. Publ.*, pages 53–63 (Cambridge Univ. Press, Cambridge).

[112] Cassidy, P.J. (1972). Differential algebraic groups, *Amer. J. Math.* **94**, 891–954.

[113] Kolchin, E.R. (1985). *Differential algebraic groups*, volume 114 of *Pure and Applied Mathematics* (Academic Press Inc., Orlando, FL).

[114] Johnson, J., Reinhart, G.M., and Rubel, L.A. (1995). Some counterexam-

ples to separation of variables, *J. Differential Equations* **121**(1), 42–66.

[115] Kovacic, J.J. (2003). The differential Galois theory of strongly normal extensions, *Trans. Amer. Math. Soc.* **355**(11), 4475–4522 (electronic).

[116] Kovacic, J.J. (2006). Geometric characterization of strongly normal extensions, *Trans. Amer. Math. Soc.* **358**(9), 4135–4157 (electronic).

[117] Umemura, H. (1990). Birational Automorphism Groups and Differential Equations, *Nagoya Mathematics Journal* **119**, 1–80.

[118] Umemura, H. (1996). Galois Theory of Algebraic and Differential Equations, *Nagoya Mathematics Journal* **144**, 1–58.

[119] Umemura, H. (1996). Differential Galois Theory of Infinite Dimension, *Nagoya Mathematics Journal* **144**, 59–135.

[120] Umemura, H. (1999). Lie-Drach-Vessiot Theory, in *CR-Geometry and Overdetermined Systems*, volume 25 of *Advanced Studies in Pure Mathematics*, pages 364–385 (North-Holland, Amsterdam).

[121] Umemura, H. (2004). Galois Theory and Painlevé Equations, Preprint.

[122] Umemura, H. (2004). Monodromy preserving deformation and differential Galois group. I, *Astérisque* **296**, 253–269 (Analyse complexe, systèmes dynamiques, sommabilité des séries divergentes et théories galoisiennes. I).

[123] Malgrange, B. (2002). On nonlinear differential Galois theory, *Chinese Ann. Math. Ser. B* **23**(2), 219–226, Dedicated to the memory of Jacques-Louis Lions.

[124] Malgrange, B. (2001). Le groupoïde de Galois d'un feuilletage, in *Essays on geometry and related topics, Vol. 1, 2*, volume 38 of *Monogr. Enseign. Math.*, pages 465–501 (Enseignement Math., Geneva).

[125] Casale, G. (2006). Feuilletages singuliers de codimension un, groupoïde de Galois et intégrales premières, *Ann. Inst. Fourier (Grenoble)* **56**(3), 735–779.

[126] Casale, G. (2006). Irréductibilité de la première équation de Painlevé, *C. R. Math. Acad. Sci. Paris* **343**(2), 95–98.

[127] Casale, G. (2007). The Galois groupoid of Picard-Painlevé VI equation, in *Algebraic, analytic and geometric aspects of complex differential equations and their deformations. Painlevé hierarchies*, RIMS Kôkyûroku Bessatsu, B2, pages 15–20 (Res. Inst. Math. Sci. (RIMS), Kyoto).

[128] Pillay, A. (1997). Differential Galois theory. II, *Ann. Pure Appl. Logic* **88**(2-3), 181–191, Joint AILA-KGS Model Theory Meeting (Florence, 1995).

[129] Pillay, A. (1998). Differential Galois theory. I, *Illinois J. Math.* **42**(4), 678–699.

[130] Marker, D. and Pillay, A. (1997). Differential Galois theory. III. Some inverse problems, *Illinois J. Math.* **41**(3), 453–461.

[131] Pillay, A. (2002). Finite-dimensional differential algebraic groups and the Picard-Vessiot theory, in *Differential Galois theory (Bedlewo, 2001)*, volume 58 of *Banach Center Publ.*, pages 189–199 (Polish Acad. Sci., Warsaw).

[132] Pillay, A. (2004). Algebraic *D*-groups and differential Galois theory, *Pacific J. Math.* **216**(2), 343–360.

[133] Bertrand, D. (2007). Schanuel's conjecture over function fields and differential Galois theory (joint with Anand Pillay), to appear Oberwohlfach Reports.

[134] Bertrand, D. (2006). Schanuel's conjecture for non-constant elliptic curves

over function fields, to appear in *Model Theory and Applications*, Cambridge University Press.

[135] Landesman, P. (2006). *Generalized Differential Galois Theory*, Graduate Center, The City University of New York Ph.D. Thesis.

[136] Franke, C.H. (1963). Picard-Vessiot theory of linear homogeneous difference equations, *Transactions of the AMS* **108**, 491–515.

[137] Bialynicki-Birula, A. (1962). On the Galois Theory of fields with operators, *Amer. J. Math.* **84**, 89–109.

[138] André, Y. (2001). Différentielles non commutatives et théorie de Galois différentielle ou aux différences, *Ann. Sci. École Norm. Sup. (4)* **34**(5), 685–739.

[139] Chatzidakis, Z. and Hrushovski, E. (1999). Model theory of difference fields, *Trans. Amer. Math. Soc.* **351**(8), 2997–3071.

[140] Kamensky, M. (2006). Definable groups of partial automorphisms, Preprint; http://arxiv.org/abs/math.LO/0607718.

[141] Chatzidakis, Z., Hardouin, C., and Singer, M.F. (2007). On the Definition of Difference Galois Groups, To appear in the Proceedings of the Newton Institute 2005 semester 'Model theory and applications to algebra and analysis' ; also arXiv:math.CA/0705.2975.

[142] Di Vizio, L. (2002). Arithmetic theory of q-difference equations: the q-analogue of Grothendieck-Katz's conjecture on p-curvatures, *Invent. Math.* **150**(3), 517–578.

[143] Di Vizio, L. (2004). Introduction to p-adic q-difference equations (weak Frobenius structure and transfer theorems), in *Geometric aspects of Dwork theory. Vol. I, II*, pages 615–675 (Walter de Gruyter GmbH & Co. KG, Berlin).

[144] André, Y. and Di Vizio, L. (2004). q-difference equations and p-adic local monodromy, *Astérisque* **296**, 55–111 (Analyse complexe, systèmes dynamiques, sommabilité des séries divergentes et théories galoisiennes. I).

[145] Etingof, P.I. (1995). Galois groups and connection matrices of q-difference equations, *Electron. Res. Announc. Amer. Math. Soc.* **1**(1), 1–9 (electronic).

[146] van der Put, M. and Reversat, M. (2007). Galois Theory of q-difference equations, *Ann. Fac. Sci. de Toulouse* **XVI**(2), 1–54.

[147] Sauloy, J. (2003). Galois theory of Fuchsian q-difference equations, *Ann. Sci. École Norm. Sup. (4)* **36**(6), 925–968 (2004).

[148] Ramis, J.P. and Sauloy, J. (2007). The q-analogue of the wild fundamental group. I, in *Algebraic, analytic and geometric aspects of complex differential equations and their deformations. Painlevé hierarchies*, RIMS Kôkyûroku Bessatsu, B2, pages 167–193 (Res. Inst. Math. Sci. (RIMS), Kyoto).

[149] Ramis, J.P., Sauloy, J., and Zhang, C. (2006). Développement asymptotique et sommabilité des solutions des équations linéaires aux q-différences, *C. R. Math. Acad. Sci. Paris* **342**(7), 515–518.

[150] Di Vizio, L., Ramis, J.P., Sauloy, J., and Zhang, C. (2003). Équations aux q-différences, *Gaz. Math.* **96**, 20–49.

[151] Ramis, J.P. (2007). Local classification of linear meromorphic q-difference equations (joint with Jacques Sauloy), to appear Oberwohlfach Reports.

[152] Hardouin, C. (2007). Iterative q-difference Galois theory, to appear Oberwohlfach Reports.

[153] Medina, R.B. (2005). *Theéorie des modèles des corps différentiellement clos avec un automorphisme géneérique*, Université Paris 7 Ph.D. Thesis.

[154] Abramov, S. and Petkovšek, M. (1994). D'Alembert Solutions of Linear Differential and Difference Equations, in *Proceedings of the 1994 International Symposium on Symbolic and Algebraic Computation (ISSAC'94)*, ed. J. von zur Gathen, pages 169–180 (ACM Press, New York).

[155] Abramov, S., Bronstein, M., and Petkovšek, M. (1995). On Polynomial Solutions of Linear Operator Equations, in *Proceedings of the 1995 International Symposium on Symbolic and Algebraic Computation (ISSAC'95)*, ed. A.H.M. Levelt, pages 290–296 (ACM Press, New York).

[156] Bronstein, M. and Petkovšek, M. (1996). An introduction to pseudo-linear algebra, *Theoretical Computer Science* **157**, 3–33.

[157] Bronstein, M. and Petkovšek, M. (1994). On Ore rings, linear operators and factorization, *Programming and Computer Software* **20**(1), 14–26.

[158] Bronstein, M., Li, Z., and Wu, M. (2005). Picard-Vessiot extensions for linear functional systems, in *Proceedings of the 2005 International Symposium on Symbolic and Algebraic Computation (ISSAC 2005)*, ed. M. Kauers, pages 68–75 (ACM Press, New York).

[159] Wu, M. (2005). *On Solutions of Linear Functional Systems and Factorization of Modules over Laurent-Ore Algebras*, Chinese Academy of Sciences and l'Université de Nice-Sophia Antipolis Ph.D. Thesis.

[160] Hardouin, C. (2006). Hypertranscendance et Groupes de Galois aux différences, arXiv:math.RT/0702846v1.

[161] Singer, M. (2007). Differential Groups and Differential Relations, to appear in Oberwohlfach Reports.

[162] Sato, M., Sturmfels, B., and Takayama, N. (2000). *Gröbner Deformations of Hypergeometric Differential Equations*, volume 6 of *Algorithms and Computation in Mathematics* (Springer-Verlag, Berlin).

[163] ed. R.P. Holzapfel, A.M. Uludağ, and M. Yoshida (2007). *Arithmetic and geometry around hypergeometric functions*, volume 260 of *Progress in Mathematics* (Birkhäuser Verlag, Basel), Lecture notes of a CIMPA Summer School held at Galatasaray University, Istanbul, June 13–25, 2005.

[164] ed. A. Adolphson, F. Baldassarri, P. Berthelot, N. Katz, and F. Loeser (2004). *Geometric aspects of Dwork theory. Vol. I, II* (Walter de Gruyter GmbH & Co. KG, Berlin), Lectures from the Dwork trimester dedicated to Bernard Dwork held at the University of Padova, Padova, May–July 2001.

2

Some methods to solve linear differential equations in closed form.

Felix Ulmer

IRMAR
Université de Rennes 1
F-35042 Rennes Cedex
ocormier@maths.univ-rennes1.fr
ulmer@univ-rennes1.fr

Jacques-Arthur Weil

XLIM
Université de Limoges
123 avenue Albert Thomas
F-87060 Limoges Cedex
jacques-arthur.weil@unilim.fr

In this article, we review various methods to find closed form solutions of linear differential equations that can for example be found in the symbolic computation package MAPLE. We focus on the presentation of the methods and not on the underlying mathematical theory which is differential Galois theory (see [1, 2]). The text contains many examples which have been computed using MAPLE†. We motivate the search of solution by an integration problem and all the algorithms are motivated via this integration problem.

2.1 Introduction

2.1.1 Integrating via linear differential equations

In 1832-3, Joseph Liouville publishes two articles on the determination of integrals which are algebraic functions [3]. His goal is to design an algorithm which decides when the integral of an algebraic function is algebraic (and compute it when it is) or prove that the integral is not algebraic. In this study, we will consider algebraic functions over $\mathbb{C}(x)$, i.e. functions which are a root of a polynomial with coefficients in $\mathbb{C}(x)$.

---

† A maple file containing all these computations and illustrating their use can be found at
http://www.unilim.fr/pages_perso/jacques-arthur.weil/atde.mws   or
http://www.unilim.fr/pages_perso/jacques-arthur.weil/atde.mpl.

Example 2.1.1 *Consider the function f given by*

$$f = \frac{1}{4} \frac{\sqrt{1 - \sqrt{\frac{1}{x}}(1 + \sqrt{x})}}{x(x-1)}.$$

It may be defined by its minimum polynomial $P(f) := \sum_{i=0}^{n} \phi_i f^i = 0$ *with $\phi_i \in C(x)$; in our example :*

$$P(f) = f^4 + \frac{1}{8} \frac{(x-1)^3}{x-2} f^2 - \frac{1}{256} \frac{(x-1)^6}{x-2}.$$

Liouville shows that if the integral of f is an algebraic function, then f must have a primitive in the algebraic field extension $K := \mathbb{C}(x, f)$ defined by the equation $P(f) = 0$, i.e $K = \mathbb{C}(x)[Y]/(P)$ (we will give a "modern" proof of this in section 5). So, if there exists an algebraic primitive of f, then it can be written in a unique way as

$$\int f dx \;\; = \;\; \alpha_0 + \alpha_1 f + \ldots + \alpha_{n-1} f^{n-1}, \quad \text{where } \alpha_i \in \mathbb{C}(x) \;\; (2.1)$$

in the basis $(1, f, \ldots, f^{n-1})$ of $K/\mathbb{C}(x)$. Formal differentiation of $P(f) = 0$ shows that the derivative of an algebraic function f is given by

$$\frac{df}{dx} = - \frac{\sum_{i=0}^{n} \phi_i' f^i}{\sum_{i=1}^{n} i \cdot \phi_i f^{i-1}}.$$

In particular $\frac{df}{dx}$ belongs to the same algebraic extension $K = \mathbb{C}(x, f)$.

Differentiating the expression 2.1 of $\int f dx$, the relation $(\int f dx)' - f = 0$ yields a differential system of order one for the α_i. Here, we obtain

$$\left(4 \frac{\alpha_1}{(x-1)^4} + \frac{3}{4} \frac{(5x-12)\alpha_3}{(x-1)(x-2)} + \alpha_3' \right) f^3 + \left(\frac{1}{2} \frac{(5x-12)\alpha_2}{(x-1)(x-2)} + \alpha_2' \right) f^2$$

$$+ \left(-1 + \alpha_1' + \frac{5}{4} \frac{\alpha_1}{x-1} \cdot + \frac{3}{64} \frac{(x-1)^2 \alpha_3}{x-2} \right) f + \alpha_0' + \frac{1}{32} \frac{(x-1)^2 \alpha_2}{x-2} = 0.$$

As P was the minimum polynomial of f, all coefficients of the above polynomial expression in f must be zero, which in turn gives us the following linear differential system :

$$(\mathcal{S}) : \begin{cases} \alpha_0' &= -\dfrac{(x-1)^2}{32(x-2)} \alpha_2(x) \\[2mm] \alpha_1' &= -\dfrac{5}{4} \dfrac{1}{x-1} \alpha_1(x) - \dfrac{3}{64} \dfrac{(x-1)^2}{x-2} \alpha_3(x) + 1 \\[2mm] \alpha_2' &= -\dfrac{(5x-12)}{2(x-1)(x-2)} \alpha_2(x) \\[2mm] \alpha_3' &= -4 \dfrac{1}{(x-1)^4} \alpha_1(x) - \dfrac{3}{4} \dfrac{(5x-12)}{(x-1)(x-2)} \alpha_3(x) \end{cases}$$

This will be our leitmotiv example. We will show in section 2.3.2 that this system admits the following *rational* solution in $(\mathbb{C}(x))^4$:

$$\begin{cases} \alpha_0 &= c \\ \alpha_1 &= \dfrac{4}{315}\,\dfrac{-102\,x^2 + 147\,x + 48 + 35\,x^3}{(x-1)^2} \\ \alpha_2 &= 0 \\ \alpha_3 &= -\dfrac{64}{315}\,\dfrac{(x-2)\left(5\,x^2 + 6\,x - 139\right)}{(x-1)^5} \end{cases}$$

where $c \in \mathbb{C}$ is a constant. We infer that :

$$\int f dx \;=\; \frac{4}{315}\,\frac{\left(-102\,x^2 + 147\,x + 48 + 35\,x^3\right)}{(x-1)^2} f \\ -\frac{64}{315}\,\frac{(x-2)\left(5\,x^2 + 6\,x - 139\right)}{(x-1)^5} f^3 + c$$

Or equivalently

$$\int f dx = \frac{\left(-5\,x^2 - 6\,x + 139 + \sqrt{x-1}\left(35\,x^2 - 62\,x + 91\right)\right)\sqrt{-1 + \sqrt{x-1}}}{315\,\sqrt{x-2}}.$$

We see that Liouville's method reduces the calculation of an integral to the (simpler) problem of computing rational solutions of a linear differential system. In what follows, we will show how to compute such solutions ; elaborating, we will in fact show that the more general problem of solving (in a sense that will be defined) a linear differential equation $L(y) = \sum_{i=0}^{n} a_i \frac{d^i y}{dx^i} = 0$ (with $a_i \in \mathbb{C}(x)$) reduces to computing rational solutions of ancillary linear differential systems.

In the two aforementioned memoirs, Liouville already considers the issue of computing algebraic solutions of $L(y) = 0$. He notes that the main theoretical problem is to bound the algebraic degree of such a solution. In 1872, H. Schwarz determines the values of parameters a, b, c of the hypergeometric equation

$$H_{a,b,c}(y) \;=\; x(x-1)\,y^{(2)} + \{c - (1 + a + b)x\}\,y' - ab\,y \;=\; 0$$

for which $H_{a,b,c}(y) = 0$ has an algebraic solution. In a long forgotten first memoir of 1862 (corrected in 1881), P. Pepin shows how to compute algebraic solutions of second order differential equations. This problem is then handled by F. Klein and L. Fuchs (1875) who introduce the use of linear groups and their invariants in this problem. In his 1878 *Mémoire sur les équations différentielles linéaires à intégrales algébriques* (J. für Math. 49), C. Jordan establishes methods for solving third order equations. Although this does not constitute a full algorithm, his ideas are a foundation of modern methods. Interested readers may consult [4, 5, 6, 7, 8] for more historical details.

After the works of Vessiot or Marotte [5], this problem seems to wither and get forgotten, probably because the corresponding calculations are infeasible by hand ; it resurfaces in 1948 with the work on E. Kolchin on algebraic groups and later with the appearance of computers and of computer algebra. In 1977, J. Kovacic gives an algorithm for solving second order equations ; in 1981, M.F. Singer gives a decision procedure for solving linear differential equations of any order. Since then, many improvements have been developed and these have been implemented in computer algebra systems so that users can now really apply them without being specialists. We will describe the state of some of these improvements in what follows. General references for this are, for example, [8, 9, 2] and references therein.

2.1.2 Solutions of Linear Differential Equations

In what follows, we consider a field k (of characteristic 0) equipped with a derivation D (typically, the reader may think of $k = \mathbb{C}(x)$ and $D = \frac{d}{dx}$). The set $C = \{a \in k \mid D(a) = 0\}$ of *constants* of k is a subfield of k ; for technical convenience, we will assume it to be algebraically closed (in general, $C = \mathbb{C}$ or $C = \overline{\mathbb{Q}}$).

Our aim is to describe some (algorithmic) methods for solving homogeneous ordinary linear differential equations

$$L(y) \;=\; y^{(n)} - a_{n-1}y^{(n-1)} - \cdots - a_0 y = 0 \quad (a_i \in k). \quad (2.2)$$

The space of solutions of such an equation is a vector space over C of dimension at most n ([10, 7]). To "solve" such an equation, we will construct n functions (in general in some field extension of the field of coefficients), linearly independent over C, which satisfy $L(y) = 0$. Similarly, to solve a non-homogeneous equation $L(y) = b$, one will construct a particular solution and adjoin a basis (a *fundamental system*) of solutions of $L(y) = 0$.

To continue with our leitmotiv example of the integration of an algebraic function, we will first show that solving linear differential equations is equivalent to solving linear differential systems.

2.2 Linear differential equations versus linear differential systems

The transformation of a linear differential equation $L(y) = b$ to a first order linear differential system in companion form

$$
\begin{pmatrix} y_1 \\ y_2 \\ \cdots \\ y_{n-1} \\ y_n \end{pmatrix}' = \begin{pmatrix} 0 & 1 & 0 & \cdots & & 0 \\ 0 & 0 & 1 & 0 & & \vdots \\ \vdots & & \ddots & \ddots & & 0 \\ 0 & \cdots & \cdots & 0 & & 1 \\ a_0 & a_1 & \cdots & a_{n-2} & a_{n-1} \end{pmatrix} \begin{pmatrix} y_1 \\ y_2 \\ \cdots \\ y_{n-1} \\ y_n \end{pmatrix} + \begin{pmatrix} 0 \\ 0 \\ \cdots \\ 0 \\ b \end{pmatrix}
$$

is well known. It is then possible to go from the solution space of one to the solution space of the other. It is however also possible to associate a linear equation $L(y) = b$ to a given arbitrary first order linear differential system $Y' = AY + B$, where $A \in \mathcal{M}_n(k)$, $B \in k^n$.

Example 2.2.1 *In the system* $(\mathcal{S})$ *obtained from our integration example, we can set* $z = \alpha_0 + \alpha_3$. *Taking derivatives (and using the relations given by the* α_i *and their derivatives), it is possible to express the derivatives* $z^{(i)}$ *as linear combinations of the four unknown functions* α_j *($j = 0, \ldots, 3$) :* $z^{(4)}$ *must therefore linearly depend on* $z, \ldots, z^{(3)}$, *so that* z *satisfies the inhomogeneous linear differential equation* $\mathcal{L}(z) = b$, *where* $\mathcal{L}(z)$ *is given by*

$$
\frac{45}{32} \frac{\left(146965\, x^4 - 1342369\, x^3 + 4634709\, x^2 - 7115419\, x + 4081618\right)}{(x - 2)\,(x - 1)^3\,(5525\, x^3 - 37260\, x^2 + 84801\, x - 64586)} z'
$$
$$
+ \frac{5}{16} \frac{\left(927095\, x^4 - 8377567\, x^3 + 28568871\, x^2 - 43330157\, x + 24580270\right)}{(x - 2)\,(x - 1)^2\,(5525\, x^3 - 37260\, x^2 + 84801\, x - 64586)} z''
$$
$$
+ \frac{160225\, x^4 - 1425785\, x^3 + 4781433\, x^2 - 7135291\, x + 3988058}{2\,(x - 2)\,(x - 1)(5525\, x^3 - 37260\, x^2 + 84801\, x - 64586)} z''' + z^{(4)},
$$

$$
b = -6 \frac{-90555\, x^3 + 284742\, x^2 - 412627\, x + 11050\, x^4 + 230430}{(5525\, x^3 - 37260\, x^2 + 84801\, x - 64586)\,(x - 1)^6\,(x - 2)}.
$$

To each solution z *of this equation, one can associate the following* *solution (α) of the first order differential system ($\mathcal{S}$):*

$$\alpha_0(x) = z - 8\frac{\left(275\,x^2 - 1213\,x + 1322\right)(x-1)(x-2)}{5525\,x^3 - 37260\,x^2 + 84801\,x - 64586}z'$$
$$- \frac{16}{3}\frac{\left(375\,x^2 - 1633\,x + 1762\right)(-1+x)^2(x-2)}{\left(5525\,x^3 - 37260\,x^2 + 84801\,x - 64586\right)}z''$$
$$- \frac{128}{3}\frac{(5\,x-11)(x-1)^3(x-2)^2}{5525\,x^3 - 37260\,x^2 + 84801\,x - 64586}z^{(3)}$$
$$+ \frac{64}{3}\frac{x(-611\,x + 145\,x^2 + 658}{\left(5525\,x^3 - 37260\,x^2 + 84801\,x - 64586\right)(x-1)^2}$$

$$\alpha_1(x) = -\frac{\left(-9219\,x^2 + 22284\,x + 1235\,x^3 - 17564\right)(x-1)^4}{2\left(5525\,x^3 - 37260\,x^2 + 84801\,x - 64586\right)}z'$$
$$- \frac{\left(-13749\,x^2 + 32644\,x + 1885\,x^3 - 25364\right)(x-1)^5}{3\left(5525\,x^3 - 37260\,x^2 + 84801\,x - 64586\right)}z''$$
$$- \frac{4}{3}\frac{\left(65\,x^2 - 323\,x + 390\right)(x-2)(x-1)^6}{5525\,x^3 - 37260\,x^2 + 84801\,x - 64586}z^{(3)}$$
$$+ \frac{4}{3}\frac{(x-1)\left(1495\,x^3 - 10197\,x^2 + 23166\,x - 17536\right)}{5525\,x^3 - 37260\,x^2 + 84801\,x - 64586}$$

$$\alpha_2(x) = -32\frac{\left(11305\,x^3 - 75012\,x^2 + 167229\,x - 124642\right)(x-2)}{(x-1)^2\left(5525\,x^3 - 37260\,x^2 + 84801\,x - 64586\right)}z'$$
$$- \frac{256}{3}\frac{\left(-12123\,x^2 + 26287\,x + 1870\,x^3 - 19034\right)(x-2)}{(x-1)\left(5525\,x^3 - 37260\,x^2 + 84801\,x - 64586\right)}z''$$
$$- \frac{512}{3}\frac{\left(85\,x^2 - 367\,x + 402\right)(x-2)^2}{5525\,x^3 - 37260\,x^2 + 84801\,x - 64586}z^{(3)}$$
$$+ \frac{2048}{3}\frac{(x-2)\left(170\,x^3 - 1047\,x^2 + 2175\,x - 1538\right)}{\left(5525\,x^3 - 37260\,x^2 + 84801\,x - 64586\right)(x-1)^5}$$

$$\alpha_3(x) = 8\frac{\left(275\,x^2 - 1213\,x + 1322\right)(x-1)(x-2)}{5525\,x^3 - 37260\,x^2 + 84801\,x - 64586}z'$$
$$+ \frac{16}{3}\frac{\left(375\,x^2 - 1633\,x + 1762\right)(-1+x)^2(x-2)}{\left(5525\,x^3 - 37260\,x^2 + 84801\,x - 64586\right)}z''$$
$$+ \frac{128}{3}\frac{(5\,x-11)(x-1)^3(x-2)^2}{5525\,x^3 - 37260\,x^2 + 84801\,x - 64586}z^{(3)}$$
$$- \frac{64}{3}\frac{x(-611\,x + 145\,x^2 + 658}{\left(5525\,x^3 - 37260\,x^2 + 84801\,x - 64586\right)(x-1)^2}$$

This shows that solving the first order differential system can be reduced to solving the linear differential equation $\mathcal{L}(z) = b$ and the above relation gives a bijection between the two solution spaces.

This approach can be turned into an algorithm as we shall see next.

2.2.1 Equivalent differential systems

In order to attach a linear differential equation to a given first order differential system $Y' = AY + B$ where $A \in \mathcal{M}_n(k)$ and $B \in k^n$, the idea is to transform the system into companion form using a basis transformation $Z = PY + \beta$ (where $P \in GL_n(k)$, $\beta \in k^n$). Such a transformation is also called a *gauge transformation*. Such a basis transformation produces a new system

$$Z' = PY' + P'Y + \beta' = P[A]Z + (PB + \beta' - P[A]\beta) \qquad (2.3)$$

where

$$P[A] = ((PAP^{-1} + P'P^{-1})).$$

Two systems $Y' = AY + B$ and $Z' = \tilde{A}Z + \tilde{B}$ are said to be *equivalent* or *of the same type* if there exists an invertible matrix $P \in GL_n(k)$ and $\beta \in k^n$ such that $\tilde{A} = P[A] := PAP^{-1} + P'P^{-1}$ and $\tilde{B} = (PB + \beta' - P[A]\beta)$. In this case the relation $Z = PY + \beta$ is a bijection between the two solution spaces.

2.2.2 From systems to scalar equation: cyclic vector approach

In order to transform the system into companion form using a basis transformation, we consider an arbitrary solution $(y_1, y_2, \cdots, y_n)^t$ of $Y' = AY + B$ and use the *Ansatz* $z_1 = \lambda_1 y_1 + \cdots + \lambda_n y_n$. Computing successively $z_2 = z_1', \ldots, z_{n+1} = z_1^{(n)}$ using the relation $Y' = AY + B$ we obtain $n+1$ relations between the n variables y_i. Therefore the variables z_i must be linearly dependent over k, showing that z_1 is a solution of the inhomogeneous linear differential equation $\mathcal{L}(z_1) = \sum_{i=0}^{n} b_i z_1^{(i)} = b$. We obtain this way $Z = PY + \beta$ and $Z' = \tilde{A}Z + \tilde{B}$.

If the matrix P is invertible, then $(\lambda_1, \ldots, \lambda_n)^t$ is called a *cyclic vector* for the initial system.

In the previous section, our choice corresponds to $[1, 0, 0, 1]$ which indeed turned out to be a cyclic vector. However $(1, 0, 0, 0)^t$ or $(0, 1, 0, 0)^t$ are example of vectors that are not cyclic for our system.

Since $z_2 = z_1', \ldots, z_{n+1} = z_1^{(n)}$ the resulting system in Z is in companion form and z_1 satisfies a linear differential equation $\mathcal{L}(z_1) = \sum_{i=0}^{n} b_i z_1^{(i)} = b$. It can be proved (e.g [11, 2]) that almost any choice of $(\lambda_1, \ldots, \lambda_n)^t$ produces a cyclic vector ("almost" means up to an algebraic sub-variety). This justifies the probabilistic approach consisting of trying random vectors in order to find a cyclic vector (in fact, non-cyclic vectors are rare but very useful because they provide a factorization of the differential system).

Some of the algorithms that we will present in the remaining sections are easier to explain for linear differential equations than for first order systems, making the cyclic vector approach an useful tool. A drawback of the transformation is the possible swell of the coefficients of the resulting equation. This motivated the search for algorithms that are directly applicable to first order systems, therefore avoiding the possibly costly transformation ([12] and the references therein).

2.3 Rational solutions

A *rational* solution is a solution in the ground field, i.e the field k to which the coefficients belong. In the following we will focus on the *classical* case $k = C(x)$ (or $C = \mathbb{C}, \overline{\mathbb{Q}}, \ldots$). Our goal in this section is to find a basis of those solutions of $L(y) = \sum_{i=0}^{n} a_i(x)y^{(i)} = 0$ (with polynomial $a_i(x) \in C[x]$) that belong to $C(x)$. To achieve this, we will use local information given by the Laurent series expansion of solutions at some point c. If we write

$$a_i(x) = a_{i,0}(x - c)^{\beta_i} + a_{i,1}(x - c)^{\beta_i + 1} + \ldots$$

and evaluate $L(y)$ at a Laurent series expansion of a solution

$$\frac{p(x)}{d(x)} = b_0(x - c)^{\lambda} + b_1(x - c)^{\lambda + 1} + \ldots,$$

then the coefficient with the lowest valuation $\gamma = min\{\lambda + \beta_i - i\}$ must be zero. This shows that the order λ of the series must be a root of the *indicial equation*

$$ind_c(\lambda) = \sum_{\{i \mid \lambda + \beta_i - i = \gamma\}} \lambda(\lambda - 1) \cdots (\lambda - (i - 1))a_{i,0} = 0$$

A root of the indicial equation is called *an exponent* of the linear differential equation at the point c, and the above shows that there are at most n exponents at each point c.

A point $c \in C$ is a regular point if the exponents at this point are $0, 1, \ldots, n - 1$ (and there is a basis of analytic solutions) and it is a singular point otherwise. It is easily checked that the singular points are the zeros of the highest coefficient $a_n(x)$ (and, possibly, ∞ if we work over the Riemann sphere).

2.3.1 Computation of the rational solutions

The following method appears also in [3]. For $k = C(x)$, a rational solution is of the form $\frac{p(x)}{d(x)}$. We start by computing the denominator $d(x)$ of the fraction. Since the roots of $d(x)$ are all singular points, those roots must also be roots of $a_n(x)$. This leads to the ansatz

$$\frac{p(x)}{d(x)} = \frac{p(x)}{\prod_{i=1}^{s}(x - c_i)^{-\lambda_i}} = p(x)\prod_{i=1}^{s}(x - c_i)^{\lambda_i}$$

where $c_i \in C$ are zeroes of $p_n(x)$ and $\lambda_i \in \mathbb{Z}$ are exponents of the linear differential equation. A necessary condition for the existence of a rational solution is that at each singular point there exists an exponent in $\mathbb{Z}$. Instead of trying all the possible combinations of exponents, we choose for λ_i the smallest exponent in $\mathbb{Z}$ at c_i. Our rational solution will be of the form

$$\frac{p(x)}{d(x)} = \frac{p(x)}{\prod_{j=1}^{s}(x - c_i)^{-\lambda_i}} = p(x)\prod_{j=1}^{s}(x - c_i)^{\lambda_i}$$

with known denominator (the other possibilities for the exponents can be put into this form by multiplying the numerator and the denominator with a common factor).

Substitution of this expression into $L(y) = 0$ leads to a new linear differential equation for the yet unknown numerator $p(x)$:

$$\tilde{\mathcal{L}}(p(x)) = L\left(p(x)\prod_{j=1}^{s}(x - c_j)^{\lambda_j}\right) = \sum_{i=0}^{n} a_i(x)\left(\frac{p(x)}{\prod_{j=1}^{s}(x - c_j)^{-\lambda_j}}\right)^{(i)}$$

$$= \sum_{i=0}^{n} \tilde{a}_i(x)\,(p(x))^{(i)} .$$

We can clear denominators in order to obtain again an equation $\tilde{\mathcal{L}}(y) = \sum_{i=0}^{n} \tilde{a}_i(x)y^{(i)}$ with $\tilde{a}_i(x) \in C[x]$. As we search for a polynomial solution, we may look at the action of $\tilde{\mathcal{L}}$ on a polynomial $p(x) = \sum_{i=0}^{N} \gamma_i x^i$ with unknown coefficients using

$$\tilde{a}_i(x) = x^{\beta_i}\left(a_{i,\infty,0} + \mathcal{O}(\frac{1}{x})\right) .$$

Since we must have $\tilde{\mathcal{L}}(p(x)) = 0$, the coefficient of the highest term must be zero. Like for the indicial equation above, this leads to a polynomial

$$ind_\infty(\lambda) = \sum_{\{i|\lambda+\beta_i-i=\gamma\}} \lambda(\lambda - 1)\cdots(\lambda - (i - 1))b_{i,\infty,0} = 0.$$

The possible degrees N of $p(x)$ are the positive integer roots of this polynomial. Again it is sufficient to consider the maximal positive integer root N, which leads to the Ansatz $p(x) = \sum_{i=0}^{N} \gamma_i x^i$ with known degree N.

From $\tilde{\mathcal{L}}(\sum_{i=0}^{N} \gamma_i x^i) = 0$ we obtain a homogeneous linear system for the $N + 1$ unknowns γ_i. Solving this linear system will produce a C-basis $\{p_1(x), \ldots, p_m(x)\}$ of the polynomial solutions of $\tilde{\mathcal{L}}(y) = 0$, and therefore a C-basis of rational solutions of $L(y) = 0$ of the form

$$\{\frac{p_1(x)}{\prod_{i=1}^{s}(x - c_i)^{-\lambda_i}}, \ldots, \frac{p_m(x)}{\prod_{i=1}^{s}(x - c_i)^{-\lambda_i}}\}$$

The computational bottleneck in this algorithm resides in the way this last linear system is handled. Algorithms exploiting the structure of this system are presented in [13, 14, 12] and an optimal (up to date) version is given in [15].

Note that the calculation of the degree N can be understood in a slightly more conceptual way as follows. We look for a solution $p(x) = x^N + p_{N-1}x^{N-1} + \cdots = (\frac{1}{x})^{-N}(1 + p_{N-1}\frac{1}{x} + \mathcal{O}(\frac{1}{x}))$ so we see that we have a pole of order N at infinity so $-N$ must be an exponent at infinity, i.e a root of the indicial equation at infinity.

Exponents are easier to compute for linear differential equations than for first order systems. However it is also possible to compute the rational solutions of $Y' = AY$ ([12] and references therein) without converting the system to a linear differential equation via a cyclic vector computation.

2.3.2 Non homogeneous equations and solution of the leitmotiv example

The previous method can be adapted to non homogeneous equations $\mathcal{L}(y) = b$ (cf. [16] and references therein) but instead we will transform the inhomogeneous equation into a homogeneous equation of higher order $L(y) = b\,(\mathcal{L}(y))' - b'\mathcal{L}(y) = 0$. Note that any rational solution of the first will also be a rational solution of the second.

In our leitmotiv example, assuming Liouville's theorem, we have reduced the integration problem of an algebraic integral to the problem of finding the rational solution of the inhomogeneous equation $\mathcal{L}(y) = b$. Using the above trick we are left with the following homogeneous equation $L(y) =$

$$\frac{45}{16}\frac{440895\,x^5 + 19124660\,x^3 - 41917722\,x^2 + 47163053\,x - 4528916\,x^4 - 21498482}{(x-2)\left(-90555\,x^3 + 284742\,x^2 - 412627\,x + 11050\,x^4 + 230430\right)(x-1)^4}\,y'$$
$$+\frac{5}{32}\frac{17478890\,x^5 - 180447867\,x^4 + 761665199\,x^3 - 1654880375\,x^2 + 1838740539\,x - 827751650}{(x-1)^3\,(x-2)\left(-90555\,x^3 + 284742\,x^2 - 412627\,x + 11050\,x^4 + 230430\right)}\,y''$$
$$+\frac{22088950\,x^5 - 228088635\,x^4 + 958764595\,x^3 - 2062761247\,x^2 + 2264278659\,x - 1007458642}{16\,(x-1)^2\,(x-2)\left(-90555\,x^3 + 284742\,x^2 - 412627\,x + 11050\,x^4 + 230430\right)}\,y^{(3)}$$
$$+\frac{453050\,x^5 - 4656315\,x^4 + 19428215\,x^3 - 41370911\,x^2 + 44911683\,x - 19779482}{2\,(x-1)\,(x-2)\left(-90555\,x^3 + 284742\,x^2 - 412627\,x + 11050\,x^4 + 230430\right)}\,y^{(4)} + y^{(5)} = 0.$$

The expressions look more and more terrifying, but are easily handled by a computer.

The (finite) singular points are $c_1 = 2$, $c_2 = 1$, and the roots c_i ($i > 2$) of the irreducible equation

$$x^4 - \frac{18111}{2210}x^3 + \frac{142371}{5525}x^2 - \frac{412627}{11050}x + \frac{23043}{1105} = 0.$$

Computing the exponents (for example using the command **gen_exp** in MAPLE), we find $(0, 1, 2, 3, \frac{3}{2})$ at $x = 2$, $(-5, -2, 0, -\frac{1}{2}, -\frac{9}{2})$ at $x = 1$, and $(0, 1, 2, 3, 5)$ at c_i (for $i > 2$). A rational solution of the homogeneous equation must be of the form $p(x)(x-2)^0(x-1)^{-5}\prod_{i>2}(x-c_i)^0$, or simply $\frac{p(x)}{(x-1)^5}$.

We compute the differential equation $\widetilde{L}$ for p with $\widetilde{L}(p(x)) = L(\frac{p(x)}{(x-1)^5})$. According to our method, the degree N of p is such that $N \in \{3, 5\}$.

The solution $p(x)$ we are looking for must be of the form

$$p(x) = \gamma_0 + \gamma_1 x + \gamma_2 x^2 + \gamma_3 x^3 + \gamma_4 x^4 + \gamma_5 x^5.$$

Plugging this ansatz into the relation $\overline{L}(p(x)) = 0$, we obtain the following linear system :

$$\begin{pmatrix} 0 & 0 & -1125 & -900 & 53959 & 267545 \\ 0 & 0 & 0 & 0 & -1 & -5 \\ 22732650 & 51866405 & 239145235 & 493746279 & 463316707 & -466026280 \\ 11443680 & 16808410 & -44942425 & -164608566 & -313297728 & -442425600 \\ 0 & 0 & 0 & 0 & 1 & 5 \\ 0 & 61880 & 3442015 & 4622388 & -39011582 & -207171040 \\ 1868670 & 13070140 & 162697775 & 420978396 & 691480368 & 811113600 \\ -3729375 & -8503865 & -47608495 & -87550269 & 34987468 & 613145030 \\ -56964465 & -140390815 & -733512410 & -1659388287 & -2308356996 & -1638036600 \end{pmatrix} \begin{pmatrix} \gamma_0 \\ \gamma_1 \\ \gamma_2 \\ \gamma_3 \\ \gamma_4 \\ \gamma_5 \end{pmatrix}$$

$$= 0$$

solving the system produces the following rational solutions of our linear differential equation :

$$y = \frac{\left(-\frac{139}{2} - \frac{5}{4}x^3 + \frac{151}{4}x + x^2\right)}{(x-1)^5}\gamma_2 + \frac{-696 - \frac{5}{2}x^3 + \frac{765}{2}x - 5x^4 + x^5}{(x-1)^5}\gamma_5.$$

where γ_2 and γ_5 are arbitrary constants in $\mathbb{C}$. Note that $\{\gamma_5 = 1, \gamma_2 = -10\}$ corresponds to the obvious solution $y = 1$.

In order to obtain the rational solutions of our inhomogeneous equation, we simply substitute y into this inhomogeneous equation and obtain

$$z = -\frac{64}{315}\frac{(x-2)\left(5x^2 + 6x - 139\right)}{(x-1)^5} + \gamma$$

where γ is an arbitrary constant. Now, to solve the integration problem of our leitmotiv example, we substitute this solution into the relations (α) page 88. The resulting solution of the system (S) is the one given in the introduction and we recover the integral given on page 84.

2.4 Factorization and Reduction of Order of a Differential Equation

One way to simplify the solving of $L(y) = 0$ is to find a *factorization* in the form of a composition $L(y) = L_1(L_2(y))$ of operators. Solutions of $L_2(y) = 0$ would then be solutions of L and we would have reduced the order of the equation that we want to solve. However, this factorization is not unique.

This factorization can be given an algebraic (and computable) meaning in an appropriate setting.

2.4.1 Differential operators

First let us write L as a differential operator

$$L(y) = (a_n(x)D^n + \ldots + a_1(x)D + a_0(x))\,(y)$$

where the symbol D represents $\frac{d}{dx}$. To endow the set of differential operators with a ring structure (of non-commutative polynomials, or *Ore polynomials*), we note that

$$(Dx)(y) = \frac{d}{dx}(xy) = x\frac{d}{dx}(y) + y = (xD + 1)\,(y)$$

Using this rule $Dx = xD + 1$, one can show that the set $\mathcal{D} := k[D]$ of differential polynomials, with the usual addition and the above multiplication rule (for any $a \in k$, $D.a := aD + a'$) is a non commutative ring. As hinted above, multiplication in this ring corresponds to the composition of differential operators. Furthermore, one can show that $\mathcal{D}$ is a left (resp. right) Euclidean ring, which means that one has notions of computable gcd (left or right), least common (left or right) multiple, etc:

2.4.2 Factorization of L

There exist several factoring algorithms; we will display them on our leitmotiv example.

Example 2.4.1 *In our leitmotiv example, we remark that D is a factor on the right of our operator L.*
Using one of the available factoring algorithms (e.g the command DFactor *in* MAPLE*), we have a "complete" factorization of L in the form*

$$L = (D + f_1)(D + f_2) L_{2,1} D, \qquad \text{with} \qquad f_1, f_2 \in \mathbb{C}(x)$$

where †

$$L_{2,1} = D^2 + \frac{(-12123\,x^2 + 1870\,x^3 + 26287\,x - 19034)}{2\,(85\,x^2 - 367\,x + 402)\,(x - 2)\,(x - 1)} D$$
$$+ \frac{3}{16}\,\frac{11305\,x^3 - 75012\,x^2 + 167229\,x - 124642}{(x - 2)\,(x - 1)^2\,(85\,x^2 - 367\,x + 402)}$$

When $L = M \cdot N$, suppose that we know a solution y of $N(y) = 0$; then $L(y) = M(N(y)) = M(0) = 0$ so solutions of N are solutions of L. There is a kind of converse to this result: if N is an *irreducible* operator of order lower than the order of L, and if there exists y such that both $N(y) = 0$ and $L(y) = 0$, then N is a right factor of L. This is a consequence of the facts that $\mathcal{D}$ is left euclidean and that an operator of some order r possesses at most r linearly independent solutions.

We now will show that L could admit many different factorizations.

Example 2.4.2 *In our example, we observe many other factorizations; they can be computed via the eigenring method introduced by Singer (see [17, 18, 19]) using the commands* DFactorLCLM *or* eigenring *in*

---

† The full expressions of f_1 and f_2 are heavy and uninteresting here ; however, the reader is encouraged to check them with her/his favorite computer algebra system.

MAPLE. *For example,*

$$L = \dot{(}D + \tilde{f}_1)(D + \tilde{f}_2)(D + \tilde{f}_3)$$

$$\left(D^2 + \frac{3}{2}\frac{(6\,x - 13)\,D}{x^2 - 3\,x + 2} + \frac{3}{16}\frac{85\,x^2 - 367\,x + 402}{(x^2 - 3\,x + 2)^2}\right),$$

with $\tilde{f}_i \in k$.

One can show (and we verify it on our examples) that the number of factors (and their order, i.e the degree in D) is unique (up to "isomorphisms" and permutation of factors)†. But there can still be infinitely many factorizations. In order to exhibit infinitely many factorizations, recall that in part 2.3.2 we had found a basis of the rational solutions to $L(y) = 0$ given by $y =$

$$\frac{(-\frac{139}{2} - \frac{5}{4}x^3 + \frac{151}{4}x + x^2)}{(x-1)^5}\gamma_2 + \frac{-696 - \frac{5}{2}x^3 + \frac{765}{2}x - 5\,x^4 + x^5}{(x-1)^5}\gamma_5$$

If we denote by $f = \frac{y'}{y}$ the logarithmic derivative of such a solution, then $D - f$ is a right-hand factor of L (because $(D - f)(y) = 0$). Now f depends on two parameters (arbitrary constants) so we have infinitely many right-hand factor which will be of the form $D-$

$$\frac{(3\,x^2 + 10\,x^3 + 1239 - 612\,x)\,(\gamma_2 + 10\,\gamma_5)}{((4\,x^5 - 20\,x^4 - 10\,x^3 + 1530\,x - 2784)\gamma_5 - (x-2)\,(5\,x^2 + 6\,x - 139)\,\gamma_2)\,(x-1)}.$$

2.4.3 Decomposition of L

We have seen that finding right-hand factors amounts to finding distinguished subspaces of the solution space of L. To study solutions of L, one may ask for something stronger than a factorization of L, namely a *decomposition* of L as the Least Left Common Multiple (LCLM) of lower-order operators L_i.

When this is the case, the solution space of L is the direct sum (as a vector space) of the solution spaces of the L_i. This simplifies greatly the operation of solving L, of course; it is roughly what the eigenring method (mentioned above) computes (when possible).

We may sketch this method as follows. The eigenring $\mathcal{E}(L)$ consists of operators R, of order less than that of L, such that there exists an

---

† it is a consequence of the Jordan-Holder theorem for $\mathcal{D}$-modules.

operator S with $L.R = S.L$. We then have that, if $L(y) = 0$ then $L(R(y)) = S(L(y)) = S(0) = 0$ so R induces an endomorphism of the solution space $V(L)$. For an eigenvalue λ of this endomorphism R, there exists an $y \in V(L)$ (i.e a solution of L) such that $(R - \lambda)(y) = 0$. This means that L and $(R - \lambda)$ have a solution in common, so the right-gcd of L and $R - \lambda$ is a non-trivial factor. The decomposition of $V(L)$ as a direct sum of characteristic subspaces of R induces a decomposition of L as the LCLM of the corresponding factors. This method is developed in [17, 18, 19].

2.4.4 Exponential Solutions

We may give an idea of how to factor by focusing on first-order factors (in fact, the search for factors of higher order may be reduced to this question, see [8, 2] for more details and references).

A first order factor is of the form $D - f$ with $f \in k$. The operator L admits such a factor on the right if and only if the equation $L(y) = 0$ admits a solution y such that $y' = fy$. Because such a y is an exponential (of an integral of f), one calls such a solution an *exponential solution* of $L(y) = 0$. These solutions will also prove useful in part 2.5.

The algorithm for computing such solutions is roughly similar to (though quite a bit longer in practice than) the algorithm for computing rational solutions: we first compute what could be the behavior at the singularities and then search for a polynomial that would "glue together" these local pieces of information.

For our leitmotiv equation $L(y) = 0$, one can show ([20, 8, 21] as our equation is fuchsian) that any exponential solution to our equation will be of the form $y = \prod_{j=1}^{s}(x-c_j)^{\lambda_j} p(x)$, where p is a polynomial, the c_j are the singularities, and the λ_j are exponents (not necessarily integers, as opposed to what happened with rational solutions) of the singularities c_j. For each combination of the λ_j, we set $y = \prod_{j=1}^{s}(x - c_j)^{\lambda_j} z$, we compute the linear differential equation satisfied by z, and look whether it admits polynomial solutions. This is hence similar to the situation for rational solutions ; however, this is much more costly because many combinations (the number is exponential in the number of singularities) of the λ_j have to be checked.

In part 2.3.2, we had already found two rational solutions to $L(y) = 0$. Studying the exponents at the singularities which we had already computed, we see that there remain three types of putative exponential solutions:

$$\frac{1}{(x-1)^5}p_1(x), \qquad \frac{1}{(x-1)^{\frac{1}{2}}}p_2(x), \qquad \frac{(x-2)^{\frac{5}{2}}}{(x-1)^{\frac{1}{2}}}p_3(x).$$

We have found a basis of two possibilities for the polynomial p_1, which gave us a vector space of rational solutions of dimension 2. For the third possibility, calculation shows that no polynomial p_3 fits. For the second possibility, calculation again shows that $p_2 = 1$ is the only polynomial solution, up to multiplication by a constant, and we obtain the exponential solution $y_3 = \frac{1}{\sqrt{x-1}}$; the corresponding right hand factor of L is $D + \frac{1}{2(x-1)}$. You could also retrieve this result using the command **expsols** in MAPLE.

A more sophisticated algorithm for exponential solutions, using a smart mix of local informations and reductions modulo primes, is given in [22].

2.5 Liouvillian solution

Since our leitmotiv example $L(y) = 0$ is of order five, we need to compute a basis of solutions of dimension five over C in order to obtain all solutions. We already computed a two dimensional space of rational solutions and one exponential solution. In order to find the remaining two solutions, we will have to look for solutions in a larger class of functions.

Those two solutions will be solutions of the second order left factor of the second factorization of L computed in the previous section :

$$L_2(y) = y''(x) + \frac{3}{2}\frac{(6\,x - 13)}{(x-2)\,(x-1)}y'(x) + \frac{3}{16}\frac{(85\,x^2 - 367\,x + 402)}{(x-2)^2\,(x-1)^2}y(x)$$

We will now turn to more general class of solutions consisting of those functions that can be *written* using the symbols $\int$, $e^\int$ and algebraic functions. The so-called *Liouvillian solutions* are formally defined in the following way

Definition 1 *A function is* Liouvillian *over a (differential) field k if the function belongs to a (differential) field extension K of k where*

$$k = K_0 \subset K_1 \subset \cdots \subset K_N = K$$

such that for $i \in \{1, \ldots, n-1\}$ we have $K_{i+1} = K_i(t_i)$ where

(i) t_i *is algebraic over K_i (algebraic extension), or*

(ii) $t_i' \in K_i$ *(extension by an integral, here $t_i = \int u$, where $u \in K_i$),*
or

(iii) $t_i'/t_i \in K_i$ *(extension by the exponential of an integral).*

The following function is liouvillian over $\mathbb{C}(x)$:

$$\frac{\sqrt{x}\,\exp(\int \sqrt{x+1}\,dx)}{\int \exp(x^2)\,dx}.$$

2.5.1 Differential Galois theory and algorithms

In classical Galois theory one shows the existence of a splitting field of a polynomial which is the smallest field extension containing all the roots. The classical Galois group sends a root into another root and therefore permutes the roots. Properties of the polynomial and its roots are mirrored in the structure of the classical Galois group. For example the polynomial is irreducible over the coefficient field if and only if the permutation action on the root is transitive and the polynomial is solvable by radicals if and only if the group is solvable.

A similar theory exists for linear differential equations as a result of work of Picard, Vessiot and Kolchin. Note that the solution space of $L(y) = 0$ is a vector space over the field of constants, and therefore the field of constants C will play an important role in the theory. It is convenient to assume that C is algebraically closed of characteristic zero and we will make this assumption throughout the remaining of this section. Another new ingredient we will have to take into account is the notion of a derivative. One shows that there exists a differential splitting field called the Picard-Vessiot extension (in short PV extension), which is the smallest differential field extension containing a full basis of solutions as well as all derivatives of those solutions, but no new constants.

Example 2.5.1 *The Airy equation $A(y) = y'' - xy$ over $k = \mathbb{C}(x)$ has a $\mathbb{C}$-basis of solutions given by*

$$y_1 = 1 + \sum_{i=1}^{\infty} \frac{x^{3i}}{\prod_{j=1}^{i} 3j(3j-1)}, \quad y_2 = x + \sum_{i=1}^{\infty} \frac{x^{3i+1}}{\prod_{j=1}^{i} 3j(3j+1)}$$

The PV extension is $\mathbb{C}(x, y_1, y_2, y_1', y_2')$. Note that higher derivatives of y_i can be expressed using the y_i and y_i', like $y_i'' = xy_i$. We have $y_1 y_2' - y_1' y_2 = 1$ as a differential relation among the solutions. ◇

Definition 2 *The differential Galois group G of the equation $L(y) = 0$*

(in fact of the PV extension K/k associated to $L(y) = 0$) is the set of automorphisms of K that leave elements of k fixed and commute with the derivation.

$$G = \{\sigma \in \text{Aut}(K/k) \,|\, \sigma D = D\sigma\}$$

Since G fixes the coefficients $a_i \in k$ of $L(y) = \sum_{i=0}^{n} a_i y^{(i)}$ and commutes with the derivation, we get for $\sigma \in G$ and any solution y_j of $L(y) = 0$ that

$$0 = \sigma(0) = \sigma\left(L(y_j)\right) = \sigma\left(\sum_{i=0}^{n} a_i y_j^{(i)}\right) = \sum_{i=0}^{n} a_i \left(\sigma(y_j)\right)^{(i)} = L(\sigma(y_j)).$$

Therefore G sends a solution of $L(y) = 0$ into a solution. If $y_1, y_2, \ldots, y_n$ is a C-basis of the solution space in the PV extension, then $\sigma(y_i) = \sum_{j=1}^{n} c_{i,j} y_j$ where $c_{i,j} \in C$. This shows that G has a faithful representation as a group of $n \times n$ matrices $(c_{i,j})$ over the field of constants C (while the classical group is a permutation group, the differential Galois group is a linear group). Like the classical Galois group, the differential Galois group mirrors the properties of the linear differential equation and its solutions.

Proposition 1 *([23]) Under our standard assumptions and notations, the following holds :*

(i) *All solutions of $L(y) = 0$ are rational (i.e. in k) if and only if $\mathcal{G} = \{id\}$.*

(ii) *All solutions of $L(y) = 0$ are algebraic over k if and only if $\mathcal{G}$ is a finite group.*

(iii) *$L(y)$ has a non trivial factorization $L_1(L_2(y))$ if and only if $\mathcal{G} \subset \text{GL}_n(C)$ is a reducible linear group (there exists a non trivial subspace W such that $\forall \sigma \in \mathcal{G}$, $\sigma(W) = W$).*

In fact the differential Galois group has an important additional property, it is a *linear algebraic group*. This means that the set of matrices $(c_{i,j})$ of the group G, where each matrix is viewed as points in the space C^{n^2}, is an algebraic variety over C. This imposes strong restrictions as we shall now see in connection with our integration problem.

Example 2.5.2 *The computation of an integral $\int a$ for $a \in k$ corresponds to the resolution of the inhomogeneous differential equation $y' = a$. With our previous trick, we can transform this to a homogeneous equation $L_f(y) = y'' - \frac{a'}{a} y'$. The C-basis of the solution space*

*of the latter is $\{1, \int a\}$, showing that the PV extension in this case is
simply $K = k(\int a)$. In order to find the matrix of an element $\sigma \in G$
in the basis $\{1, \int a\}$, we note that $\sigma(1) = 1$ (because $1 \in k$) and since σ
commutes with the derivation we obtain :*

$$\left(\sigma \left(\int a \right) - \int a \right)' = \sigma(a) - a = a - a = 0.$$

*This shows that $\sigma(\int a) - \int a$ is a constant $c_\sigma \in C$ and therefore $\sigma(\int a) =$
$\int a + c_\sigma$. The matrix of σ is thus*

$$\begin{pmatrix} 1 & c_\sigma \\ 0 & 1 \end{pmatrix}.$$

*The group $\mathcal{G}$ consists of the "points" $(c_{i,j}) \in C^4$ which are zeros of the
polynomial system $c_{1,1} = c_{2,2} = 1, c_{2,1} = 0$. This variety is isomorphic to
C and we find that $\mathcal{G}$ is a subgroup of the additive group $(C, +)$. Algebraic
sub-varieties of C must be zero sets of one polynomial and are therefore
either finite sets or C. Since an additive subgroup G of $(C, +)$ is either
$\{0\}$ or contains infinitely many elements, the only possibilities for G are
$\{0\}$ and C.
If $G = \{0\}$, then all solutions and in particular $\int a$, are rational and
therefore belong to k. Otherwise $G = (C, +)$ and not all solutions are
algebraic, which implies that $\int a$ is not algebraic (and therefore must be
transcendental over k).* ◇

The above example gives a proof of a result very similar to Liouville's
theorem on algebraic integrals stated in the introduction :

Theorem 1 *(Liouville) Let k be a differential field with algebraically
closed field of constants of characteristic zero. The primitive $\int a$ of $a \in k$
is either in k or is transcendental over k. In particular, an integral that
is algebraic over k must be in k .* ◇

Liouville's original proof of the above is easier and more linked to the
algebraic step of the Liouvillian extensions than to the *integral step* of
the Liouvillian extensions:

Example 2.5.3 *Suppose that f is algebraic over the differential field
(k, D) and that the minimal polynomial of f is $p(X) = X^m + \sum_{i=0}^{m-1} a_i X^i$.
A derivation Δ on the algebraic field extension $k(f)$ which is an exten-*

sion of D must satisfy

$$0 = p(f)' = mf'f^{m-1} + \sum_{i=0}^{m-1} a_i f' f^{i-1} + \sum_{i=0}^{m-1} a_i' f^i.$$

This shows that

$$f' = -\frac{\sum_{i=0}^{m-1} a_i' f^i}{mf^{m-1} + \sum_{i=0}^{m-1} a_i f^{i-1}}$$

and that there is at most one such derivation. One can verify that the formula indeed defines a derivation on $k(f)$. If $\sigma \in Aut(k(f)/k)$ is an automorphism, then $\sigma D \sigma^{-1}$ is also a derivation of $k(f)$ and therefore $\sigma D \sigma^{-1} = D$, showing that for any automorphism $\sigma D = D\sigma$. Thus the differential Galois group and the classical Galois group coincide for algebraic extensions.

Liouville now noted that if $f' = a$ (or $f = \int a$), then for all $\sigma \in Aut(k(f)/k)$ we have $\sigma(f)' = a$. Therefore the trace of f (the sum of all conjugates) is again an integral of a which is in k.

Another consequence of $f' \in k(f)$ is that all higher derivatives $f'', \cdots, f^{(m)}$ of f also belong to $k(f)$. Since $[k(f) : k] = m$, the $m + 1$ elements of $f, f', f'', \cdots, f^{(m)}$ must be linearly dependent over k, showing that f is the solution of a linear differential equation over k of degree at most m.

The existence of a liouvillian solution is also reflected in the differential Galois group. First we note that the set of liouvillian solutions is a G-invariant sub-space of the solution space and therefore that there is a right factor whose solutions are all liouvillian solutions of $L(y) = 0$ (cf. Example 2.4.1 and 2.4.2). For this reason most algorithms first factor (and, possibly, decompose as an LCLM) the given differential equation and then look for liouvillian solutions of irreducible right-hand factors.

For irreducible equations, the existence of a liouvillian solution is equivalent to $G \subset GL_n(C)$ having a finite subgroup H leaving a one dimensional subspace generated by a solution z of the solution space invariant ([24]). In more sophisticated terms, an irreducible equation $L(y) = 0$ has a liouvillian solution if and only if the component of the identity G° of the linear algebraic group G is solvable ([23]). Using a theorem of C. Jordan and a bound of I. Schur, it is possible to bound the index $[G : H]$ of the largest group H with the above property ([20])

$$[G : H] \leq f(n) \leq \left(\sqrt{8n} + 1\right)^{2n^2} - \left(\sqrt{8n} - 1\right)^{2n^2}$$

Note that the bound is independent of the actual equation $L(y) = 0$ and depends only on the order of the equation. Since $H \subset G$, its elements σ commute with the derivation and H leaves the line generated by z invariant ; so, we find that $\forall \sigma \in H$, $\sigma(z) = c_\sigma z$ and $\sigma(z') = c_\sigma z'$. In particular H being maximal with this property is the stabilizer of z'/z and we get that the orbit of $u = z'/z$ is finite of length $[G : H] \leq f(n)$. This shows that u is algebraic of degree at most $[G : H] \leq f(n)$ and that $L(y) = 0$ has a solution z of the form $z = e^{\int u}$ where u is algebraic of degree at most $[G : H] \leq f(n)$. We obtain that if $L(y) = 0$ has a liouvillian solution, i.e. a solution that can be written in $\int$, $e^{\int}$ and algebraic functions, then it has a solution of the form $z = e^{\int u}$ where u is algebraic of bounded degree ([23, 20]). The existing algorithms to compute liouvillian solutions will attempt to compute the minimal polynomial of u of the form

$$P(u) = u^m + b_{m-1}u^{m-1} + \ldots + b_1 u + b_0 = 0 \qquad (b_i \in k)$$

by trying degrees less than the above bound. If no polynomial is found, then there is no liouvillian solution.

A classification of the subgroups of $GL_n(C)$ allows one to give a more precise list of the possible degrees (many other degrees are possible, but this is the list of the smallest degrees):

(i) For $n = 2$, $\exists u = z'/z$ such that $[k(u) : k] \in \{1, 2, 4, 6, 12\}$ (cf. [25]).

(ii) For $n = 3$, $\exists u = z'/z$ such that $[k(u) : k] \in \{1, 3, 6, 9, 21, 36\}$ (cf. [26, 24]).

(iii) For $n = 4$, $\exists u = z'/z$ such that (cf. [27])

$$[k(u) : k] \in \{1, 4, 5, 8, 10, 12, 16, 20, 24, 40, 48, 60, 72, 120\}.$$

(iv) For $n = 5$, $\exists u = z'/z$ such that $[k(u) : k] \in \{1, 5, 6, 10, 15, 30, 40, 55\}$ (cf. [27]).

A similar finite list exists for any order.

In order to compute the minimal polynomial $P(u)$ of $u = z'/z$, we note that G permutes the roots of $P(u)$ and sends the logarithmic derivative of a solution into the logarithmic derivative of a solution. This gives the following form

$$P(u) = u^m + b_{m-1}u^{m-1} + \ldots + b_1 u + b_0$$

$$= \prod_{i=1}^{m}\left(u - \frac{z_i'}{z_i}\right)$$

$$= u^m - \frac{(\prod_{i=1}^{m} z_i)'}{\prod_{i=1}^{m} z_i}u^{m-1} + \ldots + \prod_{i=1}^{m}\frac{z_i'}{z_i}$$

where z_i are solutions of $L(y) = 0$. In particular, the coefficient b_{m-1} is the logarithmic derivative of a product of m solutions and one can show that this condition is a necessary and sufficient condition for the existence of a liouvillian solution [25, 28, 29].

In order to compute b_{m-1}, one constructs a linear differential equation $L^{\circledS m}(y)$ whose solution is spanned by the products of length m of solutions of $L(y) = 0$. We start with an arbitrary product $z = z_1 z_2 \ldots z_m$ where z_i are solutions of $L(y) = 0$. Taking derivatives of z we obtain expressions of the form $z_1^{(i_1)} z_2^{(i_2)} \ldots z_m^{(i_m)}$. Using $L(y) = 0$ as a rewrite rule, we can replace derivatives of z of order $\geq n$ by linear combinations over k of lower order derivatives of z. After $\binom{n+m-1}{n-1}$ derivations the $\binom{n+m-1}{n-1} + 1$ expressions $z^{(i)}$ are all linear combinations of at most $\binom{n+m-1}{n-1}$ unknowns and will therefore be linearly dependent over k. The linear combination involving the smallest order of derivation of z is called the m-th symmetric power of $L(y)$ and is denoted $L^{\circledS m}(y)$.

The order of the linear differential equation $L^{\circledS m}(y)$ grows very fast and it can be difficult to compute.

For second order equations $L(y)$, however, the order of $L^{\circledS m}(y)$ is always $m + 1$ and therefore maximal. In this case, $L^{\circledS m}(y)$ can be computed using a simple recurrence [30] :

Example 2.5.4 *Consider $L(y) = y'' + ry$ (where $r \in k$). Then setting*

$$\begin{cases} L_0(y) &= y, \\ L_1(y) &= y', \\ L_{i+1}(y) &= L_i(y)' + i(m - i + 1)r L_{i-1}(y). \end{cases}$$

We obtain $L_{m+1} = L^{\circledS m}$

2.5.2 Liouvillian solutions of second order equations

The solution of second order equations is simplified by the fact that symmetric products are always of maximal order and that any solution of a symmetric product is a product of solutions. A very efficient algorithm is due to Kovacic [25] and we present a slight improvement of this algorithm [28] where the computation of exponential solutions is replaced by the computation of rational solutions whenever possible:

(i) Factor $L(y)$ by computing exponential solutions. A non trivial exponential solution gives a liouvillian solution. A second independent liouvillian solution can be obtained by the method of variation of the constants.

(ii) If $L(y)$ is not of the form $y'' + ry$ (i.e. if $a_1 \neq 0$), replace y by $y \cdot e^{\int -\frac{a_1}{2}}$, in order to obtain an equation of the form $y'' + ry$ (where $r \in k$) which we denote again by $L(y)$. (The change of variable is a bijection which transforms liouvillian solutions into liouvillian solutions.)

(iii) For $m \in \{4, 6, 8, 12\}$ (in this order!) :

- Compute rational solutions f of $L^{\circledS m}(y) = 0$
- If a non trivial rational solution f exists, then

  – set $b_{m-1} = -\frac{f'}{f}$,
  – compute the other coefficients b_i of $P(u)$ using the recurrence $(\sharp_m)$:

$$\begin{cases} b_m = 1, \qquad b_{m-1} = -\frac{f'}{f} \\ b_{i-1} = \dfrac{-b_i' + b_{m-1}b_i + r(i+1)b_{i+1}}{m-i+1}, \quad m-1 \geq i \geq 0 \end{cases}$$

  – return an irreducible factor of the polynomial $P(u)$

  Otherwise try the next m

(iv) If no m produces a liouvillian solution, then there are no such solutions.

For higher order equations, the computation is more difficult and methods can for example be found in [31, 29, 32]. An important observation is that for $m \in \{6, 8, 12\}$, the solutions are in fact algebraic of known algebraic degree. In this case, it is possible to compute the minimal polynomial of a solution instead of the minimal polynomial of a logarithmic derivative [33, 34].

An alternative efficient algorithm, using the above and Klein's method for computing algebraic solutions of second order differential equations via hypergeometric functions, is given in [35, 36].

Example 2.5.5 *In the decomposition of the operator L of our leitmotiv example, the following second order equation remained to be solved :*

$$L_2(y) := y'' + \frac{3}{2}\frac{(6x-13)}{(x-2)(x-1)}y' + \frac{3}{16}\frac{85x^2 - 367x + 402}{(x-2)^2(x-1)^2}y = 0.$$

The steps in the above algorithm are

(i) $L_2(y)$ *is irreducible (it was one of the irreducible factors in section 2.4.1)*

(ii) $L_2(y)$ *is not of the form $y'' + ry$, therefore we replace y by*

$$y \cdot e^{\int -\frac{a_1}{2}} = y(x-2)^{3/4}(x-1)^{-\frac{21}{4}} \qquad (2.4)$$

and obtain the equation

$$L(y) = y'' + \frac{3}{16}\frac{x^2 - 3x + 3}{(x-2)^2(x-1)^2}y$$

(iii) *We consider the first case $m = 4$ and compute $L^{\circledS 4}(y)$:*

$$y^{(5)} + \frac{15}{4}\frac{(x^2 - 3x + 3)}{(x-2)^2(x-1)^2}y^{(3)} - \frac{45}{8}\frac{(2x^3 - 9x^2 + 17x - 12)}{(x-2)^3(x-1)^3}y''$$

$$+ \frac{45}{4}\frac{2x^4 - 12x^3 + 33x^2 - 45x + 24}{(x-2)^4(x-1)^4}y'$$

$$- \frac{45}{4}\frac{2x^5 - 15x^4 + 54x^3 - 108x^2 + 113x - 48}{(x-2)^5(x-1)^5}y$$

There is a two dimensional space of rational solutions with basis:

$$[x(x-2)(x-1), (x-2)(x-1)].$$

Taking $f = x(x-2)(x-1)$ the recurrence $(\sharp_m)$ constructs the following irreducible polynomial :

$$P_4 = u^4 - \frac{(2 - 6x + 3x^2)}{x(x-2)(x-1)}u^3 + \frac{3}{8}\frac{(9x^3 - 35x^2 + 43x - 16)}{(x-1)^2(x-2)^2 x}u^2$$

$$- \frac{27x^4 - 153x^3 + 319x^2 - 288x + 94}{16x(x-2)^3(x-1)^3}u$$

$$+ \frac{81x^5 - 594x^4 + 1723x^3 - 2466x^2 + 1737x - 480}{256(x-2)^4(x-1)^4 x}$$

Note that for any choice of (c_1, c_2) where not both are zero, the recurrence will construct a polynomial for u of degree 4 from $f = c_1 x (x - 2)(x - 1) + c_2(x - 2)(x - 1)$. For almost all values of those c_i, the resulting polynomial of degree 4 will be irreducible and define an algebraic function u. However, for 3 families of values of (c_1, c_2), the resulting polynomial will be reducible and in fact will be the square of an irreducible polynomial of degree two ([37, 38, 28]). Here, those values (computed via the discriminant of P for the variable u) are

$$\{c_1 = 0\}, \{c_2 = -2c_1\}, \{c_2 = -4c_1\}.$$

Indeed, for $c_1 = 0$, we get

$$P = \left(u^2 - \frac{(2x - 3)u}{2(x - 1)(x - 2)} + \frac{3x^2 - 9x + 7}{16(x - 2)^2(x - 1)^2} \right)^2$$

Both the degree two irreducible factor of P or the irreducible polynomial P_4 can be used to proceed. In the following we will use P_4.

In order to undo the variable change (2.4) and produce a solution for our original equation L, we replace u by $u = v + \frac{3}{4}\frac{6x - 13}{(x - 2)(x - 1)}$ in the equation $P(u) = 0$. Computing the expression $\exp(\int v)$ by computing the integral and performing the simplification, we obtain two algebraic solutions of $L(y) = 0$ which are a basis of the solution space:

$$y_4 = \frac{(x - 2)\sqrt[4]{x - 2\sqrt{-1 + x}}}{(-1 + x)^5} \quad \text{and} \quad y_5 = \frac{(x - 2)\sqrt[4]{x + 2\sqrt{-1 + x}}}{(-1 + x)^5}$$

2.6 Brief conclusion

We have presented and illustrated algorithms for rational, exponential and Liouvillian solutions of linear differential equations. Those are precisely the solutions that can be expressed using solutions of differential equations of order at most one. One may push further the class of admissible solutions, for example using solutions of equations of other orders and, eventually, compute the differential Galois group (i.e all differential relations between the solutions). But this is beyond the modest purpose of this survey; interested readers are referred to the lovely set of notes [1] and the corresponding fundamental book [2].

These methods have applications in handling holonomic functions, establishing transcendence properties of numbers, and many others ; a striking application today is the application of constructive differential Galois theory to integrability of dynamical systems [39].

It should also be mentioned that the theory extends beautifully to difference or q-difference equations; interested readers may enjoy consulting [40] or [41].

Bibliography

[1] Singer, M. (2008). Introduction to the Galois Theory of Linear Differential Equations, This volume.

[2] van der Put, M. and Singer, M. (2003). *Galois Theory of Linear Differential equations* (Springer Verlag, Heidelberg).

[3] Liouville, J. (1833). Second mémoire sur la détermination des intégrales dont la valeur est algébrique, *Journal de l'Ecole polytechnique* **14**, 149–193.

[4] Vessiot, E. (1915). *Méthodes d'intégration explicites* (Éditions Jacques Gabay, Paris).

[5] Marotte, F. (1898). *Les équations différentielles linéaires et la théorie des groupes* (Annales École normale supérieure, Gauthier-Villard, Paris).

[6] Gray, J. (1986). *Linear Differential Equations and Group Theory from Riemann to Poincaré* (Birkhäuser, Basel).

[7] Poole, E. (1936, réédition : Dover, 1960). *Introduction to the theory of linear differential equations* (Clarendon Press, Oxford).

[8] Singer, M. (1999). Direct and inverse problems in differential Galois theory, in *Selected Works of Ellis Kolchin with Commentary*, ed. C. Bass, Buium., pages 527–554 (American Mathematical Society, Providence).

[9] van der Put, M. (1999). Symbolic analysis of differential equations, in *Some tapas of computer algebra*, volume 4 of *Algorithms Comput. Math.*, pages 208–236 (Springer, Berlin).

[10] Ince, E. (1926). *Ordinary differential equations* (Dover, New York).

[11] Ramis, J. (1985). *Théorèmes d'indice Gevrey pour les équations différentielles linéaires* (AMS coll. Publications, Providence).

[12] Barkatou, M. (1999). On rational solutions of linear differential systems, *J. Symbolic Comput.* **28**(4-5), 547–567.

[13] Abramov, S. and Kvashenko, K. (1991). Fast algorithms for the rational solutions of linear differential equations with polynomial coefficients, in *Proceedings of International Symposium on Symbolic and Algebraic Computation ISSAC'91* (ACM Press, New York).

[14] Abramov, S., Bronstein, M., and Petkovšek, M. (1995). On polynomial solutions of linear operators, in *Proceedings of International Symposium on Symbolic and Algebraic Computation ISSAC'95* (ACM Press, New York).

[15] Bostan, A., Cluzeau, T., and Salvy, B. (2005). Fast algorithms for polynomial solutions of linear differential equations, in *Proceedings of International Symposium on Symbolic and Algebraic Computation ISSAC'05*, pages 45–52 (electronic) (ACM, New York).

[16] Bronstein, M. (1992). On solutions of linear differential equations in their coefficient field, *J.Symb.Comp* **13**, 413–439.

[17] Singer, M. (1996). Testing Reducibility of Linear Differential Operators: A Group Theoretic Perspective, *Applicable Algebra in Engineering* **7**, 77–104.

[18] van Hoeij, M. (1996). Rational Solutions of the Mixed Differential Equation

and its Application to Factorization of Differential Operator, in *Proceedings of International Symposium on Symbolic and Algebraic Computation ISSAC'96* (ACM Press, New-York).

[19] Barkatou, M.A. and Pflügel, E. (1998). On the equivalence problem of linear differential systems and its application for factoring completely reducible systems, in *Proceedings of International Symposium on Symbolic and Algebraic Computation ISSAC'98*, pages 268–275 (electronic) (ACM Press, New York).

[20] Singer, M. (1981). Liouvillian solutions of nth order linear differential equations, *Amer. J. Math.* **103**, 661–682.

[21] Singer, M. and Ulmer, F. (1995). Necessary Conditions for Liouvillian Solutions of (Third Order) Linear Differential Equations, *Applied Algebra in Engineering, Communication and Computing* 6, 1–22.

[22] Cluzeau, T. and van Hoeij, M. (2004). A modular algorithm for computing the exponential solutions of a linear differential operator, *J. Symbolic Comput.* **38**(3), 1043–1076.

[23] Kolchin, E. (1948). Algebraic matrix groups and the Picard-Vessiot theory of homogeneous ordinary differential equations., *Annals of Math.* 49.

[24] Ulmer, F. (1992). On liouvillian solutions of linear differential equations, *Appl. Algebra in Eng. Comm. and Comp.* **226**, 171–193.

[25] Kovacic, J. (1986). An algorithm for solving second order linear homogeneous differential equations, *J. Symb. Comp.* **2**, 3–43.

[26] Singer, M. and Ulmer, F. (1993). Galois Groups of Second and Third Order Linear Differential Equations, *J. Symb. Comp.* **16**, 9–36.

[27] Cormier, O. (2001). Résolution des équation différentielles linéaires d'ordre 4 et 5, *Thèse Université de Rennes 1* .

[28] Ulmer, F. and Weil, J. (1996). Note on Kovacic's algorithm, *J. Symb. Comp.* **22**, 179–200.

[29] Singer, M. and Ulmer, F. (1997). Linear differential equations and products of linear forms, *J. Pure Appl. Algebra* **117-118**, 549–563.

[30] Bronstein, M., Mulders, T., and Weil, J. (1997). On Symmetric Powers of Differential Operators, in *Proceedings of International Symposium on Symbolic and Algebraic Computation ISSAC'97* (ACM Press, New-York).

[31] van Hoeij, M., Ragot, J., Ulmer, F., and Weil, J. (1999). Liouvillian solutions of linear differential equations of order 3 and higher, *Journal of Symbolic Computation* **28**(4-5), 589–609.

[32] Ulmer, F. (2003). Liouvillian solutions of third order differential equations, *J. Symb. Comp.* **36**, 855–889.

[33] Fakler, W. (1997). On second order homogeneous linear differential equations with Liouvillian solutions, *Theoretical Computer Science* **187**, 27–48.

[34] Ulmer, F. (2005). Note on Algebraic solutions of differential equations with known finite Galois group, *Appl. Algebra in Eng. Comm. and Comp.* **16**, 205–218.

[35] Berkenbosch, M. (2004). *Algorithms and Moduli Spaces for Differential Equations*, Rijksuniversiteit Groningen Ph.D. thesis.

[36] van Hoeij, M. and Weil, J. (2005). Solving second order differential equations with Klein's theorem, in *Proceedings of International Symposium on Symbolic and Algebraic Computation ISSAC'05 (Beijing)* (ACM, New York).

[37] Hendriks, P. and van der Put, M. (1995). Galois action on solutions of a

differential equation, *Journal of Symbolic Computation* **19**, 559–576.

[38] Ulmer, F. (1994). Irreducible linear differential equations of prime order, *J. Symbolic Comput.* **18**(4), 385–401.

[39] Morales Ruiz, J.J. (1999). *Differential Galois theory and non-integrability of Hamiltonian systems*, volume 179 of *Progress in Mathematics* (Birkhäuser Verlag, Basel).

[40] van der Put, M. and Singer, M. (1997). *Galois theory of difference equations*, volume 1666 of *Lecture Notes in Mathematics* (Springer-Verlag, Berlin).

[41] Di Vizio, L., Ramis, J.P., Sauloy, J., and Zhang, C. (2003). Équations aux q-différences, *Gaz. Math.* **96**, 20–49.

[42] Duval, A. and Loday-Richaud, M. (1992). Kovacic's algorithm and its application to some families of special functions, *Appl. Algebra Engrg. Comm. Comput.* **3**(3), 211–246.

[43] Cluzeau, T. (2003). Factorization of differential systems in characteristic p, in *Proceedings of International Symposium on Symbolic and Algebraic Computation ISSAC'03*, pages 58–65 (electronic) (ACM, New York).

[44] Barkatou, M. and Pflügel, E. (1999). An algorithm computing the regular formal solutions of a system of linear differential equations, *J. Symbolic Comput.* **28**(4-5), 569–587, Differential algebra and differential equations.

[45] Beukers, F. (1992). Differential Galois theory, in *From Number Theory to Physics*, ed. Waldschmidt, Moussa, Luck, and Itzykson (Springer Verlag, Heidelberg).

[46] Cox, D., Little, J., and O'Shea, D. (1992). *Ideals, varieties, and Algorithms*, Undergraduate Texts in Math (Springer Verlag, Heidelberg).

[47] Geddes, K., Czapor, S., and Labahn, G. (1992). *Algorithms for computer algebra* (Kluwer Academic Publishers, Dortrecht, Holland).

[48] van Hoeij, M. and Weil, J. (1997). An algorithm for computing invariants of differential Galois groups, *J. Pure Appl. Algebra* **117-118**, 353–379.

[49] Kaplanski, I. (1957). *An introduction to differential Algebra* (Hermann, Paris).

[50] Ritt, J. (1950: or Dover, 1966). *Differential Algebra* (Hermann, Paris).

[51] Singer, M. (1990). An outline of differential Galois theory, in *Computer Algebra and Differential Equations*, ed. E. Tournier (Academic Press, New York).

[52] Singer, M. and Ulmer, F. (1993). Liouvillian and Algebraic Solutions of Second and Third Order Linear Differential Equations, *J. Symb. Comp.* **16**, 37–73.

[53] Sturmfels, B. (1993). *Algorithms in Invariant Theory*, Texts and Monographs in Symbolic Computation (Springer-Verlag, Wien and New York).

3

Factorization of Linear Differential Operators and Systems

Sergey P. Tsarev

Krasnoyarsk State Pedagogical University
Lebedevoi, 89
Krasnoyarsk, 660049,
Russia
sptsarev@mail.ru

3.1 Introduction

Factorization of systems of differential equations was first studied for the case of a single linear ordinary differential equation (LODE) with linear ordinary differential operator (LODO) of the form

$$L = f_0(x)D^n + f_1(x)D^{n-1} + \ldots + f_n(x), \quad D = d/dx, \qquad (3.1)$$

where the coefficients $f_s(x)$ belong to some differential field **K**. Factorization is a useful tool for computing a closed form solution of the corresponding linear ordinary differential equation $Ly = 0$ as well as determining its Galois group (see for example [44, 45, 52] and the paper of M. Singer in this volume). For simplicity and without loss of generality we suppose that operators (3.1) are reduced (i.e. $f_0(x) \equiv 1$) unless we explicitly state the reverse. The most popular case of the differential field $\mathbf{K} = \bar{\mathbf{Q}}(x)$ of rational functions with rational or algebraic number coefficients is a nontrivial example which is well investigated and will be considered hereafter when we discuss any constructive results.

In this paper we give a review of the current state of the theory of factorization of ordinary and partial differential operators and even more generally, of systems of linear differential equations of arbitrary type (determined as well as overdetermined). We start with the elementary algebraic theory of factorization of linear ordinary differential operators (3.1) developed in the period 1880–1930. After exposing these classical results we sketch more sophisticated algorithmic approaches developed in the last 20 years. The revival of this theory in the last two decades is motivated by the development of powerful computer algebra systems and implementation of nontrivial algebraic and differential algorithms such as factorization of polynomials and indefinite integration of elementary functions.

The main part of this paper will be devoted to modern generalizations of the factorization theory to the most general case of systems of linear partial differential equations and their relation with explicit solvability of nonlinear partial differential equations based on some constructions from the theory of rings, the theory of partially ordered sets (lattices) and that of abelian categories. Many of the results of this paper may be exposed within the framework of the Picard-Vessiot theory. But we follow a much simpler algebraic approach in order to facilitate the aforementioned generalizations.

The proper theoretical background for the simplest case—factorization of linear ordinary differential operators with rational coefficients—was known already in the end of the XIX century. Paradoxically, mathematicians of that epoch had developed even a nontrivial (theoretical) algorithm of factorization of such operators [4]! A review of this theory can be found in [39]. In contrast to the well-known property of uniqueness of factorization of usual commutative polynomials into irreducible factors, a simple example $D^2 = D \cdot D = (D + 1/(x-c)) \cdot (D - 1/(x-c))$ shows that some LODO may have essentially different factorizations with factors depending on some arbitrary parameters. Fortunately according to the results by E. Landau [26] and A. Loewy [28, 29] exposed below all possible factorizations of a given operator L over a fixed differential field have the same number of factors in different expansions $L = L_1 \cdots L_k = \overline{L}_1 \cdots \overline{L}_r$ into irreducible factors and the factors L_s, $\overline{L}_p$ are pairwise "similar". (Hereafter we always suppose the order of factors to be greater than 0: $\mathrm{ord}(L_i) > 0$, $\mathrm{ord}(\overline{L}_j) > 0$). We outline the main ideas of this classical theory in Section 3.2. For simplicity we discuss here only the case of differential operators; a generalization to the case of a general Ore ring (including difference and q-difference operators, see [6, 8]) is straightforward.

Subsequent Sections are devoted to different aspects of the theory of factorization of linear partial differential operators.

3.2 Factorization of LODO

The basics of the algebraic theory of factorization of LODO were essentially given already in [28, 29], [31]–[34]. Algebraically the main results are just an easy consequence of the fact that the ring $\mathbf{K}[D]$ of LODO with coefficients in a given differential field $\mathbf{K}$ is Euclidean: for any

LODO L, M there exist unique LODO Q, R, Q_1, R_1 such that

$$L = Q \cdot M + R, \ L = M \cdot Q_1 + R_1, \ ord(R) < ord(M), \ ord(R_1) < ord(M).$$

For any two LODO L and M using the right or left Euclidean algorithm one can determine their right greatest common divisor $rGCD(L, M) = G$, i.e. $L = L_1 \cdot G$, $M = M_1 \cdot G$ (the order of G is maximal) and their right least common multiple $rLCM(L, M) = K$, i.e. $K = \overline{M} \cdot L = \overline{L} \cdot M$ (the order of K is minimal) as well as their left analogues $lGCD$ and $lLCM$. All left and right ideals of this ring are principal and all two-sided ideals are trivial. Operator equations

$$X \cdot L + Y \cdot M = B, \qquad L \cdot Z + M \cdot T = C \qquad (3.2)$$

with unknown operators X, Y, Z, T are solvable iff $rGCD(L, M)$ divides B on the right and $lGCD(L, M)$ divides C on the left. We say that an operator L is (right) *transformed into* L_1 *by an operator (not necessarily reduced)* B, and write $L \xrightarrow{B} L_1$, if $rGCD(L, B) = 1$ and $K = rLCM(L, B) = L_1 \cdot B = B_1 \cdot L$. In this case any solution of $Ly = 0$ is mapped by B into a solution By of $L_1 y = 0$. Using (3.2) one may find with rational algebraic operations an operator B_1 such that $L_1 \xrightarrow{B_1} L$, $B_1 \cdot B = 1 (\text{mod} L)$. Operators L, L_1 will be also called *similar* or *of the same kind* (in the given differential field $\mathbf{K}$). So for similar operators the problem of solution of the corresponding LODE $Ly = 0$, $L_1 y = 0$ are equivalent. One can define also the notion of left-hand transformation of L by B into L_1: $K = lLCM(L, B) = B \cdot L_1 = L \cdot B_1$. Obviously left- and right-hand transformations are connected via the adjoint operation. Also one may prove ([33]) that two operators are left-hand similar iff they are right-hand similar. A (reduced) LODO is called *prime* or *irreducible* (in the given differential field $\mathbf{K}$) if it has no nontrivial factors aside from itself and 1. Every LODO similar to a prime LODO is also prime. Two (prime for simplicity) LODO P and Q are called *interchangeable in the product* $P \cdot Q$ and this product will be called *interchangeable* as well if $P \cdot Q = Q_1 \cdot P_1$, $Q_1 \neq P$, $P_1 \neq Q$. In this case P is similar to P_1, Q is similar to Q_1 and $P_1 \xrightarrow{Q} P$.

Theorem 1 (Landau [26] and Loewy [28]) *Any two different decompositions of a given LODO L into products of prime LODO $L = P_1 \cdots P_k = \overline{P}_1 \cdots \overline{P}_p$ have the same number of factors ($k = p$) and the factors are similar in pairs (in some transposed order). One* *decomposition may be obtained from the other through a finite sequence of interchanges of contiguous factors (in the pairs $P_i \cdot P_{i+1}$).*

All definitions here are constructive over the differential field of rational functions $\mathbf{K} = \mathbf{Q}(x)$: either using the Euclidean algorithm or finding rational solutions of LODO one can determine for example if two given LODO are similar or find all possible (parametric) factorizations of a given LODO with rational functional coefficients [1, 6, 7, 48].

The Landau-Loewy theorem also has the useful following *ring-theoretic interpretation*. Namely, every $L \in \mathbf{K}[D]$ generates the corresponding left ideal $|L\rangle$; L_1 divides L on the right iff $|L\rangle \subset |L_1\rangle$. If we have a factorization $L = L_1 \cdots L_k$ then we have a chain of ascending left principal ideals $|L\rangle \subset |L_2 \cdots L_k\rangle \subset |L_3 \cdots L_k\rangle \subset \ldots \subset |L_k\rangle \subset |1\rangle = \mathbf{K}[D]$. If the factors L_k are irreducible, the chain is maximal, i.e. it is not possible to insert some intermediate ideals between its two adjacent elements. The Landau-Loewy theorem is nothing but the **Jordan-Hölder -Dedekind chain condition:**

Theorem 2 *Any two finite maximal ascending chains of left principal ideals in the ring $\mathbf{K}[D]$ of LODO have equal length.*

Similarity of irreducible factors can also be interpreted in this approach. An even more general lattice-theoretic interpretation turned out to be fruitful for a generalization of this simple algebraic theory for the case of factorizations of *partial* differential operators [49]. Namely, let us consider the set of (left) ideals in $\mathbf{K}[D]$ as a partially ordered by inclusion set $\mathcal{M}$ (called a poset). This poset has the following two fundamental properties:

Property I) for any two elements $A, B \in \mathcal{M}$ (left ideals!) one can find a unique $C = \sup(A, B)$, i.e. such C that $C \geq A$, $C \geq B$, and C is "minimal possible". Analogously there exists a unique $D = \inf(A, B)$, $D \leq A$, $D \leq B$, D is "maximal possible".

Such posets are called *lattices* [20]. $\sup(A, B)$ and $\inf(A, B)$ correspond to the GCD and the LCM in $\mathbf{K}[D]$.

For simplicity (and following the established tradition) $\sup(A, B)$ will be hereafter denoted as $A + B$ and $\inf(A, B)$ as $A \cdot B$;

Property II) For any three $A, B, C \in \mathcal{M}$ the following *modular identity* holds:

$$(A \cdot C + B) \cdot C = A \cdot C + B \cdot C$$

Such lattices are called *modular lattices* or *Dedekind structures*.

As one can prove, modularity implies the Jordan-Hölder-Dedekind chain condition: any two finite maximal chains $L > L_1 > \cdots > L_k > 0$ and $L > M_1 > \cdots > M_r > 0$ for a given $L \in \mathcal{M}$ have equal lengths: $k = r$ (the same for ascending chains). For the interpretation of the notions of similarity, direct sums, Kurosh & Ore theorems on direct sums cf. [49].

But even more fruitful for generalizations is the following *categorical interpretation* of similarity of LODO and the Jordan-Hölder-Dedekind chain condition. Namely, let us consider the following **abelian category** $\mathcal{LODO}$ **of LODO**.

The objects of $\mathcal{LODO}$ are *reduced* operators $L = D^n + a_1(x)D^{n-1} + \ldots + a_n(x)$, $a_i \in \mathbf{K}$. One may ideally think of (finite-dimensional!) the solution spaces $Sol(L)$ of such operators in some sufficiently large Picard-Vessiot extension of the coefficient field as another way of representation of an object in $\mathcal{LODO}$. This helps one to understand the meaning of some definitions below, but one should remember that these solution spaces are *not constructive* unlike the objects-operators.

The morphisms $Hom(L, L_1)$ in this category are constructively defined as a *not necessarily reduced* LODO B such that $L_1 \cdot B = C \cdot L$ for some other LODO C. Non-constructively this B may be seen as a mapping of solutions of L into solutions of L_1. Note that all operators here have coefficients in some *fixed* differential field $\mathbf{K}$. Two operators B_1, B_2 generate the same morphism iff $B_1 = B_2(mod\ L)$. Also we should remark that this definition is *not* equivalent to the definition of a transformation of operators $L \xrightarrow{B} L_1$ introduced earlier, because for morphisms:

1) B and L may have common solutions, i.e. a nontrivial $rGCD(B, L)$. This means that the mapping of the solution space $Sol(L)$ by B may have a kernel $Sol(rGCD(B, L))$. The morphism is not injective in this case.

2) The image of the solution space $Sol(L)$ may be smaller than $Sol(L_1)$. The morphism is not surjective in this case.

Algebraically this means that $L_1 \cdot B = C \cdot L \neq rLCM(B, P)$.

Similarity of operators L and L_1 now simply means isomorphism of the objects L and L_1 in this category.

The following fact is a direct corollary of our representation of this category as a subcategory of the category of finite-dimensional vector spaces and linear mappings preserving direct sums, products etc.:

Theorem 3 *The category* $\mathcal{LODO}$ *is abelian.*

Among the many useful results for abelian categories (cf. for example [16, 18]) we need the following

Theorem 4 *Any abelian category with finite ascending chains satisfies the Jordan-Hölder property.*

This will serve us as a maximal theoretical framework for an algebraic interpretation and generalization of the Landau-Loewy theorem for the case of systems of linear partial differential equations below. Again there exist notions of direct sums, Kurosh-Ore theorems on direct sums and the powerful technique of modern homological algebra for abelian categories [16, 18]. We will see below that this rather high level of abstraction allows a very natural generalization of the definition of factorization for arbitrary systems of linear partial differential equations (LPDE).

3.3 Factorization of LPDO

In contrast to the case of ordinary operators, two main results are seemingly lost for LPDO: the Landau-Loewy theorem and the possibility to use some known solution for factorization of operators. In the case of a LODO L obviously if one has its solution $L\phi = 0$ then one can split off a first-order right factor: $L = M \cdot \left(D - \frac{\phi'}{\phi}\right)$. For a LPDO, even if one knows the *complete* set of solutions, the operator may not be factorizable, as the following classical examples show:

Example 1. The equation $Lu = \left(D_x D_y - \frac{2}{(x+y)^2}\right) u = 0$ with $D_x = \partial/\partial x$, $D_y = \partial/\partial y$ has the following complete solution:

$$Lu = 0 \Leftrightarrow u = -\frac{2(F(x) + G(y))}{x + y} + F'(x) + G'(y),$$

where $F(x)$ and $G(y)$ are two arbitrary functions of one variable each. On the other hand, as an easy calculation shows, the operator L cannot be represented as a product of two first-order operators (over *any* differential extension of the given coefficient field $\mathbf{K} = \mathbf{Q}(x, y)$).

Example 2. The equation $Lu = \left(D_x D_y - \frac{6}{(x+y)^2}\right) u = 0$ again has the following complete solution:

$$u = \frac{12(F(x) + G(y))}{(x + y)^2} - \frac{6(F'(x) + G'(y))}{x + y} + F''(x) + G''(y), \quad (3.3)$$

but the operator is again "naively irreducible". More generally, the equation

$$Lu = u_{xy} - \frac{c}{(x+y)^2}u = 0, \qquad c = \text{const} \tag{3.4}$$

has a complete solution in an "explicit" form similar to (3.3) iff $c = n(n+1)$ for $n \in \mathbf{N}$. In this case it has a complete solution in the form

$$u = c_0 F + c_1 F' + \ldots + c_n F^{(n)} + d_0 G + d_1 G' + \ldots + d_{n+1} G^{(n+1)} \tag{3.5}$$

with some definite $c_i(x,y)$, $d_i(x,y)$ and two arbitrary functions $F(x)$, $G(y)$.

Only for $c = 0$ is the corresponding operator "naively reducible": $L = D_x \cdot D_y$.

The solution technology used here is very old and can be found in [9, 19] under the name of *Laplace transformations* or *Laplace cascade method*: after a series of transformations of (3.4) one gets a *naively factorizable* LPDE!

Another unpleasant example is also ascribed in [5] to E. Landau: if

$$\begin{aligned} P &= D_x + xD_y, \quad Q = D_x + 1, \\ R &= D_x^2 + xD_xD_y + D_x + (2+x)D_y, \end{aligned} \tag{3.6}$$

then $L = Q \circ Q \circ P = R \circ Q$. On the other hand the operator R is absolutely irreducible, i.e. one cannot factor it into a product of first-order operators with coefficients in *any* extension of $\mathbf{Q}(x,y)$. So there seems to be no hope for an analogue of Landau-Loewy theorem for decomposition of LPDO into products of lower-order LPDO.

We see that a "naive" definition of a factorization of a LPDO as its representation as a product (composition) of lower-order LPDOs lacks some fundamental properties established in the previous Section for factorization of LODO.

Recently [49] an attempt to give a "good" definition of generalized factorization was undertaken. In the next subsection we only briefly sketch the ideas of this approach and describe its nontrivial relation to explicit integrability of *nonlinear* partial differential equations.

3.3.1 General theory of factorization of an arbitrary single LPDO, ring-theoretic approach

Our goal in [49] was to define a notion of factorization with "good" properties:

- Every LPDO L shall have only finite chains of ascending generalized factors. In particular D_x should be irreducible.
- Jordan-Hölder property: all possible generalized factorizations of a given operator L have the same number of "factors" in different expansions into irreducible factors and the "factors" should be pairwise "similar" in such expansions.
- Existence of *large* classes of solutions should be related to factorization.
- The classical theory of integration of LPDO using the Laplace cascade method should be an integral part of this generalized definition.

An obvious extension of the definition may be suggested if one would use ascending chains of arbitrary (not necessary principal) left ideals starting from the left ideal generated by the given operator:

$$|L\rangle \subset I_1 \subset I_2 \subset \ldots \subset I_k \subset |1\rangle. \tag{3.7}$$

Unfortunately one can easily see that even for the operator D_x we have such chains, and they have unlimited length: $|D_x\rangle \subset |D_x, D_y^m\rangle \subset |D_x', D_y^{m-1}\rangle \subset \ldots |D_x, D_y\rangle \subset |1\rangle$! So we shall take some special class of ideals, more general than the principal ideals, but much less rich then arbitrary left ideals. In [49] we gave a definition of such a suitable subclass of left ideals called *divisor ideals*.

For such special left ideals of the ring of LPDO:

- chains (3.7) will be finite and different maximal chains for a given L have the same length: if $|L\rangle \subset I_1 \subset I_2 \subset \ldots \subset I_k \subset |1\rangle$, $|L\rangle \subset J_1 \subset J_2 \subset \ldots \subset J_m \subset |1\rangle$, then $k = m$ and one can prove a natural lattice-theoretic "similarity" of "factors" in the two chains.
- Irreducible LODO will be still irreducible as LPDO.
- For $dim = 2$, $ord = 2$ (that is for operators with two independent variables of order two) a LODO is factorizable in this generalized sense (i.e. having a nontrivial chain (3.7)) iff it is integrable with the Laplace cascade method. We describe this cascade method below in subsection 3.3.3
- Algebraically, the problem is reduced from the ring $Q(x, y)[D_x, D_y]$ to factorization in rings of formal LODO with noncommutative coefficients $Q(x, y, D_x)[D_y]$ and/or $Q(x, y, D_y)[D_x]$ (Ore quotients); in these rings all left and right ideals are again principal ideals.

The details, rather involved, may be found in [49]. This approach nevertheless suffers from the following problems:

- The definition of divisor ideals given in [49] is very technical, not intuitive.
- No algorithm for such generalized factorization is known.

A generalization of this *ring-theoretic approach* to *systems* of LPDE was proposed recently by M. Singer. Another ring-theoretic approach was considered in [23].

In the next subsection we propose a different, much more intuitive definition of generalized factorization.

3.3.2 General theory of factorization of arbitrary systems of LPDE, approach of abelian categories

The Abelian category $\mathcal{SLPDE}$ of arbitrary systems of LPDE is defined by its objects which are simply systems

$$S : \begin{cases} L_{11}u_1 + \ldots + L_{1s}u_s = 0, \\ \ldots \\ L_{p1}u_1 + \ldots + L_{ps}u_s = 0, \end{cases} \quad \begin{matrix} L_{ij} \in Q(x_1,\ldots,x_n)[D_{x_1},\ldots,D_{x_n}], \\ u_k = u_k(x_1,\ldots,x_n). \end{matrix}$$

$$(3.8)$$

A morphism $P : S \to Q$ of two systems is defined as a matrix of differential operators

$$P : \begin{cases} v_1 = P_{11}u_1 + \ldots + P_{1s}u_s, \\ \ldots \\ v_m = P_{m1}u_1 + \ldots + P_{ms}u_s, \end{cases} \quad (3.9)$$

$P_{ij} \in Q(x_1,\ldots,x_n)[D_{x_1},\ldots,D_{x_n}]$ with the condition that any solution set $\{u_1,\ldots,u_s\}$ of the source system (3.8) is mapped into a subspace of the solution space $\{v_1,\ldots,v_m\}$ of the target system

$$Q : \begin{cases} M_{11}u_1 + \ldots + M_{1m}v_m = 0, \\ \ldots \\ M_{q1}v_1 + \ldots + M_{qm}v_m = 0, \end{cases} \quad \begin{matrix} M_{ij} \in Q(x_1,\ldots,x_n)[D_{x_1},\ldots,D_{x_n}], \\ v_k = v_k(x_1,\ldots,x_n). \end{matrix}$$

$$(3.10)$$

The standard differential Gröbner technique (originally developed in the beginning of the XX century as the so called Janet-Riquier theory [24, 37, 38, 40]) makes this definition constructive: (3.9) is a morphism mapping (3.8) to (3.10) iff for any i the equation $\sum_{j,k} M_{ij}P_{jk}u_k = 0$ is reducible to zero modulo the equations of the system (3.8).

Again, it is easy to see that this category is abelian: for this it is enough to check that $\mathcal{SLPDE}$ is embeddable into the category of (infinite-dimensional) vector spaces and linear morphisms and this embedding preserves direct sums, products etc.

It seems natural to refer to Theorem 4 to transfer the many properties of factorization proved for the category $\mathcal{LODO}$ in Section 3.2 to the case of the category $\mathcal{SLPDE}$. Unfortunately this is not so simple: the ascending chains of monomorphisms are infinite in general: the same example $|D_x\rangle \subset |D_x, D_y^m\rangle \subset |D_x, D_y^{m-1}\rangle \subset \ldots |D_x, D_y\rangle \subset |1\rangle$ makes this obvious. The solution to this problem is given by the standard construction of a Serre-Grothendieck factorcategory. We refer to [13, 16, 18] and especially to [17] for a detailed explanation of this important and general construction. One of the important steps of this construction is the construction of inverses of morphisms with "relatively small" kernels; the objects are not formally changed in contrast to the ring-theoretic construction of factorrings and factormodules. In our case we proceed as follows: for a given (say, determined) system of LPDE of the form (3.8) (with $s = p$), take the subcategory $\mathcal{S}_{n-2}$ of (overdetermined) systems with solution space parameterized by functions of at most $n - 2$ variables. Then the Serre-Grothendieck factorcategory $\mathcal{S}/\mathcal{S}_{n-2}$ has finite ascending chains. Another remarkable feature of this factorcategory is, as we mentioned above, the possibility to consider morphisms which have kernels defined by systems from $\mathcal{S}_{n-2}$ as *invertible morphisms*. This may lead to a more general theory of Bäcklund-type transformations (at least for the case of linear systems), for example of transformations of Moutard type ([9, 19]).

Now we can transfer all theoretical results proved in Section 3.2 to the case of the factorcategory $\mathcal{S}/\mathcal{S}_{n-2}$ and provide a theoretical foundation for the factorization theory of arbitrary linear systems of LPDE.

The obvious drawback still lies in the absence of *algorithms* for such a generalized factorization. We give an overview of currently known numerous partially algorithmic results in the next Sections 3.3.3–3.3.5.

3.3.3 *dim* = 2, *ord* = 2: *Laplace transformations and Darboux integrability of nonlinear PDEs*

Here we expose the basics of the classical theory [9, 15, 19], which is applicable to hyperbolic linear partial differential equations of order two with two independent variables. For simplicity only the case of an equation with *straight* characteristics will be discussed here:

$$Lu = u_{xy} + a(x,y)u_x + b(x,y)u_y + c(x,y)u = 0. \tag{3.11}$$

The more general case can be found in [19, 2, 50]. If one of the *Laplace invariants* of (3.11) $h = a_x + ab - c$, $k = b_y + ab - c$ vanishes, one can "naively" factorize the operator on the l.h.s. of (3.11): $k \equiv 0 \Rightarrow L = (D_y + a)(D_x + b)$; $h \equiv 0 \Rightarrow L = (D_x + b)(D_y + a)$.

If $h \neq 0$, $k \neq 0$ then (3.11) is *not factorizable* in the "naive" sense. In this case one can perform one of the two *Laplace transformations* (not to be confused with Laplace transforms!) which are invertible differential substitutions (isomorphisms in the category of $\mathcal{SLPDE}$):

$$u = \frac{1}{h}(D_x + b)u_{(1)}$$

or

$$u = \frac{1}{k}(D_y + a)u_{(-1)}.$$

In fact each of the above substitutions is the inverse of the other up to a functional factor. Each of these substitutions produces a new operator of the same form (3.11) but with different coefficients and Laplace invariants. The idea of the *Laplace cascade method* consists in application of these substitutions a few times, obtaining the (infinite in general) chain

$$\dots \leftarrow \quad L_{(-2)} \quad \leftarrow \quad L_{(-1)} \quad \leftarrow \quad L \quad \rightarrow \quad L_{(1)} \quad \rightarrow \quad L_{(2)} \quad \rightarrow \dots \quad (3.12)$$

In some cases (namely those cases considered as integrable in this approach) this gives us on some step an operator $L_{(i)}$ with vanishing $h_{(i)}$ or $k_{(i)}$. Then this chain cannot be continued further in the respective direction and one can find an explicit formula for the complete solution of the transformed equation; performing the inverse differential substitutions we obtain the complete solution of the original equation (with quadratures).

One of the main results of [49] is the following Theorems:

Theorem 5 $L = D_x \cdot D_y - a(x,y)D_x - b(x,y)D_y - c(x,y)$ *has a nontrivial generalized right divisor ideal (so is factorizable in the sense described in Section 3.3) iff the chain (3.12) of Laplace transformations is finite at least in one direction.*

Theorem 6 $L = D_x \cdot D_y - a(x,y)D_x - b(x,y)D_y - c(x,y)$ *is a lLCM of two generalized right divisor ideals iff the chain (3.12) of Laplace transformations is finite in both directions.*

This shows the meaning of the generalized definition of [49] and provides a *partial* algorithm for generalized factorization for equations of the form (3.11).

Although practically efficient for simple cases, this method has the obvious decidability problem: given an operator L, how many steps in the chain (3.12) should be tried? Currently no stopping criterion is known. As the example (3.4) shows, the number of steps in the chain (equal to n for (3.4) in the integrable case $c = n(n + 1)$) depends on some subtle arithmetic properties of the coefficients.

There exists a remarkable link of the theory of Laplace transformations to the theory of integrable *nonlinear* partial differential equations. This topic was very popular in the XIX century and led to the development of the integration methods of Lagrange, Monge, Boole and Ampere. G. Darboux [10] generalized the method of Monge (known as the method of intermediate integrals) to obtain the most powerful method for exact integration of partial differential equations known in the last century.

Recently in a series of papers [2, 46, 55] the Darboux method was cast into a more precise and efficient (although not completely algorithmic) form. For the case of a single second-order nonlinear PDE of the form

$$u_{xy} = F(x, y, u, u_x, u_y) \tag{3.13}$$

the idea consists in linearization: using the substitution $u(x, y) \to u(x, y) + \epsilon v(x, y)$ and cancelling terms with ϵ^n, $n > 1$, we obtain a LPDE

$$v_{xy} = Av_x + Bv_y + Cv \tag{3.14}$$

with coefficients depending on x, y, u, u_x, u_y. Equations of the type (3.14) are in fact amenable to the Laplace cascade method, though certainly one needs to take into consideration the original equation (3.13) while performing all the computations of the Laplace invariants and Laplace transformations: (3.13) allows us to express all the mixed derivatives of u via x, y, u and the non-mixed $u_{x \cdots x}$, $u_{y \cdots y}$). The following statement can be found in [19]; recently it was rediscovered in [2, 46]:

Theorem 7 *A second order, scalar, hyperbolic partial differential equation (3.13) is Darboux integrable if and only if the Laplace sequence (3.12) for (3.14) is finite in both directions.*

In [2, 46] this method was also generalized to the case of a general second-order nonlinear PDE

$$F(x, y, u, u_x, u_y, u_{xx}, u_{xy}, u_{yy}) = 0.$$

3.3.4 *dim = 2, ord ≥ 3:* **Generalized Laplace transformations**

In [50] we proposed a generalization of the Laplace cascade method for arbitrary strictly hyperbolic equations with two independent variables of the form

$$\hat{L}u = \sum_{i+j\le n} p_{i,j}(x,y)\hat{D}_x^i\hat{D}_y^j u = 0, \qquad (3.15)$$

as well as for $n \times n$ first-order linear systems

$$(v_i)_x = \sum_{k=1}^{n} a_{ik}(x,y)(v_k)_y + \sum_{k=1}^{n} b_{ik}(x,y)v_k \qquad (3.16)$$

with strictly hyperbolic matrix (a_{ik}).

Here we demonstrate this new method on an example of the constant-coefficient system

$$\begin{cases} D_x u_1 = u_1 + 2u_2 + u_3, \\ D_y u_2 = -6u_1 + u_2 + 2u_3, \\ (D_x + D_y)u_3 = 12u_1 + 6u_2 + u_3. \end{cases} \qquad (3.17)$$

It has the following complete explicit solution:

$$\begin{cases} u_1 = 2e^y G(x) + e^x(3F(y) + F'(y)) + \exp\frac{x+y}{2}H(x-y), \\ u_2 = e^y G'(x) + 2e^x F'(y) - 2u_1, \\ u_3 = D_x u_1 + 3u_1 - 2(e^y G'(x) + 2e^x F'(y)), \end{cases}$$

where $F(y)$, $G(x)$ and $H(x-y)$ are three arbitrary functions of one variable each.

The solution technology (cf. [50]) for the details) is again a differential substitution; in the case of the system (3.17) the transformation is given by:

$$\begin{cases} \overline{u}_1 = u_1, \\ \overline{u}_2 = u_2 + 2u_1, \\ \overline{u}_3 = ((D_x + D_y)u_1 - u_1 - 2u_2 - 4u_1). \end{cases} \qquad (3.18)$$

The transformed system has a triangular matrix and is easily integrable:

$$\begin{cases} D_x \overline{u}_3 = \overline{u}_3, \\ D_y \overline{u}_2 = 2\overline{u}_3 + \overline{u}_2, \\ (D_x + D_y)u_1 = \overline{u}_3 + 2\overline{u}_2 + u_1. \end{cases}$$

Again no stopping criterion for the sequences of generalized Laplace transformations is known in the general case. For constant coefficient systems an alternative technology was proposed by F.Schwarz (private communication, 2005): transform the system (3.17) into a Janet (Gröbner) normal form with term order: LEX, $u_3 > u_2 > u_1, x > y$:

$$u_{1,xxy} - u_{1,xx} + u_{1,xyy} - 3u_{1,xy} + 2u_{1,x} - u_{1,yy} + 2u_{1,y} - u_1 = 0,$$
$$u_{2,y} + 3u_2 - 2u_{1,x} + 8u_1 = 0,$$
$$u_{2,x} - u_2 - \tfrac{1}{2}u_{1,xx} - \tfrac{1}{2}u_{1,xy} + 3u_{1,x} + \tfrac{1}{2}u_{1,y} - \tfrac{5}{2}u_1 = 0,$$
$$u_3 + 2u_2 - u_{1,x} + u_1 = 0.$$

The first equation *factors*:

$$D_x^2 D_y - D_x^2 + D_x D_y^2 - 3D_x D_y + 2D_x - D_y^2 + 2D_y - 1$$

$$= (D_x + D_y - 1)(D_y - 1)(D_x - 1).$$

So one can find u_1 easily and then the other two functions u_2 and u_3 are obtained from the remaining equations of the Janet base producing essentially the same solution (3.18).

Conjecture: For constant-coefficient systems this Gröbner basis technology is equivalent to the generalized Laplace technology.

3.3.5 *dim* ≥ 3, *ord* $= 2$: *Dini transformations*

In [11] another simple generalization of Laplace transformations formally applicable to some second-order operators in the space of arbitrary dimension was proposed. Namely, suppose that an operator $\hat{L}$ has its principal symbol

$$Sym = \sum_{i_1 + i_2 = 2} a_{i_1 i_2}(\vec{x}) D_{x_{i_1}} D_{x_{i_2}}$$

which factors (as a formal polynomial in formal commutative variables D_{x_i}) into product of two first-order factors: $Sym = \hat{X}_1 \hat{X}_2$ ($\hat{X}_j = \sum_i b_{ij}(\vec{x}) D_{x_i}$ are first-order operators) and moreover the complete operator $\hat{L}$ may be written at least in one of the *characteristic forms*:

$$L = (\hat{X}_1 \hat{X}_2 + \alpha_1 \hat{X}_1 + \alpha_2 \hat{X}_2 + \alpha_3)$$
$$L = (\hat{X}_2 \hat{X}_1 + \overline{\alpha}_1 \hat{X}_1 + \overline{\alpha}_2 \hat{X}_2 + \alpha_3), \qquad (3.19)$$

where $\alpha_i = \alpha_i(x, y)$. Since the operators $\hat{X}_i$ do not necessarily commute we have to take into consideration in (3.19) and everywhere below the *commutation law*

$$[\hat{X}_1, \hat{X}_2] = \hat{X}_1 \hat{X}_2 - \hat{X}_2 \hat{X}_1 = P(x, y)\hat{X}_1 + Q(x, y)\hat{X}_2. \qquad (3.20)$$

. This is very restrictive since the two tangent vectors corresponding to the first-order operators $\hat{X}_i$ no longer span the complete tangent space at a generic point $(\vec{x}_0)$. (3.20) is also possible only in the case when these two vectors give an *integrable* two-dimensional distribution of the tangent subplanes in the sense of Frobenius, i.e. when one can make a change of the independent variables $(\vec{x})$ such that $\hat{X}_i$ become parallel to the coordinate plane (x_1, x_2); thus in fact we have an operator $\hat{L}$ with only D_{x_1}, D_{x_2} in it and we have got no really significant generalization of the Laplace method. If one has only (3.19) but (3.20) does not hold one cannot perform more than one step in the Laplace chain (3.12) and there is no possibility to get an operator with a zero Laplace invariant (so naively factorizable and solvable).

Below we demonstrate on an example, following an approach proposed by U. Dini in another paper [12], that one can find a better analogue of Laplace transformations for the case when the dimension of the underlying space of independent variables is greater than two. Another particular special transformation was also proposed in [3], [54]; it is applicable to systems whose order coincides with the number of independent variables. The results of [3], [54] lie beyond the scope of this paper.

Let us take the following equation:

$$Lu = (D_x D_y + x D_x D_z - D_z)u = 0. \tag{3.21}$$

It has three independent derivatives D_x, D_y, D_z, so the Laplace method is *not* applicable. On the other hand its principal symbol splits into product of two first-order factors: $\xi_1 \xi_2 + x \xi_1 \xi_3 = \xi_1(\xi_2 + x\xi_3)$. This is no longer a typical case for hyperbolic operators in dimension 3; we will use this special feature, introducing two characteristic operators $\hat{X}_1 = D_x$, $\hat{X}_2 = D_y + x D_z$. We have again a nontrivial commutator $[\hat{X}_1, \hat{X}_2] = D_z = \hat{X}_3$. The three operators $\hat{X}_i$ span the complete tangent space at every point (x, y, z). Using them one can represent the original second-order operator in one of two partially factorized forms:

$$L = \hat{X}_2 \hat{X}_1 - \hat{X}_3 = \hat{X}_1 \hat{X}_2 - 2\hat{X}_3.$$

Let us use the first one and transform the equation into a system of two first-order equations:

$$Lu = 0 \iff \begin{cases} \hat{X}_1 u = v, \\ \hat{X}_3 u = \hat{X}_2 v. \end{cases} \tag{3.22}$$

Cross-differentiating the left hand sides of (3.22) and using the obvious identity $[\hat{X}_1, \hat{X}_3] = [D_x, D_z] = 0$ we get $\hat{X}_1 \hat{X}_2 v = D_x(D_y + xD_z)v =$

$\hat{X}_3 v = D_z v$ or $0 = D_x(D_y + xD_z)v - D_z v = (D_x D_y + xD_x D_z)v = (D_y + xD_z)D_x v = \hat{X}_2\hat{X}_1 v$.

This is precisely the procedure proposed by Dini in [12]. Since it results now in another second-order equation which is "naively" factorizable we easily find its complete solution:

$$v = \int \phi(x, xy - z)\, dx + \psi(y, z)$$

where ϕ and ψ are two arbitrary functions of two variables each; they give the general solutions of the equations $\hat{X}_2\phi = 0$, $\hat{X}_1\psi = 0$.

Now we can find u:

$$u = \int \left(v\, dx + (D_y + xD_z)v\, dz \right) + \theta(y),$$

where an extra free function θ of one variable appears as a result of integration in (3.22).

So we have seen that such *Dini transformations* (3.22) in some cases may produce a complete solution in explicit form for a non-trivial three-dimensional equation (3.21). This explicit solution can be used to solve initial value problems for (3.21).

Dini did not give any general statement on the range of applicability of his trick. In [51] we have proved the following

Theorem 8 *Let $L = \sum_{i+j+k \leq 2} a_{ijk}(x, y, z) D_x^i D_y^j D_z^k$ have factorizable principal symbol: $\sum_{i+j+k=2} a_{ijk}(x, y, z) D_x^i D_y^j D_z^k = \hat{S}_1 \hat{S}_2$ (mod lower-order terms) with generic (non-commuting) first-order LPDO $\hat{S}_1$, $\hat{S}_2$. Then there exist two Dini transformations $L_{(1)}$, $L_{(-1)}$ of L.*

Proof One can represent L in two possible ways:

$$L = \hat{S}_1\hat{S}_2 + \hat{T} + a(x, y, z) = \hat{S}_2\hat{S}_1 + \hat{U} + a(x, y, z) \qquad (3.23)$$

with some first-order operators $\hat{T}$, $\hat{U}$. We will consider the first one obtaining a transformation of L into an operator $L_{(1)}$ of similar form.

In the generic case the operators $\hat{S}_1$, $\hat{S}_2$, $\hat{T}$ span the complete 3-dimensional tangent space in a generic point (x, y, z). Precisely this requirement will be assumed to hold hereafter; operators L with this property will be called *generic*.

Let us fix the coefficients in the expansions of the following commutators:

$$[\hat{S}_2, \hat{T}] = K(x, y, z)\hat{S}_1 + M(x, y, z)\hat{S}_2 + N(x, y, z)\hat{T}. \qquad (3.24)$$

$$[\hat{S}_1, \hat{S}_2] = P(x,y,z)\hat{S}_1 + Q(x,y,z)\hat{S}_2 + R(x,y,z)\hat{T}. \qquad (3.25)$$

First we try to represent the operator in a partially factorized form:
$L = (\hat{S}_1 + \alpha)(\hat{S}_2 + \beta) + \hat{V} + b(x,y,z)$ with some indefinite $\alpha = \alpha(x,y,z)$,
$\beta = \beta(x,y,z)$ and $\hat{V} = \hat{T} - \beta\hat{S}_1 - \alpha\hat{S}_2$, $b = a - \alpha\beta - \hat{S}_1(\beta)$.

Then introducing $v = (\hat{S}_2 + \beta)u$ we get the corresponding first-order system:

$$Lu = 0 \Longleftrightarrow \begin{cases} (\hat{S}_2 + \beta)u = v, \\ (\hat{V} + b)u = -(\hat{S}_1 + \alpha)v. \end{cases} \qquad (3.26)$$

Next we try to eliminate u by cross-differentiating the left hand sides, which gives

$$[(\hat{V} + b), (\hat{S}_2 + \beta)]u = (\hat{S}_2 + \beta)(\hat{S}_1 + \alpha)v + (\hat{V} + b)v. \qquad (3.27)$$

If one wants u to disappear from this new equation one should find out when $[(\hat{V} + b), (\hat{S}_2 + \beta)]u$ can be transformed into an expression involving *only* v, i.e. when this commutator is a linear combination of just two expressions $(\hat{S}_2 + \beta)$ and $(\hat{V} + b)$:

$$[(\hat{V} + b), (\hat{S}_2 + \beta)] = \mu(x,y,z)(\hat{S}_2 + \beta) + \nu(x,y,z)(\hat{V} + b). \qquad (3.28)$$

This is possible to achieve choosing the free functions $\alpha(x,y,z)$, $\beta(x,y,z)$ appropriately. In fact, expanding the left and right hand sides in (3.28) in the local basis of the initial fixed operators $\hat{S}_1$, $\hat{S}_2$, $\hat{T}$ and the zeroth-order operator 1 and collecting the coefficients of this expansion, one gets the following system for the unknown functions α, β, μ, ν:

$$\begin{cases} K + \beta P - \hat{S}_2(\beta) = \nu\beta, \\ M - \hat{S}_2(\alpha) + \beta Q = \nu\alpha - \mu, \\ N + \beta R = -\nu, \\ \beta\hat{S}_1(\beta) - \hat{T}(\beta) + \hat{S}_2(a) - \beta\hat{S}_2(\alpha) - \hat{S}_2(\hat{S}_1(\beta)) \\ \qquad = -\nu(a - \alpha\beta - \hat{S}_1(\beta)) - \mu\beta. \end{cases}$$

After elimination of ν from its first and third equations we get a first-order non-linear partial differential equation for β:

$$\hat{S}_2(\beta) = \beta^2 R + (N + P)\beta + K. \qquad (3.29)$$

This Riccati-like equation may be transformed into a second-order linear PDE via the standard substitution $\beta = \hat{S}_2(\gamma)/\gamma$. Taking any non-zero solution β of this equation and substituting $\mu = \nu\alpha + \hat{S}_2(\alpha) - \beta Q - M$ (taken from the second equation of the system) into the fourth equation of the system we obtain a first-order linear partial differential equation for α with the first-order term $\beta\hat{S}_2(\alpha)$. Any solution of this equation will give the necessary value of α. Now we can substitute $[(\hat{V}+b),(\hat{S}_2+\beta)]u = \mu(\hat{S}_2+\beta)u + \nu(\hat{V}+b)u = \mu v - \nu(\hat{S}_1+\alpha)v$ into the left hand side of (3.27) obtaining the transformed equation $L_{(1)}v = 0$.

If we started the same procedure using the second partial factorization in (3.23) we would find the other transformed equation $L_{(-1)}w = 0$. $\qquad\square$

3.4 Other results and conjectures

The theory of integration of linear and nonlinear partial differential equations was among the most popular topics in the XIX century. An enormous number of papers were devoted for example to transformations of equations to an integrable form. In particular the papers [27, 35, 36] were devoted to more general Laplace type transformations. Some of these results were obtained in the framework of classical differential geometry; cf. [14, 25] for a modern exposition of those results.

In addition to the problems studied above one should mention a class of overdetermined systems of linear partial differential equations *with finite-dimensional solution space* studied in [56, 30]. There an algorithm for factorization of such systems was proposed.

Another topic popular in the past decade was the theory of "naive" factorization, i.e. representation of a given LPDO as a product of lower-order LPDO: in [22] an algorithm for such factorization was proposed for the case of operators with symbol representable as a product of two coprime polynomials. This result was developed further in [43].

From the theory of Laplace and Dini transformations the following **conjectures** seem to be natural:

- *If a LPDO is factorizable in the generalized sense, then its principal symbol is factorizable as a multivariate commutative polynomial.*
- *If a LPDO of order n has a complete solution in a quadrature-free form (3.5) then its symbol splits into n linear factors.*

3.5 Acknowledgment

The author enjoys the occasion to thank the organizers of the LMS summer lecture course and the complete mini-program "Algebraic Theory of Differential Equations" at the International Centre for Mathematical Sciences, Edinburgh for their efforts which guaranteed the success of the mini-program as well as for partial financial support which made presentation of the results given above possible. This paper was written with partial financial support from the RFBR grant 06-01-00814.

Bibliography

[1] Abramov, S. A., Bronstein, M., and Petkovšek, M. (1995). On polynomial solutions of linear operator equations, In *Proceedings of ISSAC'95 (Montreal, Canada)*, ACM Press, 290–296.

[2] Anderson, I.M. and Kamran, N. (1997). The Variational Bicomplex for Second Order Scalar Partial Differential Equations in the Plane, *Duke Math. J.* **87**, No 2, 265–319.

[3] Athorne, C. (1995). A $\mathbf{Z}^2 \times \mathbf{R}^3$ Toda system, *Phys. Lett. A* **206**, 162–166.

[4] Beke, E. (1894). Die Irreducibilität der homogenen linearen Differentialgleichungen, *Math. Annalen* **45**, 278–300.

[5] Blumberg, H. (1912). Über algebraische Eigenschaften von linearen homogenen Differentialausdrücken, *Diss., Göttingen*.

[6] Bronstein, M., and Petkovšek, M. (1994). On Ore rings, linear operators and factorization, *Programming & Computer Software* **20**, No 1, 27–44.

[7] Bronstein, M. (1994). An improved algorithm for factoring linear ordinary differential operators, In *Proc. ISSAC'94*, 336–340.

[8] Bronstein, M. and Petkovšek, M. (1996). An introduction to pseudo–linear algebra, *Theoretical Computer Science* **157**, 3–33

[9] Darboux, G. (1887-1896), Leçons sur la théorie générale des surfaces et les applications géométriques du calcul infinitésimal, t. 2,4, Paris,

[10] Darboux, G. (1870). Sur les équations aux dérivées partielles du second ordre, *Ann. Ecole Normale Sup.* **VII**, pp. 163–173.

[11] Dini, U. (1901). Sopra una classe di equazioni a derivate parziali di second'ordine con un numero qualunque di variabili, *Atti Acc. Naz. dei Lincei. Mem. Classe fis., mat., nat.* (ser. 5) **4**, 121–178. Also *Opere* v. III, 489–566.

[12] Dini, U. (1902). Sopra una classe di equazioni a derivate parziali di second'ordine, *Atti Acc. Naz. dei Lincei. Mem. Classe fis., mat., nat.* (ser. 5) **4**, 431–467. Also *Opere* v. III, 613–660.

[13] Faith, C. (1973). *Algebra: rings, modules and categories*. V. I. Grundlehren der Mathematischen Wissenschaften, Bd. 190, Springer.

[14] Ferapontov, E.V. (1997). Laplace transformations of hydrodynamic type systems in Riemann invariants: periodic sequences, *J. Phys A: Math. Gen.* **30**, 6861–6878.

[15] Forsyth, A.R. (1906). *Theory of differential equations*. Part IV, vol. VI. Cambrudge.

[16] Freyd, P. (1984). *Abelian categories. An introduction to the theory of functors*. Harper & Row.

[17] Gabriel, P. and Zisman, M. (1967). *Calculus of fractions and homotopy theory*. Erg. der Math. und ihrer Grenzgebiete, Springer. Bd. 35.

[18] Gelfand S.I. and Manin Yu.I. (1996). *Methods of homological algebra*. Springer Verlag.

[19] Goursat, E. (1898). Leçons sur l'intégration des équations aux dérivées partielles du seconde ordre a deux variables indépendants. t. 2, Paris.

[20] Grätzer, G. (1978). *General lattice theory*. Akademie-Verlag, Berlin.

[21] Grigor'ev, D.Yu. (1990). Complexity of factoring and calculating the GCD of linear ordinary differential operators, *J. Symbolic Computation* **10**, 7–37.

[22] Grigoriev, D. and Schwarz, F. (2004). Factoring and solving linear partial differential equations, *Computing* **73**, 179–197.

[23] Grigoriev, D. and Schwarz, F. (2005). Generalized Loewy-Decomposition of D-Modules, *Proc. ISSAC'2005 (July 24–27, 2005, Beijing, China)* ACM Press, 163–170.

[24] Janet, M. (1929). *Leçons sur les systèmes d'equations aux derivées partielles*, Gautier-Villard, Paris.

[25] Kamran, N. and Tenenblat, K. (1996). Laplace transformations in higher dimensions, *Duke Math. J.* **84**, No 1, 237–266.

[26] Landau, E. (1902). Über irreduzible Differentialgleichungen. *J. für die reine und angewandte Mathematik* **124**, 115–120.

[27] Le Roux, J. (1899). Extensions de la méthode de Laplace aux équations linéaires aux derivées partielles d'ordre supérieur au second, *Bull. Soc. Math. de France* **27**, 237–262. A digitized copy is obtainable from http://www.numdam.org/

[28] Loewy, A. (1903). Über reduzible lineare homogene Differentialgleichungen, *Math. Annalen* **56**, 549–584.

[29] Loewy, A. (1906). Über vollstandig reduzible lineare homogene Differentialgleichungen, *Math. Annalen* **62**, 89–117.

[30] Min Wu. (2005). On Solutions of Linear Functional Systems and Factorization of Modules over Laurent-Ore Algebras, PhD. thesis, Beijing.

[31] Ore, O. (1932). Formale Theorie der linearen Differentialgleichungen (Erster Teil), *J. für die reine und angewandte Mathematik* **167**, 221–234.

[32] Ore, O. (1932). Formale Theorie der linearen Differentialgleichungen (Zweiter Teil), *J. für die reine und angewandte Mathematik* **168**, 233–252.

[33] Ore, O. (1933). Theory of non-commutative polynomials, *Annals of Mathematics* **34**, 480–508.

[34] Ore, O. (1931). Linear equations in non-commutative fields, *Annals of Mathematics* **32**, 463–477.

[35] Pisati, Laura (1905). Sulla estensione del metodo di Laplace alle equazioni differenziali lineari di ordine qualunque con due variabili indipendenti, *Rend. Circ. Matem. Palermo* **20**, 344–374.

[36] Petrén, L. (1911). Extension de la méthode de Laplace aux équations $\sum_{i=0}^{n-1} A_{1i} \frac{\partial^{i+1}z}{\partial x \partial y^i} + \sum_{i=0}^{n} A_{0i} \frac{\partial^i z}{\partial y^i} = 0$. *Lund Univ. Arsskrift*, Bd. 7, Nr. 3, 1–166.

[37] Reid, G. J. (1991). Algorithms for reducing a system of PDEs to standard form, determining the dimension of its solution space and calculating its Taylor series solutions, *Euro. J. Appl. Mech.* **2**, 293–318.

[38] Riquier, C. (1910). Les systèmes d'equations aux derivées partielles. *Gautier-Villard*.

[39] Schlesinger, L. (1895). *Handbuch der Theorie der linearen Differentialgleichungen*. Teubner, Leipzig.

[40] Schwarz, F. (1984). The Riquier-Janet theory and its applications to non-

linear evolution equations, *Physica D* **11**, 243–351.

[41] Schwarz, F. (1989). A factorization algorithm for linear ordinary differential equations, In *Proc. Int. Symp. on Symbolic Algebraic Comput. (ISSAC'89)*, 17–25. G. Gonnet (ed.), ACM Press.

[42] Schwarz, F. (1992). Reduction and completion algorithms for partial differential equations, In *Proc. Int. Symp. on Symbolic Algebraic Comput. (ISSAC'92)*, 49–56. P. Wang (ed.), ACM Press.

[43] Shemyakova, E., Winkler, F. Obstacles to factorizations of linear partial differential operators, Submitted, 2006.

[44] Singer, M.F. (1996). Testing reducibility of linear differential operators: a group theoretic perspective, *Applicable Algebra in Engineering, Communication and Computing* **7**, 77–104.

[45] Singer, M. F., and Ulmer, F. (1993). Galois groups of second and third order linear differential equations, *Journal of Symbolic Computation* **16**, 9–36.

[46] Sokolov, V.V., Zhiber, A.V. (1995). On the Darboux integrable hyperbolic equations, *Phys. Letters A* **208**, 303–308.

[47] Tsarev, S.P. (1994). On some problems in factorization of linear ordinary differential operators, *Programming & Comp. Software* **20**, No 1, 27–29.

[48] Tsarev, S.P. (1996). An algorithm for complete enumeration of all factorizations of a linear ordinary differential operator, *Proceedings of ISSAC'1996*, ACM Press, 226–231.

[49] Tsarev, S.P. (1998). Factorization of linear partial differential operators and Darboux integrability of nonlinear PDEs, *SIGSAM Bulletin* **32**, No. 4. 21–28. E-print http://www.arxiv.org/ cs.SC/9811002.

[50] Tsarev, S.P. (2005). Generalized Laplace Transformations and Integration of Hyperbolic Systems of Linear Partial Differential Equations, *Proc. ISSAC'2005 (July 24–27, 2005, Beijing, China)* ACM Press, 325–331. E-print: www.arxiv.org, cs.SC/0501030

[51] Tsarev, S.P. (2006). On factorization and solution of multidimensional linear partial differential equations. e-print cs.SC/0609075 at http://www.archiv.org/.

[52] van der Put, M. and Singer, M.F. (2003). Galois Theory of Linear Differential Equations, Grundlehren der Mathematischen Wissenschaften, v. 328, Springer.

[53] van Hoeij, M. (1997). Factorization of differential operators with rational functions coefficients, *J. Symbolic Comput.* **24**, 537–561.

[54] Yilmaz, H. and Athorne, C. (2002). The geometrically invariant form of evolution equations, *J. Phys. A.* **35**, 2619–2625.

[55] Zhiber, A.V. and Sokolov, V.V. (2001). Exactly integrable hyperbolic equations of Liouville type, *Russian Math. Surveys* **56**, No. 1. 61–101.

[56] Ziming Li, Schwarz, F. and Tsarev, S.P. (2003). Factoring systems of linear PDEs with finite-dimensional solution spaces, *J. Symbolic Computation* **36**, 443–471.

4

Introduction to the algorithmic D-module theory

Anton Leykin

Institute for Mathematics and its Applications
University of Minnesota
Minneapolis
MN 55455
USA
Email: leykin@ima.umn.edu
http://www.ima.umn.edu/~leykin/

The history of D-module theory starts in 1960-70s and is connected with the names of Mikio Sato and Joseph Bernstein. Many well-known mathematicians, such as Kashiwara and Malgrange, contributed to the area over the years. There is a long list of options for reading on the subject, from which one can choose according to one's liking and background. A rather elementary text [8] needs no deep knowledge of algebraic geometry, deeper treatments can be found in [5, 6, 14], more specialized introductions are provided by chapters of [24] (applications to hypergeometric equations) and [12] (computation of local cohomology modules).

What we intend to provide here is a short primer covering the basics of the *algorithmic D-module theory*. The D-modules are defined as left modules over an algebra D of linear differential operators with either polynomial coefficients or coefficients in the field of rational functions.

There are three parts: the first introduces the *Weyl algebra* and discusses its basic properties, the second is devoted to a "nice" subcategory of the category of D-modules, the *holonomic D-modules*, the third features applications of the theory and its algorithmic aspects to *hypergeometric differential equations* and *local cohomology*.

In what follows the base field k will always be assumed to have characteristic 0. Lots of geometrically interesting concepts exist for the case $k = \mathbb{C}$; for simplicity the reader may limit his or her thinking to the complex numbers. In computations, we utilize the smallest option: $k = \mathbb{Q}$. It still makes sense to talk about certain algorithms over other fields, however, the current software implementations are limited to the field of rational numbers and its finite algebraic extensions.

The ring of polynomials in n variables $\boldsymbol{x} = (x_1, \ldots, x_n)$ will be usually denoted by either R or $k[\boldsymbol{x}]$ and the field of rational functions in the same variables by $k(\boldsymbol{x})$. We shall use the multi-index notation for monomials, i.e., $\boldsymbol{x}^\alpha = x_1^{\alpha_1} \cdots x_n^{\alpha_n} \in R$ for $\alpha \in \mathbb{Z}_{\geq 0}^n$.

The n-th *Weyl algebra* $k\langle \boldsymbol{x}, \boldsymbol{\partial} \rangle$ is defined as the associative k-algebra on $2n$ symbols $\boldsymbol{x} = (x_1, \ldots, x_n)$ and $\boldsymbol{\partial} = (\partial_1, \ldots, \partial_n)$ with the relations

$$
\begin{aligned}
x_i x_j &= x_j x_i, & \text{for all } 1 \leq i, j \leq n; \\
\partial_i \partial_j &= \partial_j \partial_i & \text{for all } 1 \leq i, j \leq n; \\
\partial_i x_j &= x_j \partial_i & \text{for all } 1 \leq i, j \leq n, \ i \neq j; \\
\partial_i x_i &= x_i \partial_i + 1 & \text{for all } 1 \leq i \leq n.
\end{aligned}
$$

The Weyl algebra, which for simplicity from now on will be referred to as D, is isomorphic to the algebra of linear partial differential operators with polynomial coefficients.

We let $k(\boldsymbol{x})\langle \boldsymbol{\partial} \rangle$ stand for the algebra of linear partial differential operators with coefficients in $k(\boldsymbol{x})$. Note that $k(\boldsymbol{x})\langle \boldsymbol{\partial} \rangle = k(\boldsymbol{x}) \otimes_{k[\boldsymbol{x}]} D$, where D is considered as the left $k[\boldsymbol{x}]$-module.

We assume multi-index notation for the monomials in D as well: $\boldsymbol{x}^\alpha \boldsymbol{\partial}^\beta = x_1^{\alpha_1} \cdots x_n^{\alpha_n} \partial_1^{\beta_1} \cdots \partial_n^{\beta_n} \in D$. Note that the order of symbols in the product is important – we will always write x_i left of ∂_i. Let the set of monomials be ordered with respect to an *admissible ordering* $\leq$, which means that

(i) the ordering $\leq$ is total;

(ii) for $\alpha, \beta, \alpha', \beta', \gamma, \delta \in \mathbb{Z}_{\geq 0}^n$,

$$
\boldsymbol{x}^\alpha \boldsymbol{\partial}^\beta \leq \boldsymbol{x}^{\alpha'} \boldsymbol{\partial}^{\beta'} \ \Leftrightarrow \ \boldsymbol{x}^{\alpha+\gamma} \boldsymbol{\partial}^{\beta+\delta} \leq \boldsymbol{x}^{\alpha'+\gamma} \boldsymbol{\partial}^{\beta'+\delta};
$$

(iii) $1 = \boldsymbol{x}^0 \boldsymbol{\partial}^0$ is the lowest monomial.

Every such ordering is a refinement of a *weight ordering* $\leq_\omega$ for some weight $\omega = (\omega_x, \omega_\partial) \in \mathbb{Z}_{\geq 0}^{2n}$,

$$
\boldsymbol{x}^\alpha \boldsymbol{\partial}^\beta \leq \boldsymbol{x}^{\alpha'} \boldsymbol{\partial}^{\beta'} \ \Leftrightarrow \ \deg_\omega \boldsymbol{x}^\alpha \boldsymbol{\partial}^\beta \leq \deg_\omega \boldsymbol{x}^{\alpha'} \boldsymbol{\partial}^{\beta'},
$$

where $\deg_\omega \boldsymbol{x}^\alpha \boldsymbol{\partial}^\beta = \omega_x \cdot \alpha + \omega_\partial \cdot \beta \in \mathbb{Z}_{\geq 0}$ is the ω-degree. Note that the ordering $\leq_\omega$ is not total in general. Also to make it possible for 1 to be the minimal monomial in a refinement of $\leq_\omega$ it is required that $\omega_x + \omega_\partial \in \mathbb{Z}_{\geq 0}^n$

4.1 Weyl algebra

The main object of discussion in this part is the n-th Weyl algebra,

$$
D = k\langle \boldsymbol{x}, \boldsymbol{\partial} \rangle = k\langle x_1 \ldots x_n, \partial_1 \ldots \partial_n \rangle,
$$

defined in the introduction.

The Weyl algebra D belongs to the class of *algebras of solvable type* defined in [13]. This means that for any two generators y and z of the algebra, in our case $\{y, z\} \subset \{x_1, \ldots, x_n, \partial_1, \ldots, \partial_n\}$, there exist $c \in k$ and a (noncommutative) polynomial $p \in D$ such that $zy - cyz = p$ and the monomials of p are all $\leq yz$. Here we can take $c = 1$ for any y, z and $p = 0$ unless $y = x_i$ and $z = \partial_i$, for some $1 \leq i \leq n$, in which case $p = 1$.

The monomials of D form the so-called *PBW (Poincaré-Birkhoff-Witt) basis* meaning that every element $Q \in D$ can be written uniquely as a finite linear combination of monomials,

$$Q = \sum_{\alpha, \beta \in \mathbb{Z}_{\geq 0}^n} c_{\alpha\beta} x^\alpha \partial^\beta, \text{ with } c_{\alpha\beta} \in k \text{ and } \alpha, \beta \in \mathbb{Z}_{\geq 0}^n. \qquad (4.1)$$

4.1.1 Ideals

As we mentioned before, the Weyl algebra D is isomorphic to the algebra of linear partial differential operators and may be represented by its natural action on an appropriate function space M. For instance, we may take M to be polynomial functions $R = k[x]$, rational functions $k(x)$, or the space of smooth functions C^∞; the action of generators of D defines the action of the whole Weyl algebra:

$$x_i \cdot f = x_i f, \quad \partial_i \cdot f = \frac{\partial_i f}{\partial x_i}, \quad \text{for } f \in M \text{ and } 1 \leq i \leq n.$$

The Weyl algebra is *simple*, i.e., the only two-sided ideals of D – this makes a good exercise to show – are trivial: 0 and D itself. In view of this, in what follows we would consider only one-sided ideals of D. We denote by $(Q_1, \ldots, Q_m)$ the left ideal of D generated by the operators $Q_1, \ldots, Q_m \in D$.

In fact, the left ideal $I = (Q_1, \ldots, Q_m)$ is exactly the object one would want to examine in order to study the system of linear PDEs

$$Q_1 \cdot f = \cdots = Q_m \cdot f = 0, \qquad (4.2)$$

where f is the unknown function. Indeed, the operators of I produce equations that follow algebraically from the equations (4.2).

Here we give several examples of ODE translated into the ideals in the first Weyl algebra $D = \mathbb{C}\langle x, \partial \rangle$ accompanied by their solutions.

Example 4.1.1 Cyclic ideals represent a single linear ODE. For instance, $I_1 = (\partial^3 - \partial)$ corresponds to the equations with the space of solutions spanned by functions 1, e^x and e^{-x}. The solutions of ideal $I_2 = ((\partial - 1)^2)$ are spanned by e^x and xe^x.

Not all ideals in the first Weyl algebra are cyclic though: the ideal $I_3 = (\partial^2, x\partial - 1)$ cannot be generated by one element. Its solution space is one-dimensional and is spanned by the function x.

The following existence theorem is classical:

Theorem 9 *Let* $Q \cdot f = 0$ *be an ODE with* $Q = \sum_{i=0}^d a_i(x)\partial^i \in D$ *where* $a_i(x) \in R = \mathbb{C}[x]$ *and* $a_d(x) \neq 0$.

Then the dimension of the space of holomorphic solutions in a simply connected domain in $\mathbb{C}$ *not containing zeroes of* $a_d(x)$ *equals* d.

In case of the first Weyl algebra the *rank* of the system represented by I, i.e., the dimension of its solution space, is easy to determine: the rank is equal to the minimal order of a nonzero operator of I.

The rank can be determined algorithmically via the process of differential elimination. For example, take $I_4 = (\partial^3, x\partial^2 + x\partial - 1)$, then by an analogue of the Euclid division algorithm we get:

$$x(\partial^3) - \partial(x\partial^2 + x\partial - 1) = x\partial^2 + \partial^2,$$
$$(x\partial^2 + x\partial - 1) - (x\partial^2 + \partial^2) = -\partial^2 + x\partial - 1,$$
$$(x\partial^2 + x\partial - 1) + x(-\partial^2 + x\partial - 1) = x^2\partial + x\partial - x - 1.$$

Note that the last operator is of order 1. Moreover, by this reduction process we can obtain also ∂^2 and $x\partial - 1$ proving that $I_3 = I_4$.

In general, using such reduction steps we can always compute a set of minimal generators of the given ideal $I \subset D$ of the form

$$f_1\partial^{d_1} + Q_1, \ldots, f_m\partial^{d_m} + Q_m,$$

where $d_1 > \cdots > d_m$, $f_1, \ldots, f_m \in k[x] \setminus \{0\}$, $Q_1, \ldots, Q_m \in D$ and $\deg_\partial Q_i < d_i$, for all $1 \leq i \leq m$; the generator $f_i\partial^{d_i} + Q_i$ is minimal in the sense that there is no $g\partial^{d_i} + Q' \in I$ with $g \in k[x] \setminus \{0\}$ and $\deg_\partial Q' < d_i$ such that $\deg g < \deg f_i$.

Since it is tiring to do the computations in D by hand, the reader might find useful computer algebra systems that can handle Weyl algebras. In particular, the D-modules package [17] for *Macaulay2* [10] provides all the functionality needed for our exposition.

Example 4.1.2 Here is how one can play with a couple of examples from above:

```
i1 : loadPackage "Dmodules";

i2 : D = makeWA(QQ[x]);

i3 : I3 = ideal (dx^2, x*dx-1);

o3 : Ideal of D

i4 : I4 = ideal (dx^3, x*dx^2+x*dx-1);

o4 : Ideal of D

i5 : I3 == I4

o5 = true
```

The following is a command for computing the set of minimal generators.

```
i6 : entries gens gb I4

            2
o6 = {{dx , x*dx - 1}}
```

Of course, this computes a Gröbner basis of the ideal, which is one of the most central concepts of computational *commutative* algebra. It turns out that one can make this concept work in the "slightly" *non-commutative* case of algebras of solvable type, in particular, the n-th Weyl algebra D.

We will get back to Gröbner bases in Section 4.1.5.

There exists another extensive package for D-modules computations: *kan/sm1* [28].

4.1.2 Stafford's theorem

Now consider an example in many variables: $I = (\partial_1, \partial_2, \partial_3)$, an ideal in the third Weyl algebra

$$D = k\langle x_1, x_2, x_3, \partial_1, \partial_2, \partial_3 \rangle.$$

It turns out it is possible to generate I by only two elements.

Indeed, take $P = x_3 \partial_1 + \partial_2$ and $Q = \partial_3$, then it is easy to check that $QP - PQ = \partial_1 \in (P, Q)$. Now $\partial_2 = P - x_3 \partial_1 \in (P, Q)$, hence, $I = (P, Q)$.

In fact, a problem that is considered hard in the ring of polynomials – finding the minimal number of elements that generate an ideal – turns out to be surprisingly trivial in the view of the following.

Theorem 10 (Stafford) *Every left ideal of the Weyl algebra can be generated by two elements.*

This has been proved in [25] and algorithmic proofs have been given in [11, 16].

Therefore, a non trivial ideal of the Weyl algebra is either principal, which is easy to detect by computing a Gröbner basis, or it is possible to compute two elements that generate the whole ideal.

We have to remark, however, that such a "minimal" presentation is not very useful for practical computations, since the degrees of a pair of operators generating the whole ideal could be huge. To our knowledge, there is no known bound for these degrees except a naive one that could be derived by considering the complexity of the Gröbner bases involved in the algorithms of [11, 16].

4.1.3 Modules and homomorphisms

A k-linear space M is called a (left) D-module if there is a left action of D defined on M.

Example 4.1.3 The following are left D-modules:

- The free module D^r of rank r with D acting by the left multiplication on every component of a vector in D^r;
- Equipped with the natural action of D,

  - the ring of polynomials $k[x]$;
  - the field of rational functions $k(x)$;
  - the ring of formal power series $k[[x]]$;
  - the space of C^∞-functions;

- The quotient D/I, where $I \subset D$ is a left ideal, with the induced action on cosets.

From the computational point of view we are able to deal only with *finitely generated* D-modules. Note that several modules in Example 4.1.3 are not finitely generated: namely, $k(x)$, $k[[x]]$, and the space of C^∞-functions.

Let $M = (m_1, \ldots, m_r)$ stand for a (left) D-module (finitely) generated by $m_1, \ldots, m_r \in M$. In algorithms M is often represented as a quotient of a free D-module. To that end, one forms the module of left syzygies,

$$N = \{Q_1 e_1 + \cdots + Q_r e_r \in D^r : Q_1 m_1 + \cdots + Q_r m_r = 0\},$$

where $\{e_1, \cdots, e_r\}$ is the standard basis of D^r. Then the module D^r/N is isomorphic to M.

When D-modules are employed to study systems of linear PDE, the first question one may ask is: what are the solutions? The concept of the *solution space* is readily available in the language of D-modules.

Namely, to solve the system of represented by a cyclic D-module $M_1 = D/I$ in terms of the functional space represented by D-module M_2 means to describe the set of homomorphisms $\mathrm{Hom}_D(M_1, M_2)$.

This set has a structure of a k-linear space. Every element $\phi \in \mathrm{Hom}_D(M_1, M_2)$ is defined by its value at the coset $\bar{1} \in D/I$; since $\phi(\bar{1})$ has to be annihilated by I, it is a solution to the corresponding system of PDE.

Example 4.1.4 Let us go back to the ideals in Example 4.1.1: $I_1 = (\partial^3 - \partial)$ and $I_2 = ((\partial - 1)^2)$. According to the above-said, if M is the space of smooth functions, then

$$\mathrm{Hom}_D(D/I_1, M) = \mathrm{Span}_k\{e^x, e^{-x}, 1\},$$
$$\mathrm{Hom}_D(D/I_2, M) = \mathrm{Span}_k\{e^x, xe^x\}.$$

If we look for solutions in the ring of univariate polynomials $k[x]$, then

$$\mathrm{Hom}_D(D/I_1, k[x]) = k,$$
$$\mathrm{Hom}_D(D/I_2, k[x]) = 0.$$

Unfortunately, we cannot compute with M, since it is infinitely generated. Here is what we can do for polynomial solutions, keeping in mind that $k[x] = D/(\partial)$.

```
i7 : I1 = ideal(dx^3-dx);

i8 : I2 = ideal((dx-1)^2);

i9 : DHom(I1,ideal dx)

o9 = {| -1 |}

o9 : List

i10 : DHom(I2,ideal dx)

o10 = 0
```

Let us be adventurous and look for solutions of I_1 in the D-module D/I_2, i.e., compute $\mathrm{Hom}_D(D/I_1, D/I_2)$.

```
i11 : s = DHom(I1,I2)
o11 = {| -2dx+2 |, | -12xdx+12x-28dx+40 |}
```

The output means that the space of solutions is two-dimensional, despite us expecting only one copy of " e^x " in D/I_2.

The confusion is resolved by looking closely at the kernel of maps from D to D/I_2 defined by these two spanning solutions.

```
i12 : ker map(D^1/I2, D^1, s#0)
o12 = image | dx-1 |
                            1
o12 : D-module, submodule of D
i13 : ker map(D^1/I2, D^1, s#1)
o13 = image | -dx+1 |
                            1
o13 : D-module, submodule of D
```

Both solutions are, indeed, incarnations of " e^x " in D/I_2.

4.1.4 Filtrations and the associated graded algebra

For a weight $\omega = (\omega_x, \omega_\partial) \in \mathbb{Z}_{\geq 0}^{2n}$ the collection of k-linear subspaces $F_i \subset D$, where $i \in \mathbb{Z}_{\geq 0}$:

$$F_i = \bigoplus_{\deg_\omega x^\alpha \partial^\beta \leq i} k \cdot x^\alpha \partial^\beta,$$

is called the *filtration* by the weight ω. The union of F_i equals the whole of D and we can define $\operatorname{gr} D$, the *associated graded algebra*, in the usual way:

$$\operatorname{gr} D = \bigoplus_{i \in \mathbb{Z}_{\geq 0}} F_i/F_{i-1},$$

where $F_{-1} = 0$ is assumed.

Example 4.1.5 Two important examples of filtrations are

- the *order filtration* given by $\omega = (\mathbf{0}, \mathbf{1}) = (0, \ldots, 0, 1, \ldots, 1)$, which filters the operators in D by the differential order,
- and *Bernstein filtration* $\omega = (\mathbf{1}, \mathbf{1}) = (1, \ldots, 1, 1, \ldots, 1)$, which filters by the total degree.

In both cases the associated graded algebra is commutative, since $x_i, \partial_i \in F_1 \setminus F_0$, but their commutator $[\partial_i, x_i] \in F_0$, $1 \leq i \leq n$. This makes the noncommutative relations commutative in $\mathrm{gr}_\omega D$. In fact, $\mathrm{gr}_\omega D$ is isomorphic to the algebra of (commutative) polynomials $k[\boldsymbol{x}, \boldsymbol{\xi}]$ where the generators $\boldsymbol{\xi} = (\xi_1, \ldots, \xi_n)$ replace ∂ after the natural map $\mathrm{gr}_\omega : D \to \mathrm{gr}_\omega D$ is applied.

4.1.5 Gröbner bases

Let $\leq$ be an admissible ordering refining $\leq_\omega$ for a weight $\omega = (\omega_x, \omega_\partial) \in \mathbb{Z}_{\geq 0}^{2n}$ such that $\omega_x + \omega_\partial \in \mathbb{Z}_{>}^n$. Recall that in this case $\mathrm{gr}_\omega D$ is isomorphic to the commutative algebra $k[\boldsymbol{x}, \boldsymbol{\xi}]$. Let $\mathrm{gr}_\omega : D \to \mathrm{gr}_\omega D$ be the natural projection.

Every element Q of the Weyl algebra can be written in the form (4.1), where only finitely many monomials have nonzero coefficient. The image under the map gr_ω of the maximal such monomial with respect to the ordering $\leq$ is called the *leading monomial* and denoted $\mathrm{lm}\, Q \in k[\boldsymbol{x}, \boldsymbol{\xi}]$.

We call a subset $G \subset I$ a *Gröbner basis* of a left ideal I if

$$(\mathrm{lm}\, G) = \mathrm{lm}\, I := \{\mathrm{lm}\, Q : Q \in I\} \subset k[\boldsymbol{x}, \boldsymbol{\xi}].$$

A Gröbner basis G is called *reduced* if its elements are monic, i.e., their leading coefficients equal 1, and for every $Q \in G$ its leading monomial is not divisible by $\mathrm{lm}\, P$ for any $P \in G \setminus \{Q\}$.

Gröbner bases in the Weyl algebra possess most of the properties the usual commutative Gröbner bases have. In particular, the unique *normal form* $\mathrm{nf}_G(Q)$ of an element $Q \in D$ with respect to a Gröbner basis G is the result of the reduction (in any order) by the elements of G. At the end of the reduction we have

$$\mathrm{nf}_G(Q) = Q \mod I$$

and $\mathrm{lm}\, \mathrm{nf}_G(Q)$ not divisible by $\mathrm{lm}\, P$, for any $P \in G$.

One can look up the details of the division algorithm and a description of the basic Buchberger algorithm for computing a Gröbner basis in any introductory text on Gröbner bases in polynomial rings; the corresponding algorithms for the Weyl algebra are virtually the same.

We have already computed one Gröbner basis in Example 4.1.2. There, two seemingly different ideals I_3 and I_4 turned out to be equal, simply since they have the same reduced Gröbner basis.

The applicability of Gröbner bases is not limited to the comparison of ideals; more or less every application one can imagine in the commutative world would have a counterpart for the Weyl algebras. In particular, we may generalize the concept of Gröbner bases to the submodules of free D-modules and compute kernels of maps, syzygies, free resolutions in ways very similar to the Gröbner techniques in the case of modules over polynomial rings.

Gröbner bases methods in the Weyl algebra are the most straightforward and, therefore, the best developed in the range of differential reduction algorithms devised for dealing with the systems of PDE by means of differential algebra. The practical complexity of computing Gröbner bases in the Weyl algebras is definitely higher than that in polynomial rings with the same number of variables, simply because the basic operations, such as multiplication of an element in D by a monomial, already require more effort than in the commutative case.

In view of that, it is a very surprising fact, that the theoretical worst case complexities of Gröbner bases in the Weyl algebras and the polynomials rings are of the same order! Namely, assume that G is the reduced Gröbner basis of the ideal I of the n-th Weyl algebra D generated in degrees less than d. It has been shown in [3] that there is an upper bound on the degrees of elements in G that is doubly-exponential in the number of variables n, more precisely, the order of the bound is $O(d^{2^n})$.

4.1.6 Dimension theory

To sketch the dimension theory for D-modules we fix the *Bernstein filtration* on the Weyl algebra D introduced in Example 4.1.5. For simplicity, we let M to be a cyclic (left) D-module presented as a quotient $M = D/I$ for some (left) ideal $I \subset D$. One can generalize the dimension theory to any D-module; the reason for a rather restrictive assumption of cyclicity will become apparent in Theorem 13.

The filtration on D induces filtrations on the ideal I and the quotient $M = D/I$.

Theorem 11 *Consider the function* $t \to \dim_k M_t$, $t \in \mathbb{Z}_{\geq 0}$, *where* M_t *is the t-th component of the filtration on M.*

Then for $t \gg 0$ this function is polynomial in t with rational coefficients.

Proof For a proof of a more general statement see, for example, [12, Theorem 5.18].   □

The polynomial above is known as the *Hilbert polynomial* and denoted by $H_M(t)$. Suppose,

$$H_M(t) = c_d t^d + \text{lower terms},$$

then we define the *dimension* $\dim M$ to be d and the *multiplicity* $e(M)$ to be $d!c_d$. Both numbers are positive integers.

We remark that these definitions could have been made by passing to the gr D-module gr M and reusing the notions of dimension and multiplicity for a module over a commutative polynomial ring.

Since the dimension of the n-th Weyl algebra $\dim D = 2n$, the dimension of its module $\dim M \leq 2n$. Surprisingly, there is a lower bound different from zero!

Theorem 12 (Bernstein) *For a nonzero D-module M*

$$n \leq \dim M \leq 2n.$$

Proof For proof see, for instance, [5, 4.1].   □

It could be shown that the dimension defined by passing to the associated graded algebra with respect to the filtration by any nonnegative weight ω is the same. In particular, the dimensions with respect to the Bernstein filtration and the *order filtration* (see Example 4.1.5) are equal. Theorem 12 for the order filtration is known as the *weak fundamental theorem of algebraic analysis*.

4.2 Holonomic D-modules

Here we discuss the subcategory of D-modules that is "nice": we say that a D-module M over the n-th Weyl algebra D is *holonomic* if it is finitely generated and either $M = 0$ or $\dim M = n$, the lowest possible dimension according to Theorem 12.

A system of PDE corresponding to a holonomic module is guaranteed to have a finite-dimensional solution space. The computation of a basis of the solution space of a holonomic D-module can be carried out in the sense of the discussion preceding Example 4.1.4; algorithms for this are provided in [29].

4.2.1 Holonomic ⇒ cyclic

For the first Weyl algebra D the choice for the dimension of a quotient D-module $M = D/I$ is between 0,1, and 2, the latter corresponding to $M = D$. Therefore, if M nontrivial it is holonomic. For instance, the D-modules D/I_1 and D/I_2 with I_1 and I_2 borrowed from example 4.1.1 are holonomic ideals describing ODEs.

The following is true for the n-th Weyl algebra for arbitrary n.

Theorem 13 *Every holonomic D-module is cyclic.*

Proof For a proof of a more general statement see [5, Theorem 8.18], a constructive treatment and an algorithm to compute a cyclic generator are given in [16]. □

In the rest of our discussion we will be interested mostly in holonomic ideals, and, for convenience, the reader may think they are represented as quotients: e.g. $M = D/I$ for some *holonomic ideal I*. Theorem 13 lets us do so without loss of generality.

Example 4.2.1 Let $M = R^3 = R \oplus R \oplus R$, where $R = k[\boldsymbol{x}]$ is the ring of polynomials in n variables. M is equipped with a structure of a D-module over the n-th Weyl algebra D, for R is a D-module.

Obviously $\dim M = n$, therefore, M is holonomic. By Theorem 13 we can cook up a cyclic generator from the standard generators: $e_1 = 1 \oplus 0 \oplus 0$, $e_2 = 0 \oplus 1 \oplus 0$, and $e_3 = 0 \oplus 0 \oplus 1$.

Consider

$$v = \frac{1}{2}x_1^2 \oplus x_1 \oplus 1 = \frac{1}{2}x_1^2 e_1 + x_1 e_2 + e_3.$$

We have $\partial^2 \cdot v = e_1$, then $\partial \cdot (v - x_1^2 \partial^2 \cdot v) = e_2$, and, finally, e_3 can be obtained as a k-linear combination of v, e_1, and e_2.

Hence, the vector v is a cyclic generator of M.

4.2.2 Category of holonomic modules

Holonomic D-modules form an abelian subcategory of the category of D-modules.

Indeed, not only one can add morphisms of D-modules, but for a short exact sequence of D-modules,

$$0 \to M' \to M \to M'' \to 0,$$

M is finitely generated if and only if M' and M'' are. Moreover, M is holonomic if and only if M' and M'' are, therefore, this is a category with kernels and cokernels.

By considering the Hilbert polynomials, it is easy to prove that the multiplicity is additive, $e(M) = e(M') + e(M'')$, provided the convention $e(0) = 0$.

Theorem 14 *If M is holonomic, it is of finite length.*

Proof Consider a sequence of holonomic D-modules:

$$0 \subsetneq M_1 \subsetneq \cdots \subsetneq M_{l-1} \subsetneq M_l \subsetneq M.$$

To show that M is of finite length we need to find a bound for the length l of any such sequence.

Breaking down the sequence above into short exact sequences

$$0 \to M_{i-1} \to M_i \to M_i/M_{i-1} \to 0, \quad (i = 2, \dots, l)$$

and using the additivity of the multiplicity shows that $e(M)$ is such a bound. $\qquad\square$

The converse is false as was shown in [26]: there exist simple D-modules, i.e., of length one, that are not holonomic.

Example 4.2.2 Let $Q = \partial_1 + (1 - x_1 x_2)\partial_2 - x_2$ be an element of the second Weyl algebra D and let $M = D/(Q)$.

Note that $\dim M = 4 - 1 = 3 \neq 2$, therefore, M is not holonomic. However, by modifying [26, Theorem 1.1] to fit our notation and considering left D-modules, we can show that M is simple.

Moreover, in [4] it has been shown that, in a certain sense, almost all operators in D generate a simple D-module. To our best knowledge, the problem of determining simplicity of a D-module algorithmically is still open.

4.2.3 Holonomic rank and Weyl closure

Let I be an ideal of the Weyl algebra D and consider its extension $\bar{I} = k(\boldsymbol{x}) \otimes I$ in the algebra of differential operators with rational coefficients, $k(\boldsymbol{x})\langle\boldsymbol{\partial}\rangle = k(\boldsymbol{x}) \otimes_{k[\boldsymbol{x}]} D$.

We define the *holonomic rank* of the D-module D/I to be

$$\operatorname{rank} D/I = \dim_{k(\boldsymbol{x})} k(\boldsymbol{x})\langle\boldsymbol{\partial}\rangle/\bar{I}.$$

If $\bar{I}$ is 0-dimensional, then the holonomic rank is finite. Holonomic rank is obtained by computing a Gröbner basis of I with respect to an admissible ordering that eliminates the variables ∂ (see [24, Section 1.4] for details).

If I is a holonomic ideal then rank $D/I < \infty$: the converse is not true.

Example 4.2.3 For $I = (x_1\partial_1 + 1, x_1\partial_2)$ the rank of D/I equals 1, since $\dim_{k(x)} k(x)\langle\partial\rangle/\bar{I}$ is 1-dimensional. However, I is not holonomic, since the pair of generators above forms a Gröbner basis, but gr $D/I = k[x_1, x_2, \xi_1, \xi_2]/(x_1\xi_1, x_1\xi_2)$ has dimension 3.

```
i2 : D = makeWA(QQ[x_1,x_2]);

i3 : I = ideal(x_1*dx_1+1, x_1*dx_2);

o3 : Ideal of D

i4 : isHolonomic I

o4 = false

i5 : holonomicRank I

o5 = 1
```

The holonomic rank of a D-module $M = D/I$ provides analytic information about a system of linear PDEs corresponding to I. Namely, the dimension of the space of holomorphic solutions to the system in a neighborhood of a nonsingular point equals rank M.

One can check that in Example 4.2.3 the solution space is spanned by the function $f(x_1, x_2) = 1/x_1$.

For an ideal $I \subset D$, there is an operation that completes I with all operators annihilating the functions in its solution space. Having the natural inclusion $D \subset k(x)\langle\partial\rangle$, we define the *Weyl closure* of I as

$$\mathrm{Cl}(I) = \bar{I} \cap D.$$

Example 4.2.4 Let $I = (x\partial - 1)$ be an ideal of the first Weyl algebra, then

$$\mathrm{Cl}(I) = (\partial^2, x\partial - 1).$$

Indeed, the equation $(x\partial - 1) \cdot f = 0$ has one solution up to a multiple, $f = x$, which is annihilated by ∂^2 as well.

An algorithm for Weyl closure has been described in [30].

Going back to the open question of detecting simplicity of a D-module

D/I brought up in the previous section, since $\mathrm{Cl}(I) \supset I$, it boils down to two cases: $\mathrm{Cl}(I) = I$ and $\mathrm{Cl}(I) = D$. If $\operatorname{rank} D/I < \infty$, the former case can be effectively resolved by determining whether the $k(\boldsymbol{x})\langle\boldsymbol{\partial}\rangle$-module $k(\boldsymbol{x})\langle\boldsymbol{\partial}\rangle/\bar{I}$ is simple.

4.2.4 Localization

Consider the polynomial ring $R = k[\boldsymbol{x}]$ and its localization $R_f = k[\boldsymbol{x}, f^{-1}]$ for a nonzero $f \in R$. Let us establish the natural D-module structure on the latter:

$$x_i \cdot g f^{-j} = x_i g f^{-j},$$

$$\partial_i \cdot g f^{-j} = \left(\frac{\partial g}{\partial x_i} f - jg \right) f^{-j-1},$$

for $1 \leq i \leq n$, $g \in k[\boldsymbol{x}]$, and $j \in \mathbb{Z}_{\geq 0}$.

Theorem 15 R_f *is a holonomic D-module.*

Proof For proof see, for instance, [5, 5.5]. □

Example 4.2.5 Let $f = x \in R = k[x]$, a polynomial in one variable. Then R_f is holonomic and, by Theorem 13, cyclic.

Indeed, R_f is generated by x^{-1}; a k-multiple of every monomial x^j can be obtained from x^{-1} either via multiplication by x^{j+1}, if $j \geq -1$, or by applying a constant multiple of ∂^{-j-1} otherwise.

In general, the computation of a cyclic generator of R_f requires knowing the roots of the Bernstein-Sato polynomial of f discussed in the next subsection; algorithms for computing localization can be found in [22, 23].

4.2.5 Bernstein-Sato polynomial

Here we will consider the n-th Weyl algebra $D(s) = k(s)\langle\boldsymbol{x}, \boldsymbol{\partial}\rangle$ with the ground field of univariate rational functions $k(s)$. Also let $M = k(s)[\boldsymbol{x}, f^{-1}]f^s$ be a free module of rank one over localization of the polynomial ring $k(s)[\boldsymbol{x}]$ at a nonzero polynomial $f \in k[\boldsymbol{x}]$. The standard generator on M is denoted by the symbol f^s; the choice of the symbol is explained by the action of the Weyl algebra defined below.

We let M be a $D(s)$-module with the structure induced by the following "natural" action of the generators:

$$x_i \cdot gf^{-j}f^s = x_i gf^{-j}f^s,$$

$$\partial_i \cdot gf^{-j}f^s = \left(\frac{\partial g}{\partial x_i} f^{-j} + (s-j)gf^{-j-1} \right) f^s,$$

for $1 \le i \le n$ and $j \in \mathbb{Z}_{\ge 0}$.

Theorem 16 *M is a holonomic $D(s)$-module.*

For any nonzero $f \in k[\boldsymbol{x}]$ there exist a nonzero $b \in k[s]$ and $Q \in D(s)$ such that the functional equation

$$b \cdot f^s = Q \cdot (ff^s) \tag{4.3}$$

holds in M.

Proof See [5, 5.6,5.7]. $\qquad\qquad\square$

The nonzero monic $b \in k[s]$ of the minimal degree, for which the functional equation (4.3) exists, is called the *Bernstein-Sato polynomial* of f and is denoted by b_f.

Theorem 17 *The localization R_f at a polynomial $f \in R = k[\boldsymbol{x}]$ is generated by f^a where a is the lowest integer root of b_f.*

Proof Indeed, by specializing s to any integer $t < a$ in the functional equation (4.3) we get $b_f|_{s=t} f^t = Q|_{s=t} \cdot f^{t+1}$, where $b_f|_{s=t} \neq 0$. This means that all powers $f^t \in D \cdot f^a \subset R_f$, therefore, $D \cdot f^a = R_f$. $\qquad\square$

The study of Bernstein-Sato polynomials is a core topic of the classical D-module theory. These polynomials possess several striking properties, some of which are rather hard to prove:

- A polynomial f is regular, i.e., $(f, \partial f/\partial x_1, \ldots, \partial f/\partial x_n) = k[\boldsymbol{x}]$, if and only if $b_f = s + 1$. (Proving one direction is not hard, for the other see [7, Proposition 2.6].)
- For any $f \in k[\boldsymbol{x}]$, the roots of b_f are negative rational numbers. In particular, $b_f \in \mathbb{Q}[s]$. (See [14] for a proof.)
- For a polynomial f in n variables, the roots of b_f are $\ge -n$. (See [31].)
- The set $\{b_f : f \in k[x_1, \ldots, x_n], \text{ char } k = 0, \deg f \le d\}$ is finite. Here $d, n \in \mathbb{Z}_{>0}$ are fixed, but k ranges over *all* fields of characteristic 0. (See [18] for a proof and [15] for a constructive proof.)

Because of numerous applications, in particular one due to Theorem 17, a lot of attention has been given to the computation of Bernstein-Sato polynomials. Formulas for Bernstein-Sato polynomials exist for special cases: many interesting families are considered in [34]; for the generic hyperplane arrangements see [33].

However, it was not until the end of the 20th century that the first general algorithm was produced (see [21]). The algorithm, in practice, can be carried out for polynomials f of low degrees in a reasonably small number of variables. This is due to the dependence of this approach on Gröbner basis computations in Weyl algebras.

Here is an example of computation via the classical algorithm [21] in *D-modules for Macaulay 2* for $f = x_1^a + x_2^{a+1} + x_1 x_2^a$ for $a = 3$.

```
i7 : factorBFunction globalBFunction(x_1^3+x_2^4+x_1*x_2^3)

          /     5\/     7\/     7\/    11\/    13\/    17\
o7 = (s + 1)|s + -||s + -||s + --||s + --||s + --||s + --|
          \     6/\     6/\    12/\    12/\    12/\    12/
```

This particular computation is accomplished in less than a second, however, one may see it gets much harder for higher values of a.

The fastest software for computing Bernstein-Sato polynomials, to our knowledge, is *Risa/Asir* [20].

4.3 Applications

Two well-developed applications of the D-module theory are those to A-hypergeometric systems of differential equations and to the local cohomology of holonomic modules. The former is described in detail in [24] and is currently a very active area of research. The latter is a powerful concept used throughout commutative algebra and algebraic geometry; use [12] as an extensive "manual".

4.3.1 A-hypergeometric systems

Let $D = k\langle x, \partial \rangle$ be the first Weyl algebra and let

$$Q = x(1 - x)\partial^2 + (c - x(a + b + 1))\partial - ab \in D.$$

Then $Q \cdot f = 0$ for an indeterminate univariate function f is known as the *Gauss hypergeometric equation*. The study of this equation dates back to the 19th century. It is important, since any ODE with three regular singular points is equivalent to this equation.

Let us reformulate the problem using the so-called *Euler operator* $\theta = x\partial$:

$$xQ \cdot f = (\theta(\theta + c - 1) - x(\theta + a)(\theta + b)) \cdot f = 0. \tag{4.4}$$

It is an easy exercise to do the conversion.

It follows from Theorem 9 that locally at any point $x_0 \in \mathbb{C} \setminus \{0, 1\}$ the equation (4.4) has two linearly independent solutions. If we consider a larger space of *logarithmic solutions* of the form

$$(x - x_0)^\lambda g_0 + \left((x - x_0)^\lambda \log(x - x_0)\right) g_1$$

with functions g_0 and g_1 holomorphic in the neighborhood of x_0, then the dimension of the solution space of (4.4) equals 2 even in the neighborhoods of $x_0 = 0, 1, \infty$.

For more details see [24, Section 1.3]. In particular, [24, Theorem 1.3.4] explains how to construct solutions in the open unit disk centered at $x_0 = 0$ in the form of power series, the so-called *Gauss hypergeometric series*.

Let us reformulate the Gauss hypergeometric equation in yet another way. Consider the system of linear *PDEs* corresponding to the following ideal in the fourth Weyl algebra D:

$$I = (\partial_2 \partial_3 - \partial_1 \partial_4, \theta_1 - \theta_4 + 1 - c, \theta_2 + \theta_4 + a, \theta_3 + \theta_4 + b),$$

where $\theta_i = x_i \partial_i$, for $1 \leq i \leq 4$. A function g in four variables is annihilated by the last three generators of I above, the so-called *homogeneities*, if and only if it is of the form

$$g(\boldsymbol{x}) = x_1^{c-1} x_2^{-a} x_3^{-b} \, f\!\left(\frac{x_1 x_4}{x_2 x_3}\right), \tag{4.5}$$

for a univariate function f.

Theorem 18 *A function f satisfies the Gauss hypergeometric ODE if and only if $g(\boldsymbol{x})$ in (4.5) is annihilated by $\partial_2 \partial_3 - \partial_1 \partial_4 \in D$.*

Proof The proof is done by a routine check; see [24, Proposition 1.3.7]. $\square$

Now we can encode the ideal I in an integer matrix A and a vector β

that depends only on the parameters a, b, c:

$$A = \begin{pmatrix} 1 & 0 & 0 & -1 \\ 0 & 1 & 0 & 1 \\ 0 & 0 & 1 & 1 \end{pmatrix} \quad \text{and} \quad \beta = \begin{pmatrix} c - 1 \\ -1 \\ -b \end{pmatrix}.$$

The ideal I_A generated by all elements $\boldsymbol{\partial}^u - \boldsymbol{\partial}^v$, where $u, v \in \mathbb{Z}_{\geq 0}^4$, such that $Au^T = Av^T$, turns out to be principal and is generated by the operator $\partial_2 \partial_3 - \partial_1 \partial_4$.

Let $\boldsymbol{\theta} = (\theta_1, \theta_2, \theta_3, \theta_4)$, then the homogeneities are the entries of the vector $A\boldsymbol{\theta}^T - \beta$. The homogeneities together with I_A generate the ideal I defining a system equivalent to the Gauss hypergeometric equation.

Now we are ready for an abstract definition. Given a matrix $A \in \mathbb{Z}^{d \times n}$ of rank d and a column vector $\beta \in k^d$, let the *toric ideal* be

$$I_A = (\{\boldsymbol{\partial}^u - \boldsymbol{\partial}^v : Au^T = Av^T, \ u, v \in \mathbb{Z}_{\geq 0}^4\}) \subset k[\boldsymbol{\partial}].$$

Note that it is an ideal in a (commutative) polynomial ring.

We call the system defined by the *A-hypergeometric ideal*

$$H_A(\beta) = (I_A) + (A\boldsymbol{\theta}^T - \beta)$$

in the n-th Weyl algebra D an *A-hypergeometric system* of PDE *with parameters* β.

The class of these systems is also known as *GKZ systems*; they were introduced in the 1980s [9].

We note that a set of generators for I_A can be computed by a purely combinatorial algorithm; see [27, Algorithm 4.5], for example.

Theorem 19 *If vector* $(1, 1, \ldots, 1)$ *belongs to the* $\mathbb{Q}$-*span of the rows of A (we later refer to this as the* homogeneity condition*), then $H_A(\beta)$ is holonomic for all* $\beta \in k^n$.

Proof See the sketch of a proof in [24, Theorem 2.4.9]. $\qquad\qquad\square$

Let conv A be the convex hull of the points in $\mathbb{R}^d$ corresponding to the columns of A; this is a $(d - 1)$-dimensional convex polytope. We denote by vol A the *normalized volume* of conv A, i.e. the usual Euclidian volume multiplied by $(d - 1)!$.

From holonomicity it follows that the holonomic rank of $D/H_A(\beta)$ is finite. There is an algorithmic way to construct the *logarithmic series solutions* that account for the space of solutions of dimension vol A (see [24, Theorem 3.5.1] for definition and an algorithm).

For a long time it was believed that the rank $(D/H_A(\beta))$ is independent of β due to the following result [1]: if β lies in the complement of a certain arrangement of affine planes then rank $D/H_A(\beta) = \text{vol } A$.

This belief is shown to be false by the following example:

Example 4.3.1 Let $H_A(\beta)$ be the A-hypergeometric ideal with parameter β, where

$$A = \begin{pmatrix} 1 & 1 & 1 & 1 \\ 0 & 1 & 3 & 4 \end{pmatrix}.$$

Then $5 = \text{rank } D/H_A(\beta) \neq \text{vol } A = 4$ for $\beta = (1,2)^T$, which is confirmed by this computation:

```
i2 : A = matrix{{1,1,1,1},{0,1,3,4}};

              2          4
o2 : Matrix ZZ  <--- ZZ

i3 : holonomicRank gkz(A,{0,0})

o3 = 4

i4 : holonomicRank gkz(A,{1,2})

o4 = 5
```

In the recent paper [19] the set of exceptional parameters β, i.e., where the holonomic rank jumps, is described precisely using homological methods.

4.3.2 Local cohomology

Let $I = (f_1, ..., f_d)$ be an ideal in the ring $R = k[x]$ of polynomials in n variables. Consider the *Čech complex*

$$C^\bullet(f_1, \ldots, f_d; R): \quad 0 \to C^0 \to C^1 \to \cdots \to C^d \to 0,$$

where the chains are

$$C^k = \bigoplus_{1 \leq i_1 < \ldots < i_k \leq d} R_{f_{i_1} \cdots f_{i_k}}$$

and the maps $C^l \to C^{l+1}$ are the alternating sum of maps

$$R_{f_{i_1} \cdots f_{i_l}} \to R_{f_{j_1} \cdots f_{j_{l+1}}},$$

which are either natural inclusions of one localization into the other if $\{i_1, \ldots, i_l\} \subset \{j_1, \ldots, j_{l+1}\} \subset \{1, \ldots, d\}$, or zero maps otherwise.

It can be constructed as a tensor product of shorter complexes:

$$C^\bullet(f_1, \dots, f_d; R) = C^\bullet(f_1; R) \otimes \cdots \otimes C^\bullet(f_d; R),$$

where $C^\bullet(f_i; R)$ is

$$0 \to R \hookrightarrow R_{f_i} \to 0.$$

We define *local cohomology* modules $H_I^\bullet(R)$ as the homology of the Čech complex $C^\bullet(f_1, \dots, f_d; R)$.

Note that, since R and its localizations are holonomic D-modules for the n-th Weyl algebra D, the chains C^i of the Čech complex are holonomic as well. Now, local cohomology modules are formed by taking kernels and cokernels of D-morphisms between holonomic modules, and therefore are also holonomic.

Moreover, we can compute $H_I^\bullet(R)$ via Gröbner basis techniques; this idea was first implemented in [32].

Example 4.3.2 Consider the ideal $I = (f_1, f_2, f_3) \subset R = \mathbb{Q}[x_1, \dots, x_6]$ generated by the minors of the matrix

$$\begin{pmatrix} x_1 & x_2 & x_3 \\ x_4 & x_5 & x_6 \end{pmatrix},$$

polynomials $f_1 = x_1 x_5 - x_2 x_4$, $f_2 = x_1 x_6 - x_3 x_4$, and $f_3 = x_2 x_6 - x_3 x_5$. The Čech complex $C^\bullet(f_1, f_2, f_3; R)$ is

$$
\begin{array}{cccc}
C_0 & C_1 & C_2 & C_3 \\[4pt]
\| & \| & \| & \| \\[8pt]
0 \to R \to
\begin{bmatrix} R_{f_1} \\ \oplus \\ R_{f_2} \\ \oplus \\ R_{f_3} \end{bmatrix}
\to
\begin{bmatrix} R_{f_1 f_2} \\ \oplus \\ R_{f_1 f_3} \\ \oplus \\ R_{f_2 f_3} \end{bmatrix}
\to R_{f_1 f_2 f_3} \to 0
\end{array}
$$

We can do the computation of local cohomology in *Macaulay 2*:

```
i2 : R =QQ[x_1..x_6];

i3 : D = makeWA R;

i4 : I = minors(2, matrix{{x_1, x_2, x_3},
                          {x_4, x_5, x_6}});

i5 : H = localCohom(I, Strategy=>Walther,
                       LocStrategy=>Oaku);
```

The presentations obtained for $H_I^i(R)$, $0 \le i \le 3$, are very large: below is a "pruned" version of $H_I^3(R) = D/J$, where J is an ideal with a Gröbner basis of 163 elements:

```
i6 : presentation Dprune H#3

o6 = | x_4dx_4+x_5dx_5+x_6dx_6+6 x_1dx_4+...
     -----------------------------------------------
     ...
     -----------------------------------------------
     x_4x_6^2dx_3^2dx_5-x_4x_6^2dx_2dx_3dx_6-... |
          1     163
o6 : Matrix D  <--- D
```

Although the presentations of the local cohomology modules are hard to display, let alone analyze by hand, we can see which of them vanish.

```
i7 : scan(keys H, i->
          <<"H^"<<i<<" vanishes: "<< H#i==0 <<endl)
H^0 vanishes: true
H^1 vanishes: true
H^2 vanishes: false
H^3 vanishes: false
```

This computation establishes a highly nontrivial fact, $H_I^3(R) \ne 0$, which implies that there are no $g_1, g_2 \in R$ such that the radicals of I and $(g_1, g_2) \subset R$ are equal.

An alternative D-modules technique, that of *characteristic cycles*, was used in the algorithm of [2] to produce the same result.

Bibliography

[1] Adolphson, A. (1994). Hypergeometric functions and rings generated by monomials, *Duke Math. J.* **73(2)**, 269–290.

[2] Àlvarez Montaner, J. and Leykin, A. (2006). Computing the support of local cohomology modules, *J. Symbolic Comput.* **41(12)**, 1328–1344.

[3] Aschenbrenner, M. and Leykin, A. (2007). Degree bounds for Gröbner Bases in algebras of solvable type, Preprint.

[4] Bernstein, J. and Lunts, V. (1988). On nonholonomic irreducible D-modules, *Invent. Math.* **94(2)**, 223–243.

[5] Björk, J.-E. (1979). *Rings of differential operators*, volume 21 of *North-Holland Mathematical Library* (North-Holland Publishing Co., Amsterdam-New York).

[6] Borel, A., Grivel, P.-P, Kaup, B., Haefliger E., Malgrange, B. and Ehlers, F. (1987). *Algebraic D-modules*, volume 2 of *Perspectives in Mathematics*, (Academic Press Inc., Boston, MA).

[7] Briançon, J. and Maisonobe, Ph. (1996). Caractérisation géométrique de l'existence du polynôme de Bernstein relatif, In *Algebraic geometry and singularities (La Rábida, 1991)*, volume 134 of *Progr. Math.*, pages 215–236 (Birkhäuser, Basel).

[8] Coutinho, S. C. (1995). *A primer of algebraic D-modules*, volume 33 of *London Mathematical Society Student Texts* (Cambridge University Press, Cambridge).

[9] Gel'fand, I.M. Graev, M.I. and Zelevinskiĭ, A.V. (1987). Holonomic systems of equations and series of hypergeometric type, *Dokl. Akad. Nauk SSSR* **295(1)**, 14–19.

[10] Grayson, D.R. and Stillman, M.E. Macaulay 2, a software system for research in algebraic geometry, Available at http://www.math.uiuc.edu/Macaulay2/.

[11] Hillebrand. A. and Schmale, W. (2001). Towards an effective version of a theorem of Stafford, *J. Symbolic Comput* **32(6)**, 699–716.

[12] Iyengar, S,, Leuschke, G.J., Leykin, A., Miller, C., Miller, E., Singh, A.K. and Walther, U. (2007). *Twenty-four hours of local cohomology*, volume 87 of *Graduate Studies in Mathematics* (American Mathematical Society, Providence, RI).

[13] Kandri-Rody, A. and Weispfenning, V. (1990). Non-commutative Gröbner bases in algebras of solvable type, *J. Symbolic Computation* **9**, 1–26.

[14] Kashiwara, M. (2003) *D-modules and microlocal calculus*, volume 217 of *Translations of Mathematical Monographs*. (American Mathematical Society, Providence, RI), Translated from the 2000 Japanese original by Mutsumi Saito, Iwanami Series in Modern Mathematics.

[15] Leykin, A. (2001). Constructibility of the Set of Polynomials with a Fixed Bernstein-Sato Polynomial: an Algorithmic Approach, *Journal of Symbolic Computation* **32(6)**, 663–675.

[16] Leykin, A. (2004). Algorithmic proofs of two theorems of Stafford, *Journal of Symbolic Computation* **38(6)**, 1535–1550.

[17] Leykin, A. and Tsai, H. Software package "*D*-modules for Macaulay 2", Available at http://www.math.uic.edu/~leykin/Dmodules.

[18] Lyubeznik, G. (1997). On Bernstein-Sato Polynomials, *Proc. of the AMS* **125(7)**, 1941–1944.

[19] Matusevich, L.F. and Walther, U. (2007). Arbitrary rank jumps for *A*-hypergeometric systems through Laurent polynomials, *J. Lond. Math. Soc. (2)* **75(1)**, 213–224.

[20] Noro, M., Shimoyama, T. and Takeshima, T. Computer algebra system risa/asir, Available at http://www.math.kobe-u.ac.jp/Asir/index.html.

[21] Oaku, T. (1997). Algorithm for the b-function and *D*-modules associated with polynomial, *J. Pure Appl. Algebra* **117/118**, 495–518.

[22] Oaku, T. and Takayama N. (2001). Algorithms for *D*-modules – restriction, tensor product, localization, and local cohomology groups, *J. Pure Appl. Algebra*, 156(2-3):267–308, 2001.

[23] Oaku, T., Takayama, N. and Walther, U. A localization algorithm for *D*-modules, *J. Symbolic Comput.* **29(4-5)**, 721–728.

[24] Saito, S., Sturmfels, B. and Takayama, N. (2000). *Gröbner deformations of hypergeometric differential equations*, volume 6 of *Algorithms and Computation in Mathematics* (Springer-Verlag, Berlin).

[25] Stafford, J.T. (1978). Module structure of Weyl algebras, *J. London Math. Soc. (2)* **18(3)**, 429–442.

[26] Stafford, J.T. (1985). Nonholonomic modules over Weyl algebras and enveloping algebras, *Invent. Math.* **79(3)**, 619–638.

[27] Sturmfels, B. (2002) *Solving Systems of Polynomial Equations*, Number 97

in CBMS Regional Conference Series in Mathematics (AMS, Providence).

[28] Takayama, N. kan/sm1: a computer algebra system for algebraic analysis, www.math.sci.kobe-u.ac.jp/KAN/.

[29] Tsai, H. and Walther, U. (2001). Computing homomorphisms between holonomic *D*-modules, *J. Symbolic Comput.* **32(6)**, 597–617.

[30] Tsai, H. (2000). *Algorithms for algebraic analysis*, UC Berkeley Ph.D. thesis.

[31] Varčenko, A.N. (1981). Asymptotic Hodge structure on vanishing cohomology, *Izv. Akad. Nauk SSSR Ser. Mat.* **45(3)**, 540–591.

[32] Walther, U. (1999). Algorithmic computation of local cohomology modules and the local cohomological dimension of algebraic varieties, *J. Pure Appl. Algebra* **139(1-3)**, 303–321.

[33] Walther, U. (2005). Bernstein-Sato polynomial versus cohomology of the Milnor fiber for generic hyperplane arrangements, *Compos. Math.* **141(1)**, 121–145.

[34] Yano, T. (1978). On the theory of *b*-functions, *Publ. Res. Inst. Math. Sci.* **14(1)**, 111–202.

5

Symbolic representation and classification of integrable systems

Alexander V. Mikhailov

School of Mathematics, University of Leeds, UK

Vladimir S. Novikov

School of Mathematics, Loughborough University, UK

Jing Ping Wang

IMSAS, University of Kent, UK

5.1 Introduction

Among nonlinear partial differential equations (PDEs) there is an exceptional class of so-called *integrable* equations. Integrable equations can be studied with the same completeness as linear PDEs, at least in principle. They possess a rich set of exact solutions and many hidden properties such as infinite hierarchies of symmetries, conservation laws, etc.

There are two known classes of integrable equations. One class is linearisable equations, i.e. equations related to linear ones by differential substitutions. For example, the famous Burgers equation

$$u_t = u_{xx} + 2uu_x$$

can be linearised by a differential substitution $u = \phi_x/\phi$ (the Cole-Hopf transformation). In the new variable ϕ it takes the form of a linear heat equation:

$$\phi_t = \phi_{xx}.$$

Another class is equations solvable by the inverse scattering transform method (such as the Korteweg de–Vries equation $u_t = u_{xxx} + 6uu_x$). There is a massive literature on integrable equations, their solutions and properties (see for example the monographs [1, 2, 3, 4, 5] and references therein).

In the *symmetry approach* the existence of higher symmetries of a PDE or, more precisely, of an infinite hierarchy of higher symmetries is regarded as the definition of its integrability. Existence of a finite

156

number of symmetries of a partial differential equation may not secure its integrability [6, 7, 8, 9].

The symmetry approach based on a concept of formal recursion operator has been formulated and developed in works of Shabat and co-authors (see for example review papers [10, 11, 12, 13]). It has been shown that the existence of an infinite hierarchy of symmetries or local conservation laws or a possibility to linearise a certain equation imply the existence of a formal recursion operator. The formal recursion operator is carrying information about integrability and is not sensitive to lacunas in the infinite hierarchy of symmetries or conservation laws. The conditions of its existence give integrability conditions for the equation which can be formulated in a form of an infinite sequence of canonical densities and encode many hidden properties of the equation. If a density is non-trivial (i.e. is not a total derivative) it provides a density of a local conservation law of the equation. For linearisable equations all densities except a finite number are trivial. The sequence of canonical densities is invariant with respect to invertible (and almost invertible [14]) transformations.

It was the first approach enabling one to give a complete description of integrable evolutionary equations of the form

$$
\begin{aligned}
u_t &= f(u_{xx}, u_x, u, x, t), \\
u_t &= \partial_x^n u + f(\partial_x^{n-1} u, \ldots, u_x, u), \quad n = 3, 4, 5,
\end{aligned}
$$

where f is a smooth function of its arguments [15, 16, 17]. Integrable differential-difference equations of the form

$$
u_{n,t} = f(u_{n-1}, u_n, u_{n+1}), \qquad n \in \mathbb{Z}
$$

have been classified in [18, 19]. It can also be applied to systems of equations. In particularly a complete classification of systems of two equations of the form

$$
\mathbf{u}_t = A(\mathbf{u})\mathbf{u}_{xx} + \mathbf{F}(\mathbf{u}, \mathbf{u}_x), \quad \det(A(\mathbf{u})) \neq 0, \quad \mathbf{u} = (u, v)^T \qquad (5.1)
$$

possessing infinite hierarchies of local conservation laws has been given in [20, 21, 14]. Existence of higher conservation laws immediately implies that the trace of $A(\mathbf{u})$ vanishes. System (5.1) can still possess an infinite hierarchy of symmetries even if the trace of $A(\mathbf{u})$ is not equal to zero. In the latter case the system is linearisable by a Cole-Hopf type differential substitution. In the case of $A(\mathbf{u})$ being a unit matrix integrable systems of the type (5.1) have been listed in [22]. A complete classification of polynomial homogeneous integrable systems of the form (5.1) in the case when matrix $A(\mathbf{u})$ is constant and has two distinct eigenvalues is given in [23].

In this approach it is not assumed that equations are polynomial or rational functions of independent variables and their derivatives. A disadvantage is that for every fixed order of the equation the integrability conditions have to be derived from scratch and thus it is difficult to draw a global picture, i.e., in all orders. Also, this approach is heavily based on the concept of locality which makes it difficult to apply to integro-differential, non-evolutionary and multi-dimensional systems.

In this article we would like to give a brief account of recent development of the symmetry approach. The progress has been achieved mainly due to a symbolic representation of the ring of differential polynomials which enables us to use powerful results from algebraic geometry and number theory. Symbolic representation (an abbreviated form of the Fourier transformation) was originally applied to the theory of integrable equations by Gel'fand and Dikii† [24]. The symmetry approach in symbolic representation has been formulated and developed for the problem of the global classification of integrable evolutionary equations in [25, 26, 27]. In symbolic representation the existence of an infinite hierarchy of symmetries is linked with factorisation properties of an infinite sequence of multi-variable polynomials. Symbolic representation is a suitable tool for studying integrability of noncommutative [28], non-evolutionary [29, 30, 31], non-local (integro–differential) [32], multi–component [23, 33, 34] and multi–dimensional equations [35]. It is convenient for testing integrability of a given system, provides useful information on the structure of the symmetry hierarchy and is suitable for global classification of integrable equations. In this framework it is natural to define approximate symmetries and approximate integrability. Study and classification of approximately integrable equations is a new and unexplored area of research with a considerable potential for applications. The symmetry approach in symbolic representation has certain drawbacks due to a restriction to the ring of differential polynomials, which can be amended in some cases by suitable extensions of the ring.

In the literature one can find other attempts to describe integrable systems based on properties of solutions, such as the Painlevé property in the analytical theory (see for example [2, 3, 4]), existence of three soliton solutions for the bi-linear (Hirota) representation (see the article

† Dickey is another transliteration of the name appearing in the literature.

by Hietarinta in this book), elasticity of soliton collisions in numerical experiments, etc. Existence of one nontrivial symmetry of a prior fixed order can be used for isolation of integrable equations [36, 37]. Every method has certain advantages and disadvantages. The reader can judge the power of each approach by the results obtained, their completeness and generality.

Our article is organised as follows. In the next Section 5.2 we give basic definitions and notations. We define symmetries, approximate symmetries and the formal recursion operator. Then we introduce symbolic representation of the ring of differential polynomials. A generalisation to several dependent variables is given in the last part of this section. In Section 5.3, using approximate symmetries in symbolic representation, we study the structure of the Lie algebra of symmetries. In symbolic representation the existence of approximate symmetries can be reformulated in terms of factorisation properties of polynomials. Existence of one nontrivial symmetry enables us to constructively extend any approximate symmetry of degree 3 to any degree (Theorem 4). In symbolic representation conditions for the existence of a formal recursion operator lead to a simple test for integrability. Here a state of art result is a global classification of integrable homogeneous evolutionary equations (Section 5.3.2). The result of the classification can be accounted for as follows: *Integrable equations are symmetries (members of infinite hierarchies) of nonlinear PDEs of orders 2, 3 or 5. Thus it is sufficient to classify integrable equations of order 2, 3 or 5. There is only a finite number of such equations (namely 10) and the corresponding hierarchies of symmetries* (Theorem 9). In Section 5.4 we apply our method to non-local (Benjamin-Ono and Camassa-Holm type) equations. It requires a non-local extension of the ring of differential polynomials and symbolic representation proved to be a suitable language to tackle the problem. In Section 5.5 we present results on classification of Boussinesq type equations. In the case of even order equations we use conditions following from the existence of the formal recursion operator. Together with classification results for orders 4 and 6 we present three new integrable equations of order 10. A global classification of all integrable odd order Boussinesq type equations is given in section 5.5.2. A generalisation of the symbolic approach to 2+1-dimensional equations enables us (Section 5.6) to study the structure of symmetries of Lax integrable equations and to prove the conjecture on the structure of non-local terms. Finally we discuss progress in the problem of classification of systems of integrable equations. In particularly we have found two integrable systems of order 5 which we believe are new.

5.2 Symmetries and formal recursion operators in symbolic representation

5.2.1 Differential polynomials

We shall adopt the following notations: u_n denotes the n-th derivative $\partial_x^n u$ of the dependent variable u. In particular, u_0 denotes the function u itself (often we shall omit the zero index of u_0 and simply write u).

A u-monomial is a finite product of the form

$$u_0^{\alpha_0} u_1^{\alpha_1} \cdots u_k^{\alpha_k},$$

where all exponents $\alpha_0, \ldots, \alpha_k$ are non-negative integers ($\alpha_s \in \mathbb{Z}_{\geq 0}$), and the total degree is $|\alpha| = \alpha_0 + \alpha_1 + \cdots \alpha_k > 0$.

A finite sequence $\alpha = (\alpha_0, \ldots, \alpha_k)$ can be seen as an element of a semi-group $\mathbb{Z}_{\geq 0}^{\infty*}$ of infinite sequences of non-negative integers, such that only a finite number of entries in a sequence are non-zero and there is at least one nonzero entry. There is an obvious bijection between the set of all u-monomials and $\mathbb{Z}_{\geq 0}^{\infty*}$. We can simplify the notations as follows:

$$u^\alpha = u_0^{\alpha_1} u_1^{\alpha_2} \cdots u_k^{\alpha_k}.$$

A *differential polynomial* f in variables $u_0, u_1, \ldots$ with coefficients in $\mathbb{C}$ is a finite linear combination of u-monomials, i.e.,

$$f = \sum_{\alpha \in A} a_\alpha u^\alpha, \qquad a_\alpha \in \mathbb{C},$$

where the sum is over a finite set $A = \{\alpha \,|\, \alpha \in \mathbb{Z}_{\geq 0}^{\infty*}\}$. The set of all such differential polynomials is denoted $\mathcal{R}$. It is a ring with the usual addition and multiplication of polynomials. Moreover, it is a differential ring and $\mathbb{C} \not\subseteq \mathcal{R}$. The linear operator

$$D_x = \sum_{k \geq 0} \left(u_{k+1} \frac{\partial}{\partial u_k} \right) \tag{5.2}$$

is a derivation of $\mathcal{R}$ corresponding to the total x-derivative. The operator D_x acting on an element $f \in \mathcal{R}$ results in a finite sum depending on the choice of f. Therefore, we do not indicate the upper limit for the summation in the definition of D_x and all other operators defined in this article.

It is easy to verify that monomials u^α are eigenvectors of the following commuting linear operators

$$D_u = \sum_{k \geq 0} u_k \frac{\partial}{\partial u_k}, \qquad X_u = \sum_{k \geq 1} \left(k u_k \frac{\partial}{\partial u_k} \right) \qquad (5.3)$$

with $D_u(u^\alpha) = |\alpha| u^\alpha$, and $X_u(u^\alpha) = d_\alpha u^\alpha$, where $d_\alpha = \sum_{k \geq 1} (k \alpha_k)$.
Thus the ring $\mathcal{R}$ is graded and is a direct sum of eigenspaces

$$\mathcal{R} = \bigoplus_{n \in \mathbb{N}} \mathcal{R}^n = \bigoplus_{n, p \in \mathbb{N}} \mathcal{R}^n_{p-1},$$

$$\mathcal{R}^n = \{ f \in \mathcal{R} \mid D_u(f) = nf \}, \qquad \mathcal{R}^n_p = \{ f \in \mathcal{R}^n \mid X_u(f) = pf \}.$$

If $f \in \mathcal{R}^n_p$, $g \in \mathcal{R}^m_q$, then $f \cdot g \in \mathcal{R}^{n+m}_{p+q}$. Simply speaking, $\mathcal{R}^n_p$ is a linear subspace of homogeneous differential polynomials such that each monomial has: (i) the number of u and its derivatives being n; (ii) the total number of derivatives being p. For example,

$$u_1^2 u_7 + 2 u_2 u_3 u_4 - u_0^2 u_9 \in \mathcal{R}^3_9 \subset \mathcal{R}^3.$$

In some applications it is convenient to introduce weighted homogeneous polynomials. Let us assume that the dependent variable u has a weight λ which is a fixed rational number. We define a linear differential operator

$$W_\lambda = \lambda D_u + X_u, \qquad W_\lambda : \mathcal{R} \to \mathcal{R}.$$

Differential monomials are eigenvectors of W_λ and the spectrum of W_λ is a set $S_\lambda = \{ n\lambda + m - 1 \mid n, m \in \mathbb{N} \}$. We can decompose $\mathcal{R}$ into a direct sum of eigenspaces

$$\mathcal{R} = \bigoplus_{\mu \in S_\lambda} \mathcal{W}_\mu, \qquad \mathcal{W}_\mu = \{ f \in \mathcal{R} \mid W_\lambda(f) = \mu f \}.$$

Elements of $\mathcal{W}_\mu$ are called λ-homogeneous differential polynomials of weight μ. For example, $u_3 + 6uu_1$ is a 2-homogeneous differential polynomial of weight 5. We have $\mathcal{W}_\mu \mathcal{W}_\nu \subset \mathcal{W}_{\mu+\nu}$. Moreover, if $\lambda > 0$ then subspaces $\mathcal{W}_\mu$ are finite dimensional.

It is useful to define the little "*oh*" order symbol.

Definition 1 *Let* $f \in \mathcal{R}$. *We say that* $f = o(\mathcal{R}^n)$ *if* $f \in \bigoplus_{k > n} \mathcal{R}^k$.

For example, $f = o(\mathcal{R}^3)$ means that the differential polynomial f does not have linear, quadratic and cubic terms in u and its derivatives.

For any two elements $f, g \in \mathcal{R}$ we define a Lie bracket

$$[f, \ g] = f_*(g) - g_*(f)\,, \tag{5.4}$$

where the *Fréchet derivative* for any element $h \in \mathcal{R}$ is defined as a linear differential operator of the form

$$h_* = \sum_{k \geq 0} \frac{\partial h}{\partial u_k} D_x^k\,. \tag{5.5}$$

We say that an element $h \in \mathcal{R}$ has *order* n if the corresponding differential operator h_* is of order n.

Thus $\mathcal{R}$, treated as a linear space over $\mathbb{C}$, together with the Lie bracket (5.4) is an infinite dimensional Lie algebra of differential polynomials over $\mathbb{C}$. The bilinearity and skew-symmetry of the bracket (5.4) are obvious. The Jacobi identity can be easily verified.

The grading of $\mathcal{R}$ induces the grading of the Lie algebra of differential polynomials since we have

$$[\mathcal{R}_p^n, \mathcal{R}_q^m] \subset \mathcal{R}_{p+q}^{n+m-1}\,. \tag{5.6}$$

5.2.2 Symmetries, approximate symmetries and formal recursion operator

In this section, for the sake of simplicity, we give definitions suitable for evolutionary equations

$$u_t = F, \qquad F \in \mathcal{R}\,. \tag{5.7}$$

These definitions will later be extended to non-evolutionary equations and multi-component systems of evolutionary equations.

An evolutionary equation (5.7) defines a derivation $D_F : \mathcal{R} \mapsto \mathcal{R}$:

$$D_F(a) = a_*(F).$$

In this notation the derivative $D_x = D_{u_1}$ and it is in agreement with (5.3). Sometimes, for simplification of notations, we will denote D_F as D_t.

Definition 2 *A differential polynomial $G \in \mathcal{R}$ is said to be a symmetry (a generator of an infinitesimal symmetry) for an evolutionary partial differential equation (5.7) if the Lie bracket of F and G vanishes, i.e.,* $[F, \ G] = 0$.

If G is a symmetry, then the evolutionary equation $u_\tau = G$ is compatible with (5.7). There are many other equivalent definitions of symmetry (see for example [10, 12]). Elements of $\mathcal{R}$ do not depend on x, t explicitly. Thus our definition does not include space and time dependent symmetries such as dilatation and Galilean symmetries. In this article, when we are talking about symmetries we do mean space and time independent symmetries.

Symmetries form a subalgebra $C_\mathcal{R}(F) = \{G \in \mathcal{R} \,|\, [F,\ G] = 0\}$ which is the centraliser of F (this immediately follows from the Jacobi identity). Since $F \in \mathcal{R}$ does not depend on x, t explicitly, then any equation (5.7) possesses *trivial* symmetries $u_1, F \in C_\mathcal{R}(F)$ corresponding to translations in space and time.

Definition 3 *Equation (5.7) is said to be integrable if its algebra of symmetries $C_\mathcal{R}(F)$ is infinite dimensional.*

For nonlinear equations of the form

$$u_t = u_n + f(u_{n-1}, \dots, u) \qquad n \geq 2, \tag{5.8}$$

it is easy to show that the algebra of symmetries $C_\mathcal{R}(u_t)$ is commutative. Moreover, the symmetries $G \in \mathcal{R}$ must have a linear term (see section 5.3.1).

Having in mind the gradation (5.6) of the Lie algebra of differential polynomials we represent the right hand side of equation (5.7) and infinitesimal generators of symmetries in the form

$$F = f_1 + f_2 + \cdots, \quad G = g_1 + g_2 + \cdots, \qquad f_k, g_k \in \mathcal{R}^k$$

and study them in the sequence of terms, i.e., linear, quadratic, cubic, etc.

Definition 4 *A differential polynomial $G \in \mathcal{R}$ is said to be an approximate symmetry of degree p for an evolutionary partial differential equation (5.7) if $[F,\ G] = o(\mathcal{R}^p)$.*

Equation (5.8) possesses infinitely many approximate symmetries of degree 1. An equation may possess approximate symmetries of degree 2, but fail to possess approximate symmetries of degree 3. An integrable equation possesses infinitely many approximate symmetries of any degree. In the next section, using symbolic representation we formulate the necessary and sufficient conditions for the existence of approximate symmetries of arbitrary degree.

For example, the equation

$$u_t = u_5 + 5uu_1, \qquad (5.9)$$

has approximate symmetry of degree 2 with a generator

$$G = u_7 + 7uu_3 + 14u_1u_2.$$

Indeed,

$$[G, u_5 + 5uu_1] = 210u_1^2u_2 + 105uu_2^2 + 105uu_1u_3 = o(\mathcal{R}^2).$$

Moreover, equation (5.9) has infinitely many approximate symmetries of degree 2 (this fact will become obvious in the next section), but fails to have approximate symmetries of degree 3 and thus is not integrable.

It follows from the Jacobi identity and (5.6) that approximate symmetries of degree n form a subalgebra of $\mathcal{R}$ which we denote $C_{\mathcal{R}}^n(F)$. Obviously

$$\mathcal{R} = C_{\mathcal{R}}^1(F) \supset C_{\mathcal{R}}^2(F) \supset C_{\mathcal{R}}^3(F) \supset \cdots \supset C_{\mathcal{R}}^\infty(F) = C_{\mathcal{R}}(F).$$

Formal pseudo-differential series, which for simplicity we shall call formal series, are defined as

$$A = a_m D_x^m + a_{m-1} D_x^{m-1} + \cdots + a_0 + a_{-1} D_x^{-1} + \cdots \qquad a_k \in \mathcal{R}. \ (5.10)$$

The product of two formal series is defined by

$$aD_x^k \circ bD_x^m = a(bD_x^{m+k} + C_k^1 b_1 D_x^{k+m-1} + C_k^2 b_2 D_x^{k+m-2} + \cdots), \ (5.11)$$

where $b_j = D_x^j(b)$, $k, m \in \mathbb{Z}$ and the binomial coefficients are defined as

$$C_n^j = \frac{n(n-1)(n-2)\cdots(n-j+1)}{j!}.$$

This product is associative.

Definition 5 *A formal series*

$$\Lambda = l_m D_x^m + \cdots + l_0 + l_{-1} D_x^{-1} + \cdots, \quad l_k \in \mathcal{R} \qquad (5.12)$$

is called a formal recursion operator for equation (5.7) if

$$D_F(\Lambda) = F_* \circ \Lambda - \Lambda \circ F_*. \qquad (5.13)$$

In the literature a formal recursion operator is also called a formal symmetry of equation (5.7).

The central result of the Symmetry Approach can be represented by the following Theorem, which we attribute to Shabat:

Theorem 1 *If equation (5.7) has an infinite hierarchy of symmetries of arbitrarily high order, then a formal recursion operator exists and its coefficients can be found recursively.*

The Theorem states that for integrable equations, i.e. equations possessing an infinite hierarchy of higher symmetries, one can solve equation (5.13) and determine recursively the coefficients $l_m, l_{m-1}, \ldots$ of Λ such that all these coefficients will belong to the ring $\mathcal{R}$. The solvability conditions of equation (5.13) can be formulated in an elegant form of a canonical sequence of local conservation laws of equation (5.7). They provide powerful necessary conditions of integrability. These conditions can be used for testing for integrability for a given equation or even for a complete description of integrable equations of a particular order. A detailed description of the Symmetry Approach including the proof of the above theorem and classification results for integrable PDEs and systems of PDEs based on the concept of formal recursion operator can be found in the review papers [10, 11, 12, 13].

5.2.3 Symbolic representation

Symbolic representation transforms problems in differential algebra into ones in algebra of symmetric polynomials. This enables us to use powerful results from Diophantine equations, algebraic geometry and commutative algebra. Symbolic representation is widely used in the theory of pseudo-differential operators. For integrable systems it was first applied by Gel'fand and Dikii [24] and further developed in works of Beukers, Sanders and Wang [7, 25, 26]. Symbolic representation can be viewed as a simplified notation for a Fourier transform [29].

In order to define the symbolic representation $\hat{\mathcal{R}} = \oplus \hat{\mathcal{R}}^n$ of the ring (and Lie algebra) of differential polynomials $\mathcal{R} = \oplus \mathcal{R}^n$, we first define an isomorphism of the linear spaces $\mathcal{R}^n$ and $\hat{\mathcal{R}}^n$ and then extend it to isomorphisms of the differential ring and Lie algebra equipping $\hat{\mathcal{R}}$ with the multiplication, derivation and Lie bracket.

Symbolic transform defines a linear isomorphism between the space $\mathcal{R}^n$ of differential polynomials of degree n and the space $\mathbb{C}[\xi_1, \ldots, \xi_n]^{\mathcal{S}_n^\xi}$ of algebraic symmetric polynomials in n variables, where $\mathcal{S}_n^\xi$ is a permutation group of n variables $\xi_1, \ldots, \xi_n$. Elements of $\hat{\mathcal{R}}^n$ are denoted by $\hat{u}^n a(\xi_1, \ldots, \xi_n)$, where $a(\xi_1, \ldots, \xi_n) \in \mathbb{C}[\xi_1, \ldots, \xi_n]^{\mathcal{S}_n^\xi}$. The isomorphism of linear spaces $\mathcal{R}^n$ and $\hat{\mathcal{R}}^n$ is uniquely defined by its action on monomials.

Definition 6 *The* symbolic form *of a differential monomial is defined as*

$$u_{i_1} u_{i_2} \cdots u_{i_n} \in \mathcal{R}^n \quad \longmapsto \quad \hat{u}^n \langle \xi_1^{i_1} \xi_2^{i_2} \cdots \xi_n^{i_n} \rangle_{\mathcal{S}_n^\xi} \in \hat{\mathcal{R}}^n$$

where $\langle \cdot \rangle_{\mathcal{S}_n^\xi}$ *denotes the average over the group* $\mathcal{S}_n^\xi$ *of permutation of* n *elements* $\xi_1, \ldots, \xi_n$:

$$\langle a(\xi_1, \cdots, \xi_n) \rangle_{\mathcal{S}_n^\xi} = \frac{1}{n!} \sum_{\sigma \in \mathcal{S}_n^\xi} a(\xi_{\sigma(1)}, \cdots, \xi_{\sigma(k)}).$$

For example,

$$u_k \longmapsto \hat{u}\xi_1^k, \quad u^n \longmapsto \hat{u}^n, \quad uu_1 \longmapsto \frac{\hat{u}^2}{2}(\xi_1 + \xi_2),$$

$$uu_1^2 \longmapsto \frac{\hat{u}^3}{3}(\xi_2\xi_3 + \xi_1\xi_2 + \xi_1\xi_3), \quad u_k^3 \longmapsto \hat{u}^3 \xi_1^k \xi_2^k \xi_3^k.$$

It is easy to see that in $\hat{\mathcal{R}}$ the linear operators D_u and X_u, cf. (5.3) are represented by

$$\hat{D}_u = \hat{u}\frac{\partial}{\partial \hat{u}}, \qquad \hat{X}_u = \sum_{i=1} \xi_i \frac{\partial}{\partial \xi_i}.$$

With this isomorphism the linear spaces $\hat{\mathcal{R}}_p^n$ corresponding to $\mathcal{R}_p^n$ have the property that the coefficient functions $a(\xi_1, \ldots, \xi_n)$ of symbols

$$\hat{u}^n a(\xi_1, \ldots, \xi_n) \in \hat{\mathcal{R}}_p^n$$

are homogeneous symmetric polynomials of degree p.

One of the advantages of the symbolic representation is that the action of the operator D_x, cf. (5.2) is very simple. Indeed, let $f \in \mathcal{R}^n$ and $f \longmapsto \hat{u}^n a(\xi_1, \ldots, \xi_n)$ then

$$D_x(f) \quad \longmapsto \quad \hat{u}^n a(\xi_1, \ldots, \xi_n)(\xi_1 + \cdots + \xi_n),$$

and thus $D_x^k(f) \longmapsto \hat{u}^n a(\xi_1, \ldots, \xi_n)(\xi_1 + \cdots + \xi_n)^k$.

Let $f \in \mathcal{R}^n$, $f \mapsto \hat{u}^n a(\xi_1, \ldots, \xi_n)$ and $g \in \mathcal{R}^m$, $g \mapsto \hat{u}^m b(\xi_1, \ldots, \xi_m)$, then

(i). The product $f \cdot g$ has the following symbolic representation:

$$f \cdot g \quad \longmapsto \quad \hat{u}^n a(\xi_1, \ldots, \xi_n) \circ \hat{u}^m b(\xi_1, \ldots, \xi_m)$$

$$= \hat{u}^{n+m} \langle a(\xi_1, \ldots, \xi_n) b(\xi_{n+1}, \ldots, \xi_{n+m}) \rangle_{\mathcal{S}_{n+m}^\xi}.$$

This defines the corresponding multiplication $\circ$ in $\hat{\mathcal{R}}$. The representation of differential monomials (Definition 6) can be deduced from $u_k \longmapsto \hat{u}\xi_1^k$ and this multiplication rule.

(ii). The Lie bracket $[f, g]$, cf. (5.4) is represented by

$$[f,\ g] \longmapsto \hat{u}^{n+m-1} \tag{5.14}$$

$$\langle na(\xi_1,\ldots,\xi_{n-1},\xi_n+\cdots+\xi_{n+m-1})b(\xi_n,\ldots,\xi_{n+m-1})-$$
$$mb(\xi_1,\ldots,\xi_{m-1},\xi_m+\cdots+\xi_{n+m-1})a(\xi_m,\ldots,\xi_{n+m-1})\rangle_{\mathcal{S}^\xi_{n+m-1}}$$

For example, if $f \in \mathcal{R}^1$, $f \mapsto \hat{u}\omega(\xi_1)$ and $g \in \mathcal{R}^n$, $g \mapsto \hat{u}^n a(\xi_1,\ldots,\xi_n)$, then

$$[f,g] \longmapsto (\omega(\xi_1+\cdots+\xi_n) - \omega(\xi_1) - \cdots - \omega(\xi_n))\ \hat{u}^n a(\xi_1,\ldots,\xi_n).$$

In particularly, for $f = u_1$ we have $\omega(\xi_1) = \xi_1$ and $[u_1, g] = 0$ for any $g \in \mathcal{R}$. Thus u_1 is a symmetry for any evolutionary equation $u_t = g$.

Symbolic representation of differential operators (such as the Fréchet derivative (5.5) and formal series (5.10)) is motivated by the theory of linear pseudo-differential operators in Fourier representation. To operator D_x (5.2) we shall assign a special symbol η and the following rules of action on symbols:

$$\eta(\hat{u}^n a(\xi_1,\ldots,\xi_n)) = \hat{u}^n a(\xi_1,\ldots,\xi_n) \sum_{j=1}^n \xi_j$$

and the composition rule

$$\eta \circ \hat{u}^n a(\xi_1,\ldots,\xi_n) = \hat{u}^n a(\xi_1,\ldots,\xi_n)(\sum_{j=1}^n \xi_j + \eta).$$

The latter corresponds to the Leibnitz rule $D_x \circ f = D_x(f) + f D_x$. Now it can be shown that the composition rule (5.11) can be represented as follows. Let us have two operators fD_x^q and gD_x^s such that f and g have symbols $\hat{u}^i a(\xi_1,\ldots,\xi_i)$ and $\hat{u}^j b(\xi_1,\ldots,\xi_j)$ respectively. Then

$$fD_x^q \longmapsto \hat{u}^i a(\xi_1,\ldots,\xi_i)\eta^q, \qquad gD_x^s \longmapsto \hat{u}^j b(\xi_1,\ldots,\xi_j)\eta^s$$

and

$$fD_x^q \circ gD_x^s \longmapsto \hat{u}^{i+j}\langle a(\xi_1,\ldots,\xi_i)(\eta+\sum_{m=i+1}^{i+j}\xi_m)^q b(\xi_{i+1},\ldots,\xi_{i+j})\eta^s\rangle_{\mathcal{S}^\xi_{i+j}}.$$

$$\tag{5.15}$$

Here the symmetrisation is taken over the group of permutations of all $i + j$ arguments $\xi_1,\ldots\xi_{i+j}$; the symbol η is not included in this set. In particular, it follows from (5.15) that $D_x^q \circ D_x^s \mapsto \eta^{q+s}$. The composition rule (5.15) is valid for both positive and negative exponents. In the case of positive exponents it is a polynomial in η and the result is a Fourier image of a differential operator. In the case of negative exponents one can expand the result on η at $\eta \to \infty$ in order to identify it with (5.11). In the symbolic representation instead of formal series (5.10) it is natural to consider a more general object, namely formal series of the form

$$B = b(\eta) + \hat{u}b_1(\xi_1, \eta) + \hat{u}^2 b_2(\xi_1, \xi_2, \eta) + \hat{u}^3 b_3(\xi_1, \xi_2, \xi_3, \eta) + \cdots,$$

where the coefficients $b(\eta) \neq 0, b_k(\xi_1, \ldots, \xi_k, \eta)$ are rational functions of their arguments (with certain restrictions which will be discussed in the next section).

The symbolic representation of the Fréchet derivative of the element $f \longmapsto \hat{u}^n a(\xi_1, \ldots, \xi_n)$ is

$$f_* \longmapsto n\hat{u}^{n-1} a(\xi_1, \ldots, \xi_{n-1}, \eta).$$

For example, let $F = u_3 + 6uu_1$, then $F \mapsto \hat{u}\xi_1^3 + 3\hat{u}^2(\xi_1 + \xi_2)$ and

$$F_* \mapsto \eta^3 + 6\hat{u}(\xi_1 + \eta).$$

It is interesting to notice that the symbol of the Fréchet derivative is always symmetric with respect to all permutations of arguments, including the argument η. Moreover, the following obvious but useful Proposition holds [29]:

Proposition 1 *A differential operator is a Fréchet derivative of an element of* $\mathcal{R}$ *if and only if its symbol is invariant with respect to all permutations of its argument, including the argument* η.

The symbolic representation has been extended and proved to be very useful in the case of noncommutative differential rings [28]. In the next sections symbolic representation will be extended to the cases of many dependent variables, suitable for study of system of equations and further to the cases of non-local and multidimensional equations.

5.2.4 Generalisation to several dependent variables

The definitions and most of the statements formulated in the previous sections 5.2.1–5.3.1 can be easily extended to several dependents, i.e., to systems of equations. In this section we will give a brief account of definitions and some results concerning two dependent variables. A generalisation for N dependent variables is straightforward. For details see [11, 12, 13, 30, 31, 33].

Similarly to u–monomials (Section 5.2.1), we define v-monomials

$$v^\beta = v_0^{\beta_0} v_1^{\beta_1} \cdots v_s^{\beta_s}, \quad \beta \in \mathbb{Z}_{\geq 0}^\infty.$$

A differential polynomial f in variables $u_0, v_0, u_1, v_1, \ldots$ is a finite linear combination of the form

$$f = \sum_{(\alpha,\beta) \in A} a_{\alpha,\beta} u^\alpha v^\beta, \quad a_{\alpha,\beta} \in \mathbb{C},$$

where the sum is taken over a finite set

$$A = \{(\alpha,\beta) \mid \alpha, \beta \in \mathbb{Z}_{\geq 0}^\infty, \ |\alpha| + |\beta| > 0\}.$$

It is a differential ring. We again denote it $\mathcal{R}$. The derivation D_x, (cf. (5.2)) is now replaced by

$$D_x = \sum_{k \geq 0} \left(u_{k+1} \frac{\partial}{\partial u_k} + v_{k+1} \frac{\partial}{\partial v_k} \right).$$

For any $f \in \mathcal{R}$ the Fréchet derivative f_* is a (row) vector operator

$$f_* = (f_{*u}, f_{*v}) = \left(\sum_{k \geq 0} \frac{\partial f}{\partial u_k} D_x^k, \ \sum_{k \geq 0} \frac{\partial f}{\partial v_k} D_x^k \right). \tag{5.16}$$

Systems of two evolutionary equations we will write in vector form

$$\mathbf{u}_t = \mathbf{F}(\mathbf{u}, \mathbf{u}_1, \ldots, \mathbf{u}_n), \tag{5.17}$$

where $\mathbf{u}_k = (u_k, v_k)^T$ and $\mathbf{F} = (F_1, F_2)^T$ are vector-columns where $F_1, F_2 \in \mathcal{R}$ (the upper index T stands for the transposition).

Let us introduce an infinite dimensional linear space over $\mathbb{C}$

$$\mathcal{L} = \{(H_1, H_2)^T \mid H_1, H_2 \in \mathcal{R}\}.$$

We equip $\mathcal{L}$ with a Lie bracket

$$[\mathbf{F}, \mathbf{G}] = \mathbf{F}_*(\mathbf{G}) - \mathbf{G}_*(\mathbf{F}) \in \mathcal{L},$$

where the Fréchet derivative $\mathbf{H}_*$ for any $\mathbf{H} \in \mathcal{L}$ is defined as

$$\mathbf{H}_* = \begin{pmatrix} H_{1*u} & H_{1*v} \\ H_{2*u} & H_{2*v} \end{pmatrix}.$$

Thus $\mathcal{L}$ has a structure of an infinite dimensional Lie algebra over $\mathbb{C}$. The subalgebra of symmetries of equation (5.17) is the centraliser $C_\mathcal{L}(\mathbf{F})$ of $\mathbf{F}$ in $\mathcal{L}$ (cf. section 5.2.1).

The evolutionary system (5.17) defines a derivation $D_{\mathbf{F}} : \mathcal{R} \mapsto \mathcal{R}$

$$D_{\mathbf{F}}(a) = a_{*u}(F_1) + a_{*v}(F_2),$$

which is also a derivation of the Lie algebra $\mathcal{L}$. This derivation we often denote as D_t. In this notation the derivation D_x coincides with $D_{\mathbf{u}_1}$ and $D_u + D_v$ with $D_{\mathbf{u}}$.

The ring $\mathcal{R}$ has several gradings. Here we define a *monomial degree* grading

$$\mathcal{R} = \bigoplus_{k \in \mathbb{N}} \mathcal{R}^k, \quad \mathcal{R}^k = \{a \in \mathcal{R} \,|\, D_{\mathbf{u}}a = ka\}.$$

We say that $a = o(\mathcal{R}^n)$ if $a \in \bigoplus_{k > n} \mathcal{R}^k$.

The Lie algebra $\mathcal{L}$ inherits the gradings of $\mathcal{R}$. A *monomial degree* grading

$$\mathcal{L} = \bigoplus_{k \in \mathbb{Z}_{\geq 0}} \mathcal{L}^k, \quad \mathcal{L}^k = \{\mathbf{H} \in \mathcal{L} \,|\, D_{\mathbf{u}}\mathbf{H} = (k+1)\mathbf{H}\},$$

and thus $[\mathcal{L}^p, \mathcal{L}^q] \subset \mathcal{L}^{p+q}$ is convenient for the definition of approximate symmetries. For $\mathbf{H} \in \mathcal{L}$ we say that $\mathbf{H} = o(\mathcal{L}^n)$ if $\mathbf{H} \in \bigoplus_{k > n} \mathcal{L}^k$. Approximate symmetries of equation (5.17) of degree n are defined as elements of the approximate centraliser

$$C_{\mathcal{L}}^n(\mathbf{F}) = \{\mathbf{G} \in \mathcal{L} \,|\, [\mathbf{F}, \mathbf{G}] = o(\mathcal{L}^{n-1})\},$$

which is a subalgebra of $\mathcal{L}$.

The *weighted gradation* is useful for the study of homogeneous systems. We assign weights $\mathbf{w} = (w_u, w_v)$ with rational entries to the vector variable $\mathbf{u}$ and define a linear operator

$$W = (w_u D_u + w_v D_v + X_u + X_v) \begin{pmatrix} 1 & 0 \\ 0 & 1 \end{pmatrix} - \begin{pmatrix} w_u & 0 \\ 0 & w_v \end{pmatrix}$$

with spectrum $S_W = \{(p-1)w_u + (q-1)w_v + r \,|\, p, q, r \in \mathbb{Z}_{\geq 0}, \, p+q > 0\}$. Then the linear subspaces $\mathcal{L}_\mu$ in the decomposition

$$\mathcal{L} = \bigoplus_{\mu \in S_W} \mathcal{L}_\mu, \quad \mathcal{L}_\mu = \{\mathbf{H} \in \mathcal{L} \,|\, W\mathbf{H} = \mu\mathbf{H}\}$$

satisfy

$$[\mathcal{L}_\mu, \mathcal{L}_\nu] \subset \mathcal{L}_{\mu+\nu}. \tag{5.18}$$

The elements of $\mathcal{L}_\mu$ we call $\mathbf{w}$–homogeneous differential polynomial vectors of weight μ.

For example, if the weight vector of variables (u, v) is $\mathbf{w} = (1/2, 1)$, then

$$\mathbf{F} = \begin{pmatrix} v_1 \\ u_2 + 3uv_1 + vu_1 - 3u^2u_1 \end{pmatrix}$$

is a $\mathbf{w}$–homogeneous element of weight $3/2$, indeed $\mathbf{F} \in \mathcal{L}_{3/2}$.

If $\mathbf{F} \in \mathcal{L}_\mu$ is a homogeneous vector, then $D_\mathbf{F}$ is a homogeneous derivation of weight μ:

$$D_\mathbf{F}\mathcal{L}_\nu \subset \mathcal{L}_{\nu+\mu}\,, \qquad \mathbf{F} \in \mathcal{L}_\mu.$$

From (5.18) there immediately follows

Proposition 2 *Let* $\mathbf{G} = \mathbf{G}_{\nu_1} + \cdots + \mathbf{G}_{\nu_m}$, *and* $\mathbf{G}_\gamma \in \mathcal{L}_\gamma$ *be a generator of a symmetry of a homogeneous equation, then each* $\mathbf{w}$*–homogeneous component* $\mathbf{G}_{\nu_k}$ *is a generator of a symmetry.*

For the evolutionary system (5.17) a recursion operator (a formal recursion operator) Λ can be defined as a differential or pseudo-differential operator (or a formal series)

$$\Lambda = \Lambda_k D_x^k + \Lambda_{k-1}D_x^{k-1} + \cdots, \qquad \Lambda_s \in \mathrm{Mat}_{2\times 2}(\mathcal{R})\,,$$

which satisfies the following operator equation

$$D_\mathbf{F}(\Lambda) = \mathbf{F}_* \circ \Lambda - \Lambda \circ \mathbf{F}_* \tag{5.19}$$

(compare with Definition 5). If the action of Λ is well defined on a symmetry $\mathbf{G}_1$, i.e. $\mathbf{G}_2 = \Lambda(\mathbf{G}_1) \in \mathcal{L}$, then $\mathbf{G}_2$ is a new symmetry of the evolutionary system (5.17). Starting from a "seed" symmetry $\mathbf{G}_1$, one can build up an infinite hierarchy of symmetries $\mathbf{G}_{n+1} = \Lambda^n(\mathbf{G}_1)$, provided that each action of Λ produces an element of $\mathcal{L}$.

The symbolic representation of the ring $\mathcal{R}$ generated by two independent variables u, v is quite straightforward. It is a $\mathbb{C}$–linear isomorphism which it is sufficient to define for the monomials. Suppose we have a monomial $u^\alpha v^\beta$. Let the symbolic representation for the monomial u^α be $\hat{u}^{|\alpha|}a(\xi_1, \ldots, \xi_{|\alpha|})$ where $a(\xi_1, \ldots, \xi_{|\alpha|})$ is a symmetrical polynomial (see Definition 6). Acting by the same rule, but reserving a set of variables $\zeta_1, \zeta_2, \ldots$ (instead of $\xi_1, \xi_2, \ldots$) for the symbolic representation of v–monomials we get $v^\beta \longmapsto \hat{v}^{|\beta|}b(\zeta_1, \ldots, \zeta_{|\beta|})$. Then

$$u^\alpha v^\beta \longmapsto \hat{u}^{|\alpha|}\hat{v}^{|\beta|}a(\xi_1, \ldots, \xi_{|\alpha|})b(\zeta_1, \ldots, \zeta_{|\beta|})\,.$$

Note that the symbol obtained is invariant with respect to the direct product of two permutation groups $\mathcal{S}_{|\alpha|}^\xi \times \mathcal{S}_{|\beta|}^\zeta$.

To the product of two elements $f, g \in \mathcal{R}$ with symbols

$$f \mapsto \hat{u}^n \hat{v}^m a(\xi_1, \ldots, \xi_n, \zeta_1, \ldots, \zeta_m) \text{ and } g \mapsto \hat{u}^p \hat{v}^q b(\xi_1, \ldots, \xi_p, \zeta_1, \ldots, \zeta_q)$$

corresponds:

$$
\begin{aligned}
fg \;\longmapsto\; & \hat{u}^{n+p} \hat{v}^{m+q} \langle\!\langle a(\xi_1, \ldots, \xi_n, \zeta_1, \ldots, \zeta_m) \\
& b(\xi_{n+1}, \ldots, \xi_{n+p}, \zeta_{m+1}, \ldots, \zeta_{m+q}) \rangle\!\rangle_{S_{n+p}^\xi} \rangle_{S_{m+q}^\zeta},
\end{aligned}
\qquad (5.20)
$$

where the symmetrisation operation is taken with respect to permutations of all arguments ξ and then ζ (the symmetrisation can be made in any order).

If $f \in \mathcal{R}$ has a symbol $f \mapsto \hat{u}^n \hat{v}^m a(\xi_1, \ldots, \xi_n, \zeta_1, \ldots, \zeta_m)$, then the symbolic representation for the derivative $D_x(f)$ is:

$$D_x(f) \mapsto \hat{u}^n \hat{v}^m (\xi_1 + \cdots + \xi_n + \zeta_1 + \cdots + \zeta_m) a(\xi_1, \ldots, \xi_n, \zeta_1, \ldots, \zeta_m).$$

To the operator D_x we shall assign a special symbol η satisfying the following composition rule (the Leibnitz rule)

$$
\begin{aligned}
& \eta \circ \hat{u}^n \hat{v}^m a(\xi_1, \ldots, \xi_n, \zeta_1, \ldots, \zeta_m) \\
= \; & \hat{u}^n \hat{v}^m a(\xi_1, \ldots, \xi_n, \zeta_1, \ldots, \zeta_m)(\xi_1 + \cdots + \xi_n + \zeta_1 + \cdots + \zeta_m + \eta).
\end{aligned}
$$

For $f \in \mathcal{R}$ with symbol $f \mapsto \hat{u}^n \hat{v}^m a(\xi_1, \ldots, \xi_n, \zeta_1, \ldots, \zeta_m)$ the symbolic representation of the Fréchet derivative (5.16) is

$$
\begin{aligned}
f_{*u} &\longmapsto \hat{u}^{n-1} \hat{v}^m n\, a(\xi_1, \ldots, \xi_{n-1}, \eta, \zeta_1, \ldots, \zeta_m), \\
f_{*v} &\longmapsto \hat{u}^n \hat{v}^{m-1} m\, a(\xi_1, \ldots, \xi_n, \zeta_1, \ldots, \zeta_{m-1}, \eta).
\end{aligned}
$$

5.3 Integrability of evolutionary equations

5.3.1 Study of symmetries of evolutionary equations in symbolic representation

Using the above symbolic representation of the Lie bracket we can study the properties of symmetries in great detail.

Theorem 2 *Let the right hand side of evolutionary equation (5.7) have symbolic representation*

$$F \longmapsto \hat{u}\omega(\xi_1) + \hat{u}^2 a_1(\xi_1, \xi_2) + \hat{u}^3 a_2(\xi_1, \xi_2, \xi_3) + \cdots$$

and the degree of polynomial $\omega(\xi_1)$ be greater than 1. If

$$G \longmapsto \hat{u}\Omega(\xi_1) + \hat{u}^2 A_1(\xi_1, \xi_2) + \hat{u}^3 A_2(\xi_1, \xi_2, \xi_3) + \cdots \qquad (5.21)$$

is a symmetry, then its coefficients can be found recursively

$$A_1(\xi_1, \xi_2) = \frac{G^\Omega(\xi_1, \xi_2)}{G^\omega(\xi_1, \xi_2)} a_1(\xi_1, \xi_2) \tag{5.22}$$

$$A_{m-1}(\xi_1, ..., \xi_m) = \frac{1}{G^\omega(\xi_1, ..., \xi_m)} \bigg(G^\Omega(\xi_1, ..., \xi_m) a_{m-1}(\xi_1, ..., \xi_m)$$

$$+ \sum_{j=1}^{m-2} \bigg\langle (j+1) A_j(\xi_1, ..., \xi_j, \sum_{l=j}^{m-1} \xi_{l+1}) a_{m-1-j}(\xi_{j+1}, ..., \xi_m) \tag{5.23}$$

$$- (m-j) a_{m-1-j}(\xi_1, ..., \xi_{m-1-j}, \sum_{l=0}^{j} \xi_{m-l}) A_j(\xi_{m-j}, ..., \xi_m) \bigg\rangle_{S_m^\xi} \bigg),$$

where

$$G^\omega(\xi_1, ..., \xi_m) = \omega(\sum_{n=1}^{m} \xi_n) - \sum_{n=1}^{m} \omega(\xi_n). \tag{5.24}$$

Proof: The proof of the Theorem is straightforward (see, for example [29]). Using (5.14) we can compute the Lie bracket between F and G. When it vanishes up to $\hat{\mathcal{R}}^2$, we express $A_1(\xi_1, \xi_2)$ from the result, which leads to formula (5.22). The Lie bracket vanishing up to $\hat{\mathcal{R}}^m$ is equivalent to formula (5.23). □

Corollary 1 *For the equation stated in Theorem 2, (i) any symmetry has a linear part, that is, $\Omega(\xi_1) \neq 0$; (ii) the algebra of symmetries is commutative.*

Proof: (i). Let us assume that $\Omega(\xi_1) = 0$. Then it follows from (5.22) that $A_1(\xi_1, \xi_2) = 0$. Assuming that $A_k = 0$ for all $1 < k < m-1$, we get from (5.23) that $A_{m-1} = 0$. Thus by induction, we get $G = 0$.

(ii). The commutator of two symmetries is a symmetry due to the Jacobi identity, but it does not contain a linear part. Thus it must vanish. □

Theorem 2 states that a symmetry of an equation is uniquely determined by its linear part (i.e. dispersion). For fixed $\Omega(\xi_1)$ all coefficients in the series (5.21) can be found recursively. Theorem 2 does not mean that any evolutionary equation has a symmetry. The right hand side of (5.21) must represent a valid symbol, i.e., an element of $\hat{\mathcal{R}}$. Thus:

    a. all coefficients $A_m(\xi_1, ..., \xi_{m+1})$ must be polynomial,

    b. there should be a finite number of non-vanishing coefficients A_m.

In general, as follows from (5.22) and (5.23), the coefficients A_k are rational functions – they have denominators G^ω. In order to define a symbol of a symmetry, these denominators must cancel with appropriate factors in the numerators. Thus factorisation properties of the polynomials G^ω and G^Ω are crucial for the structure of the symmetry algebra of the equation.

Proposition 3 (F. Beukers [38]) *For any positive integer $m \geq 2$ the polynomial*

$$h_{c,m} = (\xi_1 + \xi_2 + \xi_3 + \xi_4)^m - c_1^{m-1}\xi_1^m - c_2^{m-1}\xi_2^m - c_3^{m-1}\xi_3^m - c_4^{m-1}\xi_4^m,$$

where $\Pi_{i=1}^4 c_i \neq 0$, is irreducible over $\mathbb{C}$.

Proof: Suppose that $h_{c,m} = A \cdot B$ with A, and B two polynomials of positive degree. Then the projective hypersurface Σ given by $h_{c,m} = 0$ consists of two components Σ_A, Σ_B given by $A = 0, B = 0$ respectively. The intersection $\Sigma_A \bigcap \Sigma_B$ consists of an infinite number of points, which should be singularities of Σ since

$$\frac{dh_{c,m}}{d\xi_i} = \frac{dA}{d\xi_i} \cdot B \bigg|_{\Sigma_A \bigcap \Sigma_B} + A \cdot \frac{dB}{d\xi_i} \bigg|_{\Sigma_A \bigcap \Sigma_B} = 0.$$

Thus it suffices to show that Σ has finitely many singular points.

We compute the singular points by setting the partial derivatives of $h_{c,m}$ equal to zero, i.e.,

$$\begin{cases} (\xi_1 + \xi_2 + \xi_3 + \xi_4)^{m-1} - (c_1\xi_1)^{m-1} = 0 \\ (\xi_1 + \xi_2 + \xi_3 + \xi_4)^{m-1} - (c_2\xi_2)^{m-1} = 0 \\ (\xi_1 + \xi_2 + \xi_3 + \xi_4)^{m-1} - (c_3\xi_3)^{m-1} = 0 \\ (\xi_1 + \xi_2 + \xi_3 + \xi_4)^{m-1} - (c_4\xi_4)^{m-1} = 0 \end{cases}$$

From these equations it follows in particular that

$$\xi_1 = \zeta_1/c_1, \ \xi_2 = \zeta_2/c_2, \ \xi_3 = \zeta_3/c_3, \ \xi_4 = \zeta_4/c_4,$$

where $\zeta_i^{m-1} = 1$ and $\zeta_1/c_1 + \zeta_2/c_2 + \zeta_3/c_3 + \zeta_4/c_4 = 1$. For given $c_i, i = 1 \cdots 4$, we get finitely many singular points. □

Corollary 2 *The polynomials $G^\omega(\xi_1, \ldots, \xi_n)$, cf. (5.24) are irreducible for $n \geq 4$.*

Proof: Let $\omega(\xi) = \alpha^m \xi^m + \cdots + \alpha_0$. If $0 \leq m \leq 1$ then G^ω is a constant and therefore irreducible. If $m \geq 2$ polynomial G^ω has the form

$$G^\omega(\xi_1, \ldots, \xi_n) = \alpha_m G^{(m)}(\xi_1, \ldots, \xi_n) + g^\omega,$$

where $\deg(g^\omega) < m$ and

$$G^{(m)}(\xi_1, \ldots, \xi_n) = (\xi_1 + \cdots + \xi_n)^m - \xi_1^m - \cdots - \xi_n^m, \qquad (5.25)$$

which is irreducible according to Proposition 3. $\qquad\square$

Theorem 3 *The algebra of symmetries of the evolutionary equation*

$$u_t = \sum_{k=0}^{n} \alpha_k u_k + f(u_{n-1}, \ldots, u) = F, \quad n \geq 2, \quad \alpha_n \neq 0 \qquad (5.26)$$

where $f(u_{n-1}, \ldots, u) \neq 0$ *and*

$$f(u_{n-1}, \ldots, u) \in \bigoplus_{m>3} \bigoplus_{p<n} \mathcal{R}_p^m \qquad (5.27)$$

is trivial, i.e., $C_\mathcal{R}(F) = Span_\mathbb{C}\{u_1, F\}.$

Proof: In symbolic representation

$$F \longmapsto \hat{u}\omega(\xi_1) + \hat{u}^{m+1}a_m(\xi_1, \ldots, \xi_{m+1}) + \hat{u}^{m+2}a_{m+1}(\xi_1, \ldots, \xi_{m+2}) + \cdots,$$

where $\omega(\xi_1) = \alpha_n\xi_1^n + \cdots + \alpha_1\xi_1 + \alpha_0$. The condition (5.27) implies $m \geq 3$ and $\deg(a_m(\xi_1, \ldots, \xi_{m+1})) < n$.

A symmetry of (5.26) is of the form

$$G \longmapsto \hat{u}\Omega(\xi_1) + \hat{u}^2 A_1(\xi_1, \xi_2) + \hat{u}^3 A_2(\xi_1, \xi_2, \xi_3) + \hat{u}^4 A_3(\xi_1, \xi_2, \xi_3, \xi_4) + \cdots$$

if it exists. We know that its linear part $\Omega(\xi_1) \neq 0$ from Corollary 1. It follows from (5.23) that $A_k(\xi_1, \ldots, \xi_{k+1}) = 0$ for $k < m$ and

$$A_m(\xi_1, \ldots, \xi_{m+1}) = \frac{G^\Omega(\xi_1, \ldots, \xi_{m+1})}{G^\omega(\xi_1, \ldots, \xi_{m+1})} a_m(\xi_1, \ldots, \xi_{m+1}). \qquad (5.28)$$

Suppose $\Omega(\xi) \neq \alpha\xi + \beta\omega(\xi)$ for any $\alpha, \beta \in \mathbb{C}$. From Corollary 2, we have that polynomials $G^\Omega(\xi_1, \ldots, \xi_{m+1})$ and $G^\omega(\xi_1, \ldots, \xi_{m+1})$ are irreducible and therefore they are co-prime. Since

$$\deg(G^\omega(\xi_1, \ldots, \xi_{m+1})) = n > \deg(a_m(\xi_1, \ldots, \xi_{m+1})),$$

the right hand side of (5.28) is a rational function (not a polynomial). Thus there are no symmetries under the assumption.

When $\Omega(\xi) = \alpha\xi + \beta\omega(\xi)$ for some $\alpha, \beta \in \mathbb{C}$, it follows from (5.23) that $G = \alpha u_1 + \beta F \in Span_\mathbb{C}\{u_1, F\}$. $\qquad\square$

According to Definition 3, equation (5.26) is not integrable. In (5.27) condition $p < n$ is essential. Indeed, the equation

$$u_t = u_2 + u^m u_1^2$$

$(p = n = 2)$ is integrable for any m.

If an evolutionary equation (5.7) with linear part of order 2 or higher has a nontrivial symmetry, then any approximate symmetry of degree 3 is amendable to any degree. Thus, if we have infinitely many approximate symmetries of degree 3, then we have infinitely many approximate symmetries of arbitrarily high degree.

Theorem 4 *Let $\omega(\xi_1)$ be a polynomial of degree greater than 1. Assume that the evolutionary equation (5.7) with linear terms $\hat{u}\omega(\xi_1)$ has a nontrivial symmetry. Then for an approximate symmetry $\sum_{j=1}^{3} h_j$, $h_j \in \mathcal{R}^j$, of degree 3, there exists a unique $H = \sum_{j\geq 1} h_j$, $h_j \in \mathcal{R}^j$, such that H is an approximate symmetry of any degree.*

This Theorem is the direct consequence of a more general Theorem 2.3 in [26] (see also Theorem 2.76 p.27 [25]) formulated in the context of filtered Lie modules. According to Theorem 2.3, in application to an evolutionary equation (5.7) with linear terms $\hat{u}\omega(\xi_1)$, we should require that the polynomials $G^{\omega}(\xi_1, ..., \xi_{m+1})$ and polynomials $G^{\Omega}(\xi_1, ..., \xi_{m+1})$ defined by (5.24) have no common factors for some $m > 1$. This is the case for $m = 3$ as follows from Corollary 2.

The result of Theorem 4 confirms the remark made in [36]:

Another interesting fact regarding the symmetry structure of evolution equations is that in all known cases the existence of one generalised symmetry implies the existence of infinitely many. (However, this has not been proved in general.)

For systems of equations and for non-evolutionary equations the conjecture that *the existence of one symmetry implies the existence of (infinitely many) others* has been disproved. In [7] it was shown that the example given in [6] is indeed a counterexample to the conjecture (see also [9]). Even a rectified conjecture [39] that *for N-component equations one needs N symmetries* is incorrect. An example of a system of two equations possessing exactly two nontrivial symmetries is given in [8].

These examples do not contradict the spirit of our Theorem 4 since they are based upon the nonexistence of approximate symmetries of degree 2, which is one of the conditions in the theorem.

As we have already mentioned above, the existence of a formal recursion operator Λ (5.12) for an evolutionary equation is a necessary condition for the existence of an infinite hierarchy of symmetries. A similar, but not equivalent, theorem can be stated in the symbolic representation. The difference is in the natural ordering. In the standard representation the coefficients l_k are ordered by the power of D_x^k in the formal series Λ (5.12). In the symbolic representation the natural ordering is by the power of symbol $\hat{u}$. The fact that l_k must be local, i.e. $l_k \in \mathcal{R}$ in the symbolic representation, suggests the following definition:

Definition 7 *We say that a function* $b_m(\xi_1, ..., \xi_m, \eta)$, $m \geq 1$ *is k-local if in the expansion as* $\eta \to \infty$

$$b_m(\xi_1, ..., \xi_m, \eta) = \beta_{m1}(\xi_1, ..., \xi_m)\eta^{n_m} + \beta_{m2}(\xi_1, ..., \xi_m)\eta^{n_m - 1} + \cdots$$

the first k coefficients $\beta_{ms}(\xi_1, ..., \xi_m)$, $s = 1, ..., k$ *are symmetric polynomials in* $\xi_1, ..., \xi_m$. *We say that* $b_m(\xi_1, ..., \xi_m, \eta)$ *is local if it is k-local for any k.*

The existence of an infinite hierarchy of symmetries implies the existence of a formal recursion operator with local coefficients (Proposition 3 in [29]). The existence of an infinite hierarchy of approximate symmetries of degree N implies that the first $N - 1$ coefficients of the formal recursion operator are local. The details of the proof of the following Theorem can be found in [29] (Proposition 3).

Theorem 5 *Suppose equation (5.7) has an infinite hierarchy of approximate symmetries of degree* N

$$u_{t_i} = \hat{u}\Omega_i(\xi_1) + \sum_{j \geq 1} \hat{u}^{j+1} A_{ij}(\xi_1, \ldots, \xi_{j+1}) = G_i, \quad i = 1, 2, \ldots$$

where $\Omega_i(\xi_1)$ *are polynomials of degree* m_i *and* $m_1 < m_2 < \cdots < m_i < \cdots$. *Then the coefficients* $\phi_m(\xi_1, ..., \xi_m, \eta)$, $m = 1, ..., N-1$ *of the formal recursion operator*

$$\Lambda = \eta + \hat{u}\phi_1(\xi_1, \eta) + \hat{u}^2\phi_2(\xi_1, \xi_2, \eta) + \cdots$$

are local.

In symbolic representation equation (5.13) can be solved [29] in the sense that the coefficients of a formal series Λ can be found recursively for any evolutionary equation (5.7):

Theorem 6 *Let* $\phi(\eta)$ *be an arbitrary function and the formal series*

$$\Lambda = \phi(\eta) + \hat{u}\phi_1(\xi_1, \eta) + \hat{u}^2\phi_2(\xi_1, \xi_2, \eta) + \hat{u}^3\phi_3(\xi_1, \xi_2, \xi_3, \eta) + \cdots$$

be a solution of equation (5.13), then its coefficients $\phi_m(\xi_1,...,\xi_m,\eta)$ can be found recursively

$$\phi_1(\xi_1,\eta) = \frac{2(\phi(\eta+\xi_1)-\phi(\eta))}{G^\omega(\xi_1,\eta)}a_1(\xi_1,\eta)$$

$$\phi_m(\xi_1,...,\xi_m,\eta) = \frac{1}{G^\omega(\xi_1,...,\xi_m,\eta)}\Big((m+1)(\phi(\eta+\xi_1+...+\xi_m)$$

$$-\phi(\eta))a_m(\xi_1,...,\xi_m,\eta)$$

$$+\sum_{n=1}^{m-1}\langle n\phi_n(\xi_1,..,\xi_{n-1},\xi_n+\cdots+\xi_m,\eta)a_{m-n}(\xi_n,..,\xi_m)$$

$$+(m-n+1)\phi_n(\xi_1,..,\xi_n,\eta+\sum_{l=n+1}^{m}\xi_l)a_{m-n}(\xi_{n+1},..,\xi_m,\eta)$$

$$-(m-n+1)a_{m-n}(\xi_{n+1},..,\xi_m,\eta+\sum_{l=1}^{n}\xi_l)\phi_n(\xi_1,..,\xi_n,\eta)\rangle_{S_m^\xi}\Big).$$

The existence of a formal recursion operator with local coefficients is a necessary condition for the existence of an infinite hierarchy of symmetries. It suggests the following test for integrability of equations (5.7):

- Find a first few coefficients $\phi_n(\xi_1,...,\xi_n,\eta)$ (the first three nontrivial coefficients ϕ_n were sufficient for the analysis in all cases known to us).
- Expand these coefficients in series of $1/\eta$

$$\phi_n(\xi_1,...,\xi_n,\eta) = \sum_{s=s_n}\Phi_{ns}(\xi_1,...,\xi_n)\eta^{-s}$$

and check that the functions $\Phi_{ns}(\xi_1,...,\xi_n)$ are polynomials (not rational functions).

This test will be extended and used for non-local and non-evolutionary equations in sections 5.4.1 and 5.4.2.

5.3.2 Global classification of integrable homogeneous evolutionary equations

In this section, we give the ultimate global classification of integrable equations of the form

$$u_t = u_n + f(u,\cdots,u_{n-1}), \qquad n \geq 2 \tag{5.29}$$

where $u_n + f(u, \cdots, u_{n-1})$ is a λ–homogeneous differential polynomial and $\lambda \geq 0$. We give a complete description of integrable equations for all n.

Theorem 4 implies that if equation (5.29) possesses one higher symmetry and infinitely many approximate symmetries of degree 3, then it possesses infinitely many approximate symmetries of any degree. Therefore to classify integrable equations (5.29) it suffices to classify equations which possess infinitely many degree 3 approximate symmetries and then impose the condition of existence of at least one exact symmetry. The classification has been done in the case of λ-homogeneous equations with $\lambda \geq 0$. In the case $\lambda > 0$ see the details in [26], while in the case $\lambda = 0$ the details can be found in [27].

Now we sketch the results for the case $\lambda > 0$ without the detailed proofs. The following statement on factorisation properties of the polynomials $G^{(k)}(\xi_1, \ldots, \xi_n)$ (5.25) plays an important role in the classification of integrable equations:

Theorem 7 $G^{(k)}(\xi_1, \ldots, \xi_n) = t^{(k)} g^{(k)}$, where $(g^{(k)}, g^{(l)}) = 1$ for all $k < l$, and $t^{(k)}$ is one of the following cases.

- $n = 2$:
  - $k = 0 \pmod 2$: $\xi_1 \xi_2$
  - $k = 3 \pmod 6$: $\xi_1 \xi_2 (\xi_1 + \xi_2)$
  - $k = 5 \pmod 6$: $\xi_1 \xi_2 (\xi_1 + \xi_2)(\xi_1^2 + \xi_1 \xi_2 + \xi_2^2)$
  - $k = 1 \pmod 6$: $\xi_1 \xi_2 (\xi_1 + \xi_2)(\xi_1^2 + \xi_1 \xi_2 + \xi_2^2)^2$

- $n = 3$:
  - $k = 0 \pmod 2$: 1
  - $k = 1 \pmod 2$: $(\xi_1 + \xi_2)(\xi_1 + \xi_3)(\xi_2 + \xi_3)$

- $n > 3$: 1

For $n > 3$ the statement follows from the more general Theorem 3. For $n = 3$, it has been proven by Beukers and was published in [25, 26] with his kind permission. The case $n = 2$ has a quite remarkable history. In affine co–ordinates $x = \xi_1/\xi_2$ we have $G^{(k)}(\xi_1, \xi_2) = \xi_2^k P_k(x)$ and the problem reduces to factorisation properties of the Cauchy-Liouville-Mirimanoff polynomials

$$P_k(x) = (1 + x)^k - x^k - 1.$$

The common factors $P_k = x(1 + x)^\alpha (1 + x + x^2)^\beta g^{(k)}(x)$ and their periodicity were established in the joint report of Cauchy and Liouville

[40]. Using Diophantine approximation theory Beukers has shown that the factors $g^{(k)}, g^{(m)}$ are co-prime for $k \neq m$ [41]. Beukers also conjectured that the factors $g^{(p)}$ are irreducible over $\mathbb{Q}$. For prime p the irreducibility of $g^{(p)}$ over $\mathbb{Q}$ was earlier conjectured by Mirimanoff [42]. Remarkable progress towards the proof of the Mirimanoff conjecture has been recently achieved in [43].

We now consider λ-homogeneous equations of the form

$$u_t = u_n + f_2 + f_3 + \cdots, \qquad f_i \in \mathcal{R}^i \tag{5.30}$$
$$\longmapsto \quad \hat{u}a_0(\xi_1) + \hat{u}^2 a_1(\xi_1, \xi_2) + \hat{u}^3 a_2(\xi_1, \xi_2, \xi_3) + \cdots,$$

where $n \geq 2$, $\lambda > 0$ and the degree of a polynomial a_j is $n - j\lambda$. Note that if λ is not integer and $i\lambda \notin \mathbb{N}$, then $a_i = 0$. This reduces the number of relevant λ to a finite set.

Let $G \in \mathcal{R}$ be a nontrivial symmetry of (5.30). Then it is of the form

$$G = u_m + g_2 + g_3 + \cdots, \qquad g_i \in \mathcal{R}^i$$
$$\longmapsto \quad \hat{u}A_0(\xi_1) + \hat{u}^2 A_1(\xi_1, \xi_2) + \hat{u}^3 A_2(\xi_1, \xi_2, \xi_3) + \cdots,$$

where $2 \leq m \neq n$ and the degree of the polynomial A_j is $m - j\lambda$. For all integers $r \geq 0$ the following formula holds

$$\sum_{i=0}^{r} [\hat{u}^{i+1} a_i, \hat{u}^{r-i+1} A_{r-i}] = 0. \tag{5.31}$$

Clearly we have $[\hat{u}a_0, \hat{u}A_0] = 0$. The next equation to be solved is $[\hat{u}a_0, \hat{u}^2 A_1] + [\hat{u}^2 a_1, \hat{u}A_0] = 0$, which is trivially satisfied if equation (5.30) has no quadratic terms: $f_2 = 0$. Let us concentrate on the case $f_2 \neq 0$. In this case, using Theorem 4, we see that the existence of a symmetry is uniquely determined by the existence of its quadratic term [25, 26].

We now make a very interesting observation. Assume n and q are both odd. Let us compute the symmetry of equation (5.30) with linear term u_q. Its quadratic terms, cf. (5.31), have the following symbolic expression

$$\frac{a_1 \, (\xi_1^2 + \xi_1 \xi_2 + \xi_2^2)^{s-s'} \, g^{(q)}(\xi_1, \xi_2)}{g^{(n)}(\xi_1, \xi_2)}. \tag{5.32}$$

Proposition 7 implies that $\lambda \leq 3 + 2\min(s, s')$, where $s' = \frac{n+3}{2}$ (mod 3) and $s = \frac{q+3}{2}$ (mod 3). We see that if expression (5.32) is a polynomial, then it defines a symmetry $Q = u_q + Q_2 + \cdots$ since Q is determined by its quadratic term Q_2. The evolutionary equation defined by Q has the same symmetries as equation (5.30). So instead of (5.30) we may consider the equation given by Q. The lowest possible q is $2s + 3$ for $s = 0, 1, 2$. Therefore we only need to consider λ-homogeneous equations with $\lambda \leq 7$ of orders not greater than 7.

A similar observation can be made for even $n > 2$. Suppose we have found a nontrivial symmetry with quadratic term

$$\frac{a_1 \, G^{(q)}}{\xi_1 \xi_2 \, g^{(n)}}$$

This immediately implies $\lambda \leq 2$. Then the quadratic term $2\frac{a_1}{g^{(n)}}$ defines a symmetry Q starting with u_2. Therefore, we only need to find the symmetries of equations of order 2 to get the complete classification of symmetries of λ-homogeneous scalar polynomial equations (with $\lambda \leq 2$) starting with an even linear term.

Finally, we analyse the case when equation (5.30) has no quadratic terms. If $a_i = 0$ for $i = 1, \cdots, j - 1$, then we look at the equation $[\hat{u}a_0, \hat{u}^{j+1}A_j] + [\hat{u}^{j+1}a_j, \hat{u}A_0] = 0$, i.e.

$$A_j = \frac{G^{(m)}(\xi_1, \cdots, \xi_{j+1}) \, a_j}{G^{(n)}(\xi_1, \cdots, \xi_{j+1})}.$$

From Proposition 7 and the proof of Theorem 3, we know there are no symmetries for the equation when $j \geq 3$, or when $j = 2$ and n is even. When $j = 2$ and n is odd, it can only have odd order symmetries. In this case one can remark that if the equation possesses symmetries for any m then it must possess a symmetry of order 3.

By now, we have proved the following

Theorem 8 *A nontrivial symmetry of a λ-homogeneous equation with $\lambda > 0$ is part of a hierarchy starting at order 2, 3, 5 or 7.*

Only an equation with nonzero quadratic or cubic terms can have a nontrivial symmetry. For each possible $\lambda > 0$, we must find a third order symmetry for a second order equation, a fifth order symmetry for a third order equation, a seventh order symmetry for a fifth order equation with quadratic terms, and a thirteenth order symmetry for a seventh order equation with quadratic terms. The last case can be easily reduced to the case of fifth order equations by determining the quadratic terms of the equation. The details of this final computation are described in [44].

Theorem 9 *Let $\lambda > 0$. Suppose that a λ-homogeneous polynomial* *evolutionary equation*

$$u_t = u_n + f(u, \cdots, u_{n-1}), \qquad n \geq 2$$

possesses nontrivial symmetries. Then it is a symmetry of one of the following equations up to a transformation $u \mapsto \alpha u$, $\alpha \in \mathbb{C}$:

Burgers equation

$$u_t = u_2 + uu_1$$

Korteweg–de Vries equation

$$u_t = u_3 + uu_1$$

Potential Korteweg–de Vries equation

$$u_t = u_3 + u_1^2$$

Modified Korteweg–de Vries equation

$$u_t = u_3 + u^2 u_1$$

Ibragimov-Shabat equation

$$u_t = u_3 + 3u^2 u_2 + 9uu_1^2 + 3u^4 u_1$$

Kaup-Kupershmidt equation

$$u_t = u_5 + 10uu_3 + 25u_1 u_2 + 20u^2 u_1$$

Potential Kaup-Kupershmidt equation

$$u_t = u_5 + 10u_1 u_3 + \frac{15}{2}u_2^2 + \frac{20}{3}u_1^3$$

Sawada-Kotera equation

$$u_t = u_5 + 10uu_3 + 10u_1 u_2 + 20u^2 u_1$$

Potential Sawada-Kotera equation

$$u_t = u_5 + 10u_1 u_3 + \frac{20}{3}u_1^3$$

Kupershmidt Equation

$$u_t = u_5 + 5u_1 u_3 + 5u_2^2 - 5u^2 u_3 - 20\,uu_1 u_2 - 5u_1^3 + 5u^4 u_1$$

Finally we note that all the considerations in this section can be extended to the case when the dependent variable u takes its values in some associative non-commutative algebra (such as matrix, operator, Clifford, and group algebras). A complete classification for $\lambda > 0$ homogeneous equations with linear leading term in the case of a non-commutative free associative algebra was carried out in [28].

5.4 Classification results for Non-local integrable equations

The perturbative symmetry approach in the symbolic representation allows one to derive integrability conditions for certain types of non-local equations. In this section we consider two types of such equations: the Benjamin-Ono type and the Camassa-Holm type [29, 32].

The Benjamin-Ono equation reads

$$u_t = H(u_2) + 2uu_1, \tag{5.33}$$

where H denotes the Hilbert transform

$$H(f) = \frac{1}{\pi} \int_{-\infty}^{\infty} \frac{f(y)}{y - x} dy.$$

It is well known that the higher symmetries and conservation laws of the Benjamin-Ono equation contain nested Hilbert transforms and thus an appropriate extension of the differential ring $\mathcal{R}$ is required. The construction of such an extension is similar to the one proposed by Mikhailov and Yamilov in [45] for $2+1$ dimensional equations (see also section 5.6).

The second example is the Camassa-Holm type equation:

$$m_t = cmu_1 + um_1, \quad m = u - u_2, \quad c \in \mathbb{C} \setminus \{0\}. \tag{5.34}$$

This equation is known to be integrable for $c = 2$ [46] and for $c = 3$ [47]. Equation (5.34) is not in the evolutionary form, but if we exclude one of the dependent variables (say u) then we obtain a non-local equation

$$m_t = cm\Delta(m_1) + \Delta(m)m_1, \quad \Delta = (1 - D_x^2)^{-1} \tag{5.35}$$

and again the ring extension is required.

The symbolic representation and the concept of quasi-locality [45] are the key ideas in the extension of the symmetry approach to non-local equations. In the definitions of all basic objects such as symmetries, formal recursion operators, conservation laws etc., we replace the ring of differential polynomials $\mathcal{R}$ by an appropriately extended ring. The elements of this extended ring we call quasi-local polynomials (see details in the next section). Symbolic representation gives us a simple criteria to decide if a given expression belongs to the extended ring. In this extended setting Theorem 2 and most of the results of Section 5.3.1 hold if we just replace "local" by "quasi-local" in the conditions and statements.

5.4.1 Benjamin-Ono type equations and Intermediate long wave equation

Let us consider the following sequence of ring extensions:

$$\mathcal{R}_{H^0} = \mathcal{R}, \quad \mathcal{R}_{H^{n+1}} = \overline{\mathcal{R}_{H^n} \bigcup H(\mathcal{R}_{H^n})},$$

where the set $H(\mathcal{R}_{H^n})$ is defined as $H(\mathcal{R}_{H^n}) = \{H(a); a \in \mathcal{R}_{H^n}\}$ and the horizontal line denotes the ring closure. Each $\mathcal{R}_{H^n}$ is a ring and the index n indicates the nesting depth of the operator H:

$$\mathcal{R}_{H^0} \subset \mathcal{R}_{H^1} \subset \mathcal{R}_{H^2} \subset \cdots \subset \mathcal{R}_{H^n} \subset \cdots \subset \mathcal{R}_{H^\infty} = \mathcal{R}_H.$$

The elements of $\mathcal{R}_{H^n}$, $n \geq 1$ we call *quasi–local polynomials*. The right hand side of equation (5.33), its symmetries and densities of conservation laws are quasi–local polynomials.

We now consider scalar evolutionary equations, whose right hand side is a quasi–local polynomial

$$u_t = F, \quad F \in \mathcal{R}_H. \tag{5.36}$$

For the definition of its symmetry we replace $\mathcal{R}$ by $\mathcal{R}_H$ in Definition 2. Actual computations in $\mathcal{R}_H$ lead to quite cumbersome calculations. On the other hand, in the symbolic representation computations simplify drastically and results can be neatly formulated.

In the symbolic representation the operator H is represented by $i \operatorname{sign}(\eta)$. So the symbolic representation of the ring extensions is obvious. Suppose $f \in \mathcal{R}_{H^0}$ and

$$f \mapsto \hat{u}^n a(\xi_1, \ldots, \xi_n).$$

Then

$$H(f) \mapsto \hat{u}^n i \operatorname{sign}(\xi_1 + \cdots + \xi_n) a(\xi_1, \ldots, \xi_n).$$

In the extended ring all the definitions, such as the Fréchet derivative, Lie bracket and approximate symmetries, are exactly the same as in the local case. However, the symbols of elements of the extended ring are symmetric sign-polynomials instead of symmetric polynomials. For example, the symbolic representation of $H(u_n)$ and $H(uH(u_1))$ is:

$$H(u_n) \mapsto \hat{u}\, i \operatorname{sign}(\xi_1)\xi_1^n,$$

$$H(uH(u_1)) \mapsto -\frac{\hat{u}^2}{2} \operatorname{sign}(\xi_1 + \xi_2)\left(\xi_1 \operatorname{sign}(\xi_1) + \xi_2 \operatorname{sign}(\xi_2)\right).$$

The symbolic representation of the Benjamin-Ono equation (5.33) is

$$u_t = i\hat{u} \operatorname{sign}(\xi_1)\xi_1^2 + \hat{u}^2(\xi_1 + \xi_2).$$

Counting the degrees of sign-polynomials we assume that $\deg(\,\text{sign}(\xi_1 + \cdots + \xi_k)) = 0$.

Theorem 2 for evolutionary equations and symmetries in $\mathcal{R}_H$, in the symbolic representation, remains the same with the only amendment that all symbols now are sign–polynomials. To introduce a formal recursion operator for equation (5.36) we introduce a notion of asymptotically local functions, which generalises the notion of local functions, cf. Definition 7:

Definition 8 *A function $a_n(\xi_1, \ldots, \xi_n, \eta)$ is called asymptotically local if the coefficients $a_{np}(\xi_1, \ldots, \xi_n)$ and $\tilde{a}_{np}(\xi_1, \ldots, \xi_n)$ of its expansion at $\eta \to \infty$:*

$$a_n(\xi_1, .., \xi_n, \eta) = \sum_{p=s_n}^{\infty} a_{np}(\xi_1, .., \xi_n)\eta^{-p} + \sum_{p=\tilde{s}_n}^{\infty} \tilde{a}_{np}(\xi_1, .., \xi_n)\, \text{sign}(\eta)\eta^{-p}$$

are sign–polynomials, i.e. represent elements from the extended ring $\mathcal{R}_H$.

In the above expansion we take into account $\text{sign}(\eta + \sum_j \xi_j) = \text{sign}(\eta)$ as $\eta \to \infty$.

Definition 9 *A formal series*

$$\Lambda = \phi(\eta) + \hat{u}\phi_1(\xi_1, \eta) + \hat{u}^2\phi_2(\xi_1, \xi_2, \eta) + \hat{u}^3\phi_3(\xi_1, \xi_2, \xi_3, \eta) + \cdots$$

is called a formal recursion operator of equation (5.36) if it satisfies equation (5.13) and all its coefficients are asymptotically local.

Without loss of generality the function $\phi(\eta)$ can be chosen as either $\phi(\eta) = \eta$ or $\phi(\eta) = \eta\,\text{sign}(\eta)$.

As in the local case, we can solve equation (5.13) with respect to coefficients of the formal recursion operator and Proposition 6 holds. The generalisation of Theorem 5 is straightforward, we just replace "local" by asymptotically local.

For the Benjamin-Ono equation (5.33), the first coefficient $\phi_1(\xi_1, \eta)$ of the corresponding formal recursion operator

$$\Lambda = \eta + u\phi_1(\xi_1, \eta) + u^2\phi_2(\xi_1, \xi_2, \eta) + \cdots$$

looks like

$$\phi_1(\xi_1, \eta) = \text{sign}(\eta) + \frac{\xi_1(\,\text{sign}(\xi_1) + \text{sign}(\eta))}{2\eta} + O(\frac{1}{\eta^7})$$

and it is asymptotically local. One may easily check asymptotic locality of other coefficients $\phi_2(\xi_1, \xi_2, \eta)$, $\phi_3(\xi_1, \xi_2, \xi_3, \eta)$, ...

In this setting we can classify the generalisation of Benjamin-Ono type equations. Consider an equation of the form

$$u_t = H(u_2) + c_1 u u_1 + c_2 H(uu_1) + c_3 u H(u_1) + c_4 u_1 H(u) +$$
$$+ c_5 H(uH(u_1)) + c_6 H(u)H(u_1), \tag{5.37}$$

where c_j are complex constants. The linear term of this equation coincides with the linear term of the Benjamin-Ono equation and all possible homogeneous terms are included if we suppose that H is a zero-weighted operator $W(H(f)) = W(f)$ and the weight of the variable u equals 2. We also take into account that $H^2 = -1$ and the Hilbert-Leibnitz rule

$$H(fg) = fH(g) + gH(f) + H(H(f)H(g)).$$

The following theorem holds (see the details and proof in [32]).

Theorem 10 *An equation of the form (5.37) possesses an infinite hierarchy of higher symmetries if and only if it is, up to the point transformation $u \mapsto au + bH(u)$, $a^2 + b^2 \neq 0$ and re-scalings $x \mapsto \alpha x$, $t \mapsto \beta t$, $a, b, \alpha, \beta \in \mathbb{C}$, one of the list*

$$u_t = H(u_2) + D_x(\frac{1}{2}c_1 u^2 + c_2 uH(u) - \frac{1}{2}c_1(u)^2); \tag{5.38}$$

$$u_t = H(u_2) + D_x(\frac{1}{2}c_1 u^2 + \frac{1}{2}c_2 H(u^2) - c_2 uH(u)); \tag{5.39}$$

$$u_t = H(u_2) + iuu_1 \pm H(uu_1) \mp uH(u_1) \mp 2u_1 H(u)$$
$$- iH(uH(u_1)); \tag{5.40}$$

$$u_t = H(u_2) + H(uu_1) + u_1 H(u) \pm iH(uH(u_1))$$
$$\pm iH(u)H(u_1). \tag{5.41}$$

The proof of this theorem requires checking the quasi–locality of the first three coefficients of the corresponding formal recursion operators. Equations (5.38), (5.40) and (5.41) can be reduced to the Burgers equation. In the case $c_1^2 + c_2^2 \neq 0$ equation (5.39) can be transformed into the Benjamin-Ono equation (5.33). When $c_1^2 + c_2^2 = 0$, it is equivalent to the Burgers equation. The explicit forms of the transformations are in [32]. The properties of the Benjamin-Ono equation have been studied in [2].

Finally we draw attention to the intermediate long wave equation

$$u_t = -\delta^{-1} u_1 + 2uu_1 + T(u_2),$$

where δ is a real constant parameter and

$$T(u(x)) := \frac{1}{2\delta} \int_{-\infty}^{\infty} \coth\left(\frac{\pi}{2\delta}(x-y)\right) u(y)dy \ .$$

This equation was derived by Joseph [48] as the equation describing propagation of non-linear waves in a fluid of finite depth. The intermediate long wave equation is an intermediate between the Benjamin-Ono and the Korteweg–de Vries equations in the sense that the limit $\delta \to \infty$ yields the Benjamin-Ono equation, while $\delta \to 0$ gives the KdV equation. The intermediate long wave equation possesses an infinite hierarchy of higher symmetries and is integrable by the inverse scattering method [49]. As in the case of the Benjamin-Ono equation, all its higher symmetries contain nested T operators.

We consider the general non-linear equation of the intermediate long wave form with some linear operator T

$$u_t = T(u_2) + 2uu_1 \tag{5.42}$$

and address the question: for which linear operators T does this equation possess an infinite hierarchy of higher symmetries/conservation laws? In [49] Ablowitz *et al* have shown that if equation (5.42) possesses infinitely many conservation laws then the conditions

$$T(uTv + vTu) = (Tu)(Tv) - uv, \tag{5.43}$$

$$\int_{-\infty}^{\infty} (uTv + vTu)dx = 0$$

must be satisfied.

The perturbative symmetry approach allows us to derive conditions for the operator T necessary for the existence of an infinite hierarchy of higher symmetries. All the steps are similar to the case of Benjamin-Ono type equations:

- Extend the differential ring by the operator T exactly in the same way as we did with H and define $\mathcal{R}_T$ and its symbolic representation.
- Define higher symmetries in the extended ring $\mathcal{R}_T$.
- Introduce a formal recursion operator and asymptotic locality of its coefficients.

Without going into the details (see [29] and [50]) we present the following statement:

Theorem 11 *Assume that the operator T has the symbolic representation $if(k)$:*

$$T(u(x)) = i \int_{-\infty}^{\infty} f(k)\hat{u}(k)e^{ikx}dk$$

and that $f(k) \to 1$, faster than any power of k^{-1}, as $k \to +\infty$. Then if equation (5.42) possesses a formal recursion operator with the first three coefficients asymptotically local then $f(k)$ satisfies the functional equation:

$$f(x+y)(f(x)+f(y)) = f(x)f(y) + 1 \qquad (5.44)$$

Formula (5.44) is equivalent to (5.43) in the symbolic representation. Its general odd solution, smooth on the real line except the origin, is given by

$$f(k) = \coth(\delta k)$$

which corresponds to the intermediate long wave equation, and the limiting case $\delta \to +\infty$ corresponds to the Benjamin-Ono equation

$$f(k) = \text{sign}(k).$$

These are the only such equations possessing infinitely many conservation laws. The only even solution of (5.44) is

$$f(k) = const$$

leading to the Burgers equation (up to a re-scaling), which has no non-trivial conservation laws.

5.4.2 Camassa-Holm type equations

We now apply the perturbative symmetry approach to determine integrable cases of the Camassa-Holm type equation (5.35). It contains the operator $\Delta = (1 - D_x^2)^{-1}$ and therefore we extend the differential ring $\mathcal{R}$ by the operator Δ and define the Δ–extended ring $\mathcal{R}_\Delta$ as we did for $\mathcal{R}_H$ in section 5.4.1.

The symbolic representation of the operator Δ is $\Delta \mapsto \frac{1}{1-\eta^2}$. Therefore if $f \in \mathcal{R}$ with symbol $\hat{u}^n a(\xi_1, \ldots, \xi_n)$ then $\Delta(f)$ has the symbol $\hat{u}^n \frac{a(\xi_1, \ldots, \xi_n)}{1-(\xi_1 + \cdots + \xi_n)^2}$.

All the definitions remain the same as in the local case with the amendment $\mathcal{R} \to \mathcal{R}_\Delta$.

We consider a formal recursion operator for the equation (5.35). First of all we introduce a linear term to the equation (5.35) by the change of variable $m \mapsto m + 1$ (note that $\Delta(1) = 1$) and consider the equation:

$$m_t = c\Delta(m_1) + cm\Delta(m_1) + \Delta(m)m_1 + m_1 := F. \qquad (5.45)$$

Its symbolic representation reads:

$$F \mapsto \hat{m}\omega(\xi_1) + \hat{m}^2 a(\xi_1, \xi_2),$$

where $\omega(\xi_1) = \frac{c\xi_1}{1-\xi_1^2} + \xi_1$, $a(\xi_1, \xi_2) = \frac{c\xi_1 + \xi_2}{2(1-\xi_1^2)} + \frac{c\xi_2 + \xi_1}{2(1-\xi_2^2)}$. The following statement holds [32]:

Theorem 12 *The first two coefficients of the formal recursion operator*

$$\Lambda = \eta + \hat{m}\phi_1(\xi_1, \eta) + \hat{m}^2\phi_2(\xi_1, \xi_2, \eta) + \cdots$$

for the equation (5.45) are quasi-local if and only if $c = 2$ or $c = 3$.

The case $c = 2$ corresponds to the Camassa-Holm equation, while the case $c = 3$ corresponds to the Degasperis-Processi equation. In fact, one can show [51] that equation (5.45) (or equation (5.35)) with $c = 2$ or $c = 3$ possesses an infinite dimensional algebra of *local* higher symmetries in variable m even if the equation is non-local.

5.5 Integrable Boussinesq type equations

In this section we give a brief account of our results (see details in [30, 31]) on integrable systems of the form:

$$\begin{cases} u_t = v_r, \\ v_t = \alpha u_{p-r} + \beta v_q + F(u, u_1, ..., u_{p-r-1}, v, v_1, ..., v_{q-1}), \end{cases} \quad (5.46)$$

where $p > q \geq r \geq 0$, $\alpha, \beta \in \mathbb{C}$. System (5.46) can be reduced to a single second order (in time) non-evolutionary equation of order p:

$$w_{tt} = \alpha w_p + \beta w_{q,t} + F(w_r, w_{r+1}, \ldots, w_{p-1}, w_t, w_{1,t}, \ldots, w_{q-1,t})$$

in the variable w, such that $v = \partial_t w$, $u = w_r$ and w_k denotes $\partial_x^k w$. If the function F does not depend on $v, v_1, \ldots, v_{r-1}$ then one can eliminate v from the second equation and rewrite the system in the form

$$u_{tt} = \alpha u_p + \beta u_{t,q} + K(u, u_1, u_2, ..., u_{p-1}, u_t, u_{t,1}, u_{t,2}, ..., u_{t,q-1}), \quad (5.47)$$

where $K = D_x^r(F)$.

The famous integrable Boussinesq equation [52]

$$u_{tt} = u_{xxxx} + (u^2)_{xx} \quad (5.48)$$

belongs to this class. Recently all integrable equations of the form

$$u_{tt} = u_{xxx} + F(u, u_x, u_{xx}, u_t, u_{tx})$$

have been classified and comprehensively studied in [53]. In particular, it has been shown that equation

$$w_{tt} = w_{xxx} + 3w_x w_{tx} + w_{xx} w_t - 3w_x^2 w_{xx}.$$

is integrable.

Sixth order ($p = 6$) integrable equations of the form (5.47) can be obtained as reductions of the Sato hierarchies corresponding to KP, BKP and CKP equations [54, 55] as well as derived from equations studied by Drinfeld and Sokolov [56, 57].

A non-evolutionary equation

$$u_{tt} = K(u, u_x, u_{xx}, \cdots, \partial_x^n u, u_t, u_{tx}, u_{txx}, \cdots, \partial_x^m u_t), \qquad (5.49)$$

can always be replaced by a system of two evolutionary equations

$$\begin{cases} u_t = v, \\ v_t = K(u, u_x, u_{xx}, ..., \partial_x^n u, v, v_x, v_{xx}, ..., \partial_x^m v). \end{cases}$$

A non-evolutionary equation (5.49) may have other representations in the evolutionary form. If $K = D_x(G)$, then the system of evolutionary equations

$$u_t = v_x, \qquad v_t = G.$$

also represents (5.49).

For example, eliminating variable v from the following systems
Case 1:

$$u_t = v, \qquad v_t = u_{xxxx} + (u^2)_{xx}, \qquad (5.50)$$

Case 2:

$$u_t = v_x, \qquad v_t = u_{xxx} + (u^2)_x, \qquad (5.51)$$

Case 3:

$$u_t = v_{xx}, \qquad v_t = u_{xx} + u^2, \qquad (5.52)$$

we obtain the same Boussinesq equation (5.48) on variable u.

There is a subtle, but important difference between the above representations of the Boussinesq equation (5.48). The system (5.50) has only a finite number of local infinitesimal symmetries (i.e. symmetries whose generators can be expressed in terms of u, v and a finite number of their derivatives) while equations (5.51) and (5.52) have infinite hierarchies of local symmetries. The reason is simple – infinitesimal symmetries of equation (5.51), which depend on the variable v, cannot be expressed in terms of u and its derivatives, since formally $v = D_x^{-1} u_t$. Equation

(5.51) possesses only a finite number of symmetries that do not depend explicitly on the variable v.

Homogeneous equations play a central role in the theory of integrable equations.

Proposition 4 *Let system (5.46) be **w**–homogeneous of weight μ, then* $\mathbf{w} = (w_u, w_u + \mu - r)$ *and*

i. if p is even, $p = 2n$, $n \in \mathbb{N}$, then $\mu = q = n$,

ii. if p is odd, $p = 2n + 1$, $n \in \mathbb{N}$, then $\beta = 0$, $\mu = p/2$, $q = n + 1$.

The system (5.46) we will write in the vector form

$$\mathbf{u}_t = \mathbf{F}, \quad \mathbf{F} = (v_r, f)^T. \tag{5.53}$$

Suppose $\mathbf{G} = (g, h)^T$ is a generator of a symmetry, then

$$[\mathbf{F}, \mathbf{G}] = 0 \iff D_x^r(h) = D_{\mathbf{F}}(g), \quad D_{\mathbf{G}}(f) = D_{\mathbf{F}}(h).$$

It follows from the first equation that $D_{\mathbf{F}}(g)$ belongs to the image of D_x^r and thus the second component of the symmetry generator $\mathbf{G}$ can be expressed as $h = D_x^{-r}(D_{\mathbf{F}}(g))$ and the symmetry is completely defined by its first component. Substitution into the second equation yields

$$D_{\mathbf{F}}^2(g) - D_x^r D_{\mathbf{G}}(f) = 0.$$

For approximate symmetry of degree n we obviously get

$$D_{\mathbf{F}}^2(g) - D_x^r D_{\mathbf{G}}(f) = o(\mathcal{R}^n).$$

Using the above observation for systems (5.46) we can restrict the action of the recursion operator to the first component of symmetries.

For example, if we represent the Boussinesq equation (5.48) in the form of the evolutionary system (5.51), then

$$\mathbf{F}_* = \begin{pmatrix} 0 & D_x \\ D_x^3 + 2uD_x + 2u_1 & 0 \end{pmatrix}$$

and it is easy to verify that a pseudo-differential operator

$$\Lambda = \begin{pmatrix} 3v + 2v_1 D_x^{-1} & 4D_x^2 + 2u + u_1 D_x^{-1} \\ \Lambda_{21} & 3v + v_1 D_x^{-1} \end{pmatrix},$$

where $\Lambda_{21} = 4D_x^4 + 10uD_x^2 + 15u_1 D_x + 9u_2 + 4u^2 + (2u_3 + 4uu_1)D_x^{-1}$, satisfies equation (5.19) and therefore is a recursion operator. We can restrict the action of the recursion operator to the first component. If g_1

is the first component of a symmetry of the Boussinesq equation (5.51), then

$$g_2 = \Re(S) = \left(3v + 2v_1 D_x^{-1} + (4D_x^2 + 2u + u_1 D_x^{-1})D_t D_x^{-1}\right)(g_1)$$

is the first component of the next symmetry in the hierarchy. We call $\Re$ a restricted recursion operator.

A space-shift, generated by $\mathbf{u}_1 = (u_1, v_1)^{\mathrm{tr}}$, is a symmetry of the Boussinesq equation (5.51). Taking u_1 as a seed, we can construct an infinite hierarchy $g_{3k+1} = \Re^k(u_1)$ of symmetries of weights $3k+1, k = 0, 1, 2, \ldots$. For example,

$$g_4 = 4v_3 + 4v_1 u + 4vu_1 = 4D_x(v_2 + vu).$$

We see that g_4 is a total derivative and therefore $g_7 = \Re(g_4) \in \mathcal{R}$ is the next symmetry in the hierarchy, etc. The Boussinesq equation itself is not a member of this hierarchy. If we take a seed symmetry, corresponding to the time-translation $(v_1, u_3 + 2uu_1)^{\mathrm{tr}}$ we obtain another infinite hierarchy of symmetries $g_{3k+2} = \Re^k(v_1), k = 0, 1, 2, \ldots$. The Boussinesq equation does not have symmetries of weight $3k, k \in \mathbb{N}$. One can show that g_{3k+1} and g_{3k+2} are elements of the ring $\mathcal{R}$ for any $k \in \mathbb{N}$ and therefore $\Re$ generates two infinite hierarchies of symmetries of the Boussinesq equation. Moreover, all symmetries from both hierarchies commute with each other.

In general, for the system (5.53) a recursion operator is completely determined by its two entries Λ_{11} and Λ_{12} (the first row of Λ). The restricted recursion operator $\Re$ for system (5.53) can be represented as

$$\Re = \Lambda_{11} + \Lambda_{12} D_x^{-r} D_t.$$

In symbolic representation the system (5.46) takes the form

$$\begin{cases} \hat{u}_t = \hat{v}\zeta_1^r, \\ \hat{v}_t = \alpha\hat{u}\xi_1^{p-r} + \beta\hat{v}\zeta_1^q \\ \quad + \sum_{k\geq 2}\sum_{i=0}^k \hat{u}^i\hat{v}^{k-i}a_{i,k-i}(\xi_1, .., \xi_i, \zeta_1, .., \zeta_{k-i}). \end{cases} \quad (5.54)$$

It follows from Proposition 4 that for $\mathbf{w}$–homogeneous systems (5.46) $q = p/2$ (for even p) and coefficients $a_{i,j}(\xi_1, \ldots, \xi_i, \zeta_1, \ldots, \zeta_j)$ are homogeneous polynomials of degree

$$\deg(a_{i,j}) = p + (j-1)r - (i+j-1)w_u + (j-1)r - \frac{1}{2}jp,$$

in variables $\xi_1, \ldots, \xi_i, \zeta_1, \ldots, \zeta_j$, and they are symmetric in $\xi_1, \ldots, \xi_i$ and in $\zeta_1, \ldots, \zeta_j$. If $\deg(a_{i,j})$ is not a non-negative integer, then $a_{i,j} = 0$.

We shall assume that the weight $w_u > 0$ and thus the sum in (5.54) is finite due to only a finite number of non-zero coefficients a_{ij}.

5.5.1 Even order equations

In this section we study homogeneous (with a positive weight $w_u > 0$) even order ($p = 2n$) equations (5.46) assuming $r = 1$.

In symbolic representation such equations take the form

$$\begin{cases} \hat{u}_t = \hat{v}\zeta_1 \\ \hat{v}_t = \hat{u}\omega_1(\xi_1) + \hat{v}\omega_2(\zeta_1) + \hat{u}^2 a_{20}(\xi_1, \xi_2) + \hat{u}\hat{v}a_{11}(\xi_1, \zeta_1) + \\ \quad + \hat{v}^2 a_{02}(\zeta_1, \zeta_2) + \hat{u}^3 a_{30}(\xi_1, \xi_2, \xi_3) + \cdots = \hat{f}, \end{cases} \quad (5.55)$$

where $\omega_1(\xi_1) = \alpha\xi_1^{2n-1}$ and $\omega_2(\zeta_1) = \beta\zeta_1^n$ and without loss of generality we shall represent α in the form

$$\alpha = \frac{\mu^2 - \beta^2}{4}$$

and use parameters μ, β instead of α, β.

Symmetries of the system (5.55) are determined by their first component, which in symbolic representation can be written in the form

$$\hat{u}_\tau = \hat{u}\Omega_1(\xi_1) + \hat{v}\Omega_2(\zeta_1) + \hat{u}^2 A_{20}(\xi_1, \xi_2) + \hat{u}\hat{v}A_{11}(\xi_1, \zeta_1) + \\ + \hat{v}^2 A_{02}(\zeta_1, \zeta_2) + \hat{u}^3 A_{30}(\xi_1, \xi_2, \xi_3) + \cdots,$$

where $\Omega_1(\xi_1)$, $\Omega_2(\zeta_1)$, A_{ij} are polynomials. For fixed $\Omega_1(\xi_1)$, $\Omega_2(\zeta_1)$ the coefficients A_{ij} can be found recursively (there is a generalisation of Theorem 2 to the case of many dependent variables [31]).

The Fréchet derivative $\mathbf{F}_*$ of the system (5.53) with $r = 1$ in the symbolic representation has the form

$$\hat{\mathbf{F}}_* = \begin{pmatrix} 0 & \eta \\ \hat{f}_{*,u} & \hat{f}_{*,v} \end{pmatrix}$$

where

$$\hat{f}_{*,u} = \omega_1(\eta) + 2\hat{u}a_{20}(\xi_1, \eta) + \hat{v}a_{11}(\eta, \zeta_1) + 3\hat{u}^2 a_{30}(\xi_1, \xi_2, \eta) + \cdots$$
$$\hat{f}_{*,v} = \omega_2(\eta) + \hat{u}a_{11}(\xi_1, \eta) + 2\hat{v}a_{02}(\zeta_1, \eta) + \hat{u}^2 a_{21}(\xi_1, \xi_2, \eta) + \cdots$$

A formal recursion operator can be defined as a 2×2 matrix $\hat{R}$ whose entries are formal series

$$\hat{R}_{11} = \phi_{00}(\eta) + \hat{u}\phi_{10}(\xi_1, \eta) + \frac{1}{2}\hat{v}\phi_{01}(\zeta_1, \eta) + \hat{u}^2\phi_{20}(\xi_1, \xi_2, \eta) +$$
$$+ \hat{u}\hat{v}\phi_{11}(\xi_1, \zeta_1, \eta) + \hat{v}^2\phi_{02}(\zeta_1, \zeta_2, \eta) + \cdots$$
$$\hat{R}_{12} = \psi_{00}(\eta) + \frac{1}{2}\hat{u}\psi_{10}(\xi_1, \eta) + \hat{v}\psi_{01}(\zeta_1, \eta) + \hat{u}^2\psi_{20}(\xi_1, \xi_2, \eta) +$$
$$+ \hat{u}\hat{v}\psi_{11}(\xi_1, \zeta_1, \eta) + \hat{v}^2\psi_{02}(\zeta_1, \zeta_2, \eta) + \cdots,$$

and

$$\hat{R}_{21} = \eta^{-1} \circ (\hat{R}_{11,t} + \hat{R}_{12} \circ \hat{f}_{*,u}), \quad \hat{R}_{22} = \eta^{-1} \circ (\hat{R}_{12,t} + \hat{R}_{12} \circ \hat{f}_{*,v} + \hat{R}_{11} \circ \eta).$$

satisfying equation

$$\hat{R}_t = [\hat{\mathbf{F}}_*, \hat{R}]$$

and all the coefficients $\phi_{ij}, \psi_{ij}, i, j = 0, 1, 2, \ldots$ of the formal series $\hat{R}_{11}$ and $\hat{R}_{12}$ are local. An approximate recursion operator of degree k can be viewed as a truncation of $\hat{R}$ (terms with $\hat{u}^i\hat{v}^j, i + j > k$ are omitted or ignored), so the existence of an approximate recursion operator is a necessary condition for the existence of a formal recursion operator and also a necessary condition for the existence of an infinite hierarchy of symmetries of the equation (5.55). For fixed $\phi_{00}(\eta), \psi_{00}(\eta)$ the coefficients of a formal recursion operator can be found recursively, similar to the case of one dependent variable (Theorem 6). The property of locality of the coefficients imposes constraints on the coefficients of the equation (5.55) and eventually leads to the isolation of integrable systems (see details in [30]). The latter test was applied in [30] to the problem of classification of even order $\mathbf{w}$–homogeneous integrable systems of the form (5.55). The results obtained can be summarised as follows:

The 4th order equations

It is easy to see that a homogeneous system (5.46) with $p = 4, r = 1$ is linear if $w_u > 3$. In the case $w_u = 3$ the only possibility is $F = \gamma u^2, \gamma \neq 0$ which leads to a non-integrable equation for any choice of α, β and γ. Thus the weight w_u can be equal to 2 or 1.

Case 1: The case $w_u = 2$

The most general nonlinear homogeneous system of equations (5.46) with $p = 4, r = 1, w_u = 2$ (correspondingly $w_v = 3$) is of the form:

$$\begin{cases} u_t = v_1 \\ v_t = \alpha u_3 + \beta v_2 + c_1 u u_1 + c_2 u v, \end{cases} \tag{5.56}$$

where c_1, c_2 are arbitrary constants and at least one of them is not zero.

Without loss of generality we need to consider the following three types of system (5.56):

$$\begin{cases} u_t = v_1 \\ v_t = \frac{\mu^2 - \beta^2}{4} u_3 + \beta v_2 + c_1 u u_1 + c_2 u v, \ \mu \notin \{0, \pm\beta\} \end{cases} \quad (5.57)$$

$$\begin{cases} u_t = v_1 \\ v_t = v_2 + c_1 u u_1 + c_2 u v \end{cases} \quad (5.58)$$

$$\begin{cases} u_t = v_1 \\ v_t = -\frac{1}{4} u_3 + v_2 + c_1 u u_1 + c_2 u v \end{cases} \quad (5.59)$$

The first system represents the generic case $\alpha \notin \{0, -\frac{\beta^2}{4}\}$, while the other two represent the degenerate cases.

Theorem 13 *The system (5.57) possesses two formal recursion operators with $\phi_{00}(\eta) = \eta$, $\psi_{00}(\eta) = 0$ and $\phi_{00}(\eta) = 0$, $\psi_{00}(\eta) = \eta$ if and only if $\beta = c_2 = 0$. By re-scalings it can be put in the form*

$$\begin{cases} u_t = v_1 \\ v_t = u_3 + 2uu_1 \end{cases} \quad (5.60)$$

Systems (5.58) and (5.59) are not integrable; they do not possess a formal recursion operator with $\phi_{00} = 0$, $\psi_{00} = \eta$ unless $c_1 = c_2 = 0$.

System (5.60) represents the Boussinesq equation (5.48), which is known to be integrable. In the Proposition 13 and below by re-scalings we mean an invertible change of variable of the form:

$$u \mapsto \alpha_1 u, \quad v \mapsto \alpha_2 v, \quad x \mapsto \alpha_3 x, \quad t \mapsto \alpha_4 t, \quad \alpha_i \in \mathbb{C}.$$

Case 2: The case $w_u = 1$

The most general homogeneous system of equations (5.46) with $p = 4, r = 1, w_u = 1$ is of the form

$$\begin{cases} u_t = v_1 \\ v_t = \alpha u_3 + \beta v_2 + c_1 u u_2 + c_2 u_1^2 + c_3 u_1 v + c_4 u v_1 + c_5 v^2 \\ \quad + c_6 u^2 u_1 + c_7 u^2 v + c_8 u^4 \end{cases} \quad (5.61)$$

where c_i, $i = 1 \ldots 8$ are arbitrary constants and we assume that at least one of the coefficients $c_1, \ldots, c_5$ is not zero. Without loss of generality we consider the following three types of the system (5.61):

$$\begin{cases} u_t = v_1 \\ v_t = \frac{\mu^2 - \beta^2}{4} u_3 + \beta v_2 + c_1 u u_2 + c_2 u_1^2 + c_3 u_1 v + c_4 u v_1 + c_5 v^2 \\ \quad + c_6 u^2 u_1 + c_7 u^2 v + c_8 u^4, \end{cases} \quad (5.62)$$

where $\mu \notin \{0, \pm\beta\}$, $\mu, \beta \in \mathbb{C}$ and

$$\begin{cases} u_t = v_1 \\ v_t = v_2 + c_1 u u_2 + c_2 u_1^2 + c_3 u_1 v + c_4 u v_1 + c_5 v^2 \\ \quad + c_6 u^2 u_1 + c_7 u^2 v + c_8 u^4 \end{cases} \quad (5.63)$$

$$\begin{cases} u_t = v_1 \\ v_t = -\frac{1}{4}u_3 + v_2 + c_1 u u_2 + c_2 u_1^2 + c_3 u_1 v + c_4 u v_1 \\ \quad + c_5 v^2 + c_6 u^2 u_1 + c_7 u^2 v + c_8 u^4 \end{cases} \quad (5.64)$$

Theorem 14 *The system (5.62) possesses two formal recursion operators with $\phi_{00}(\eta) = \eta$, $\psi_{00}(\eta) = 0$ and $\phi_{00}(\eta) = 0$, $\psi_{00}(\eta) = \eta$ if and only if (up to re-scalings) it is one of the list*

$$\begin{cases} u_t = v_1 \\ v_t = u_3 + u_1^2 \end{cases} \quad (5.65)$$

$$\begin{cases} u_t = v_1 \\ v_t = u_3 + 2u_1 v + 2u^2 u_1 \end{cases} \quad (5.66)$$

$$\begin{cases} u_t = v_1 \\ v_t = u_3 + 2u_1 v + 4u v_1 - 6u^2 u_1 \end{cases} \quad (5.67)$$

$$\begin{cases} u_t = v_1 \\ v_t = u_3 + 4u u_2 + 3u_1^2 - v^2 + 6u^2 u_1 + u^4 \end{cases} \quad (5.68)$$

$$\begin{cases} u_t = v_1 \\ v_t = \alpha u_3 + v_2 + 4\alpha u u_2 + 3\alpha u_1^2 + u_1 v + 2u v_1 - v^2 \\ \quad + 6\alpha u^2 u_1 + u^2 v + \alpha u^4, \ \alpha \neq -\frac{1}{4} \end{cases} \quad (5.69)$$

The system (5.63) possesses a formal recursion operator with $\phi_{00}(\eta) = 0$, $\psi_{00}(\eta) = \eta$ if and only if (up to re-scalings) it is one of the list

$$\begin{cases} u_t = v_1 \\ v_t = v_2 + 2u v_1 \end{cases} \quad (5.70)$$

$$\begin{cases} u_t = v_1 \\ v_t = v_2 - u_1^2 + 2u_1 v - v^2 \end{cases} \quad (5.71)$$

$$\begin{cases} u_t = v_1 \\ v_t = v_2 - 2u u_2 - 2u_1^2 + 2u_1 v + 6u v_1 - 12u^2 u_1 \end{cases} \quad (5.72)$$

The system (5.64) possesses two formal recursion operators with $\phi_{00}(\eta) = \eta$, $\psi_{00}(\eta) = 0$ and $\phi_{00}(\eta) = 0$, $\psi_{00}(\eta) = \eta$ if and only if (up to re-

scalings) it is

$$
\begin{cases}
u_t = v_1 \\
v_t = -\frac{1}{4}u_3 + v_2 - uu_2 - \frac{3}{4}u_1^2 + u_1v + 2uv_1 - v^2 \\
\quad -\frac{3}{2}u^2u_1 + u^2v - \frac{1}{4}u^4.
\end{cases} \tag{5.73}
$$

Equation (5.65) is a potential version of the Boussinesq equation. Equations (5.66) and (5.67) are known to be integrable. Corresponding Lax representations and references can be found in [4].

Equations (5.68) and (5.69) can be mapped into linear equations

$$
w_{tt} = w_4 , \qquad w_{tt} = \alpha w_4 + w_{2t}
$$

respectively by the Cole-Hopf transformation $u = (\log w)_x$.

Equation (5.70) can be reduced to the Burgers equation with time independent forcing

$$
u_t = u_2 + 2uu_1 + w_1 , \qquad w_t = 0
$$

by the invertible transformation $v = w + u_1 + u^2$. The latter can be linearised by the Cole-Hopf transformation.

Equation (5.71) can be reduced to the system

$$
u_t = u_2 + w_1 , \qquad w_t = -w^2
$$

by a simple invertible change of the variable, $v = w + u_1$. The system (5.71) provides an example of an equation that possesses neither higher symmetries nor a recursion operator. However, a formal recursion operator does exist and therefore it is in the list of the Proposition. Its integration can be reduced to the integration of a linear nonhomogeneous heat equation with a source term of a special form.

By a simple shift of the variable $v = w + u_1 + 2u^2$, system (5.72) can be transformed to the form

$$
u_t = u_2 + 4uu_1 + w_1 , \qquad w_t = 2D_x(uw) . \tag{5.74}
$$

The system (5.72) possesses an infinite hierarchy of symmetries of all orders generated by a recursion operator

$$
\Re = -2u + 2u_1 D_x^{-1} + D_t D_x^{-1}
$$

starting from the seed u_1.

Equation (5.73) is a particular case of (5.69) corresponding to the exceptional case $\alpha = -\frac{1}{4}$.

The 6th order equations

Homogeneous 6th order ($p = 6$) equations (5.46) with non-zero quadratic terms correspond to $w_u \leq 5$. We restrict ourselves to the case $w_u > 0$. The weights $3, 4, 5$ do not lead to integrable equations:

Proposition 5 *For weights $w_u = 3, 4, 5$ there are no equations possessing a formal recursion operator.*

Case 1: The case $w_u = 2$

The most general homogeneous system (5.46) with $p = 6, r = 1, w_u = 2$ can be written as

$$\begin{cases} u_t = v_1 \\ v_t = \alpha u_5 + \beta v_3 + D_x(c_1 u u_2 + c_2 u_1^2 + c_5 u^3) \\ \qquad + c_3 u v_1 + c_4 v u_1, \end{cases} \qquad (5.75)$$

where $\alpha, \beta, c_i, i = 1, \ldots, 4$ are arbitrary constants and we assume that at least one of $c_1, \ldots, c_4$ is not zero.

Without loss of generality we consider the following three types of the above system:

$$\begin{cases} u_t = v_1 \\ v_t = \frac{\mu^2 - \beta^2}{4} u_5 + \beta v_3 + D_x(c_1 u u_2 + c_2 u_1^2 + c_5 u^3) \\ \qquad + c_3 u v_1 + c_4 v u_1, \quad \mu \notin \{0, \pm\beta\}, \quad \mu, \beta \in \mathbb{C} \end{cases} \qquad (5.76)$$

$$\begin{cases} u_t = v_1 \\ v_t = v_3 + D_x(c_1 u u_2 + c_2 u_1^2 + c_5 u^3) + c_3 u v_1 + c_4 v u_1 \end{cases} \qquad (5.77)$$

$$\begin{cases} u_t = v_1 \\ v_t = -\frac{1}{4} u_5 + v_3 + D_x(c_1 u u_2 + c_2 u_1^2 + c_5 u^3) \\ \qquad + c_3 u v_1 + c_4 v u_1 \end{cases} \qquad (5.78)$$

Theorem 15 *If system (5.76) possesses two formal recursion operators with $\phi_{00}(\eta) = \eta$, $\psi_{00}(\eta) = 0$ and $\phi_{00}(\eta) = 0$, $\psi_{00}(\eta) = \eta$ then, up to re-scalings, it is one of the list*

$$\begin{cases} u_t = v_1 \\ v_t = 2u_5 + v_3 + D_x(2u u_2 + u_1^2 + \frac{4}{27} u^3) \end{cases} \qquad (5.79)$$

$$\begin{cases} u_t = v_1 \\ v_t = \frac{1}{5} u_5 + v_3 + D_x(u u_2 + u v + \frac{1}{3} u^3) \end{cases} \qquad (5.80)$$

$$\begin{cases} u_t = v_1 \\ v_t = \frac{1}{5} u_5 + v_3 + D_x(2u u_2 + \frac{3}{2} u_1^2 + 2u v + \frac{4}{3} u^3) \end{cases} \qquad (5.81)$$

If system (5.77) possesses a formal recursion operator with $\phi_{00}(\eta) = 0$, $\psi_{00}(\eta) = \eta$ then, up to re-scalings, it is one of the list

$$\begin{cases} u_t = v_1 \\ v_t = v_3 + uv_1 + u_1 v \end{cases} \tag{5.82}$$

$$\begin{cases} u_t = v_1 \\ v_t = v_3 + 2uu_3 + 4u_1 u_2 - 4u_1 v - 8uv_1 - 24u^2 u_1 \end{cases} \tag{5.83}$$

The system (5.78) does not possess a formal recursion operator for any non-trivial choice of c_i.

Recursion operators and bi-Hamiltonian structures for these equations can be found in [30]. Lax representations can be found in [56, 57, 58, 30, 59].

Case 2: The case $w_u = 1$

Homogeneous systems of equations (5.46) with $p = 6, r = 1, w_u = 1$ can be written in the form:

$$\begin{cases} u_t = v_1 \\ v_t = \alpha u_5 + \beta v_3 + c_1 u_2^2 + c_2 u_1 u_3 + c_3 u u_4 + c_4 u_2 v + c_5 u_1 v_1 \\ \quad + c_6 u v_2 + c_7 v^2 + c_8 u_1^3 + c_9 u u_1 u_2 + c_{10} u^2 u_3 + c_{11} u^2 v_1 + \\ \quad + c_{12} u u_1 v + c_{13} u^2 u_1^2 + c_{14} u^3 u_2 + c_{15} u^3 v + c_{16} u^4 u_1 + c_{17} u^6, \end{cases} \tag{5.84}$$

where $\alpha, \beta \in \mathbb{C}$, all $c_i, i = 1, \ldots, 17$ are arbitrary constants and at least one of $c_1, \ldots c_7$ is not zero.

We need to consider the following cases of the system (5.84):

$$\begin{cases} u_t = v_1 \\ v_t = \frac{\mu^2 - \beta^2}{4} u_5 + \beta v_3 + c_1 u_2^2 + c_2 u_1 u_3 + c_3 u u_4 + c_4 u_2 v \\ \quad + c_5 u_1 v_1 + c_6 u v_2 + c_7 v^2 + c_8 u_1^3 + + c_9 u u_1 u_2 + c_{10} u^2 u_3 \\ \quad + c_{11} u^2 v_1 + c_{12} u u_1 v + c_{13} u^2 u_1^2 + c_{14} u^3 u_2 \\ \quad + c_{15} u^3 v + c_{16} u^4 u_1 + c_{17} u^6 \end{cases} \tag{5.85}$$

where $\mu \notin \{0, \pm\beta\}$, $\mu, \beta \in \mathbb{C}$, and

$$\begin{cases} u_t = v_1 \\ v_t = v_3 + c_1 u_2^2 + c_2 u_1 u_3 + c_3 u u_4 + c_4 u_2 v + c_5 u_1 v_1 \\ \quad + c_6 u v_2 + c_7 v^2 + c_8 u_1^3 + c_9 u u_1 u_2 + c_{10} u^2 u_3 + c_{11} u^2 v_1 \\ \quad + c_{12} u u_1 v + c_{13} u^2 u_1^2 + c_{14} u^3 u_2 + c_{15} u^3 v + c_{16} u^4 u_1 + c_{17} u^6 \end{cases} \tag{5.86}$$

$$\begin{cases} u_t = v_1 \\ v_t = -\frac{1}{4} u_5 + v_3 + c_1 u_2^2 + c_2 u_1 u_3 + c_3 u u_4 + c_4 u_2 v + c_5 u_1 v_1 \\ \quad + c_6 u v_2 + c_7 v^2 + c_8 u_1^3 + c_9 u u_1 u_2 + c_{10} u^2 u_3 + c_{11} u^2 v_1 \\ \quad + c_{12} u u_1 v + c_{13} u^2 u_1^2 + c_{14} u^3 u_2 + c_{15} u^3 v + c_{16} u^4 u_1 + c_{17} u^6 \end{cases} \tag{5.87}$$

Theorem 16 *If system (5.85) possesses two formal recursion operators with $\phi_{00}(\eta) = \eta$, $\psi_{00}(\eta) = 0$, and $\phi_{00}(\eta) = 0$, $\psi_{00}(\eta) = \eta$, up to re-scalings, it is one of the equations in the following list*

$$\begin{cases} u_t = v_1 \\ v_t = 2u_5 + v_3 + u_2^2 + 2u_1u_3 + \frac{4}{27}u_1^3 \end{cases} \tag{5.88}$$

$$\begin{cases} u_t = v_1 \\ v_t = \frac{1}{5}u_5 + v_3 + u_1u_3 + u_1v_1 + \frac{1}{3}u_1^3 \end{cases} \tag{5.89}$$

$$\begin{cases} u_t = v_1 \\ v_t = \frac{1}{5}u_5 + v_3 + 2u_1u_3 + \frac{3}{2}u_2^2 + 2u_1v_1 + \frac{4}{3}u_1^3 \end{cases} \tag{5.90}$$

$$\begin{cases} u_t = v_1 \\ v_t = \alpha u_5 + v_3 + 10\alpha u_2^2 + 15\alpha u_1u_3 + 6\alpha uu_4 + vu_2 \\ \quad +3u_1v_1 + 3uv_2 - v^2 + 15\alpha u_1^3 + 15\alpha u^2u_3 + \\ \quad +60\alpha uu_1u_2 + 3uu_1v + 3u^2v_1 + 45\alpha u^2u_1^2 \\ \quad +20\alpha u^3u_2 + u^3v + 15\alpha u^4u_1 + \alpha u^6, \ \alpha \neq -\frac{1}{4} \end{cases} \tag{5.91}$$

$$\begin{cases} u_t = v_1 \\ v_t = u_5 + 6uu_4 + 15u_1u_3 + 10u_2^2 - v^2 + 15u^2u_3 + \\ \quad + 15u_1^3 + 60uu_1u_2 + 45u^2u_1^2 + 20u^3u_2 + 15u^4u_1 + u^6 \end{cases} \tag{5.92}$$

If system (5.86) possesses a formal recursion operator with $\phi_{00}(\eta) = 0$, $\psi_{00}(\eta) = \eta$, up to re-scalings, it is one of the list

$$\begin{cases} u_t = v_1 \\ v_t = v_3 + u_1v_1 \end{cases} \tag{5.93}$$

$$\begin{cases} u_t = v_1 \\ v_t = v_3 + 3u_1v_1 + 3uv_2 + 3u^2v_1 \end{cases} \tag{5.94}$$

$$\begin{cases} u_t = v_1 \\ v_t = v_3 - u_2^2 + 2u_2v - v^2 \end{cases} \tag{5.95}$$

If system (5.87) possesses two formal recursion operators with $\phi_{00}(\eta) = \eta$, $\psi_{00}(\eta) = 0$, and $\phi_{00}(\eta) = 0$, $\psi_{00}(\eta) = \eta$, up to re-scalings, then it is

$$\begin{cases} u_t = v_1 \\ v_t = -\frac{1}{4}u_5 + v_3 - \frac{5}{2}u_2^2 - \frac{15}{4}u_1u_3 - \frac{3}{2}uu_4 + vu_2 \\ \quad +3u_1v_1 + 3uv_2 - v^2 - \frac{15}{4}u_1^3 - \frac{15}{4}u^2u_3 - 15uu_1u_2 + 3uu_1v \\ \quad +3u^2v_1 - \frac{45}{4}u^2u_1^2 - 5u^3u_2 + u^3v - \frac{15}{4}u^4u_1 - \frac{1}{4}u^6, \end{cases} \tag{5.96}$$

Recursion operators, bi-Hamiltonian structures and Lax representations of equations (5.88), (5.89), (5.90) and (5.93) are discussed in detail in [30].

Equations (5.91),(5.92),(5.94) and (5.96) can be linearised by a Cole-Hopf type transformation [30].

Equation (5.95) has similar properties to system (5.71). It can also be reduced to a triangular system

$$u_t = u_3 + w_1, \qquad w_t = -w^2$$

in variables u and $w = v - u_2$.

10th order equations

In this section we present three examples of 10th order integrable non-evolutionary equations.

Proposition 6 *The following systems possess infinite hierarchies of higher symmetries:*

$$\begin{cases} u_t = v_1 \\ v_t = \frac{9}{64}u_9 + v_5 + D_x \left(3uu_6 + 9u_1u_5 + \frac{65}{4}u_2u_4 + \frac{35}{4}u_3^2 + \right. \\ \quad + 2u_1v_1 + 4uv_2 + 20u^2u_4 + 80uu_1u_3 + 60uu_2^2 + 88u_1^2u_2 \\ \quad \left. + \frac{256}{5}u^3u_2 + \frac{384}{5}u^2u_1^2 + \frac{1024}{125}u^5 \right) \end{cases} \tag{5.97}$$

$$\begin{cases} u_t = v_1 \\ v_t = -\frac{1}{54}u_9 + v_5 + \frac{5}{6}u_7u_1 + \frac{5}{3}u_6u_2 + \frac{5}{2}u_5u_3 + \frac{25}{12}u_4^2 \\ \quad -5u_3v_1 - \frac{15}{2}u_2v_2 - 10u_1v_3 - \frac{45}{4}u_5u_1^2 - \frac{75}{2}u_1u_2u_4 \\ \quad -\frac{75}{4}u_3^2u_1 - \frac{75}{4}u_2^2u_3 + \frac{45}{2}u_1^2v_1 + \frac{225}{4}u_3u_1^3 + \frac{675}{8}u_2^2u_1^2 - \frac{405}{16}u_1^5 \end{cases} \tag{5.98}$$

$$\begin{cases} u_t = v_1 \\ v_t = v_5 + 2u_2u_5 + 6u_3u_4 - 6u_3v - 22u_2v_1 - 30u_1v_2 \\ \quad -20uv_3 + 96uu_1v + 96u^2v_1 + 120D_x(4u^3u_2 + 6u^2u_1^2) \\ \quad -2D_x(8u^2u_4 + 32uu_1u_3 + 13u_1^2u_2 + 24uu_2^2) - 3840u^4u_1 \end{cases} \tag{5.99}$$

Bi-Hamiltonian structures and recursion operators for equations (5.97), (5.98) and (5.99) as well as the Lax representation for equation (5.97) can be found in [30]. The system (5.99) has been considered independently by the authors of [59, 60] and [61], where the corresponding Lax representation has been obtained. The Lax representation for equation (5.98) is still not known.

5.5.2 Odd order equations

In this section we formulate the diagonalisation method and globally classify homogeneous (with a positive weight $w_u > 0$) odd order ($p = 2n + 1$) equations (5.46) in the same spirit as we did for the scalar homogeneous evolution equations in Section 5.3. For the details, we refer the reader to the recent paper [31].

A symmetry of an odd order homogeneous equation (5.46) always starts with linear terms. A homogeneous symmetry in the symbolic representation starts either with $u\xi_1^m$ or with $v\zeta_1^{m+r}$. Without loss of generality we have

$$u_\tau = G = u\xi_1^m + \sum_{s \geq 2} \sum_{j=0}^{s} u^j v^{s-j} A_{j,s-j}(\xi_1, .., \xi_j, \zeta_1, .., \zeta_{s-j}), \quad m > 1$$

(5.100)

or

$$u_\tau = G = v\zeta_1^{m+r} + \sum_{s \geq 2} \sum_{j=0}^{s} u^j v^{s-j} A_{j,s-j}(\xi_1, .., \xi_j, \zeta_1, .., \zeta_{s-j}), \quad m > 0$$

(5.101)

The functions $A_{i,j}(\xi_1, \ldots, \xi_i, \zeta_1, \ldots, \zeta_j)$ in (5.100) and (5.101) are homogeneous polynomials in their variables, symmetric with respect to arguments $\xi_1, \ldots, \xi_i$ and $\zeta_1, \ldots, \zeta_j$. These functions can be explicitly determined in the terms of the system (5.54), the symbolic representation of system (5.46), from the compatibility conditions.

Let us first concentrate on how to compute the Lie bracket between the linear part of system (5.54) denoted by K^1, i.e.

$$K^1 = \begin{pmatrix} v\zeta_1^r \\ u\xi_1^{2n+1-r} \end{pmatrix} = L \begin{pmatrix} u \\ v \end{pmatrix}, \quad L = \begin{pmatrix} 0 & \eta^r \\ \eta^{2n+1-r} & 0 \end{pmatrix}$$

and any pair of differential polynomials. We know its symbolic representation takes a simple and elegant form if the matrix L is diagonal [23]. Inspired by this, we shall diagonalise matrix L, produce the required formula in new variables and then transform back to the original variables.

Notice that matrix L has two eigenvalues $\pm\eta^{n+\frac{1}{2}}$. Therefore, there exists a linear transformation

$$T = \begin{pmatrix} 1 & 1 \\ \eta^{n+\frac{1}{2}-r} & -\eta^{n+\frac{1}{2}-r} \end{pmatrix}$$

such that

$$T^{-1}LT = \text{diag}(\eta^{n+\frac{1}{2}}, -\eta^{n+\frac{1}{2}}).$$

Let us introduce new variables $\hat{u}$ and $\hat{v}$

$$\begin{pmatrix} u \\ v \end{pmatrix} = T \begin{pmatrix} \hat{u} \\ \hat{v} \end{pmatrix} = \begin{pmatrix} \hat{u} + \hat{v} \\ \hat{u}\xi_1^{n+\frac{1}{2}-r} - \hat{v}\zeta_1^{n+\frac{1}{2}-r} \end{pmatrix}.$$

Equally, we have

$$
\begin{pmatrix} \hat{u} \\ \hat{v} \end{pmatrix} = T^{-1} \begin{pmatrix} u \\ v \end{pmatrix} = \frac{1}{2} \begin{pmatrix} u + \eta^{-n-\frac{1}{2}+r}(v) \\ u - \eta^{-n-\frac{1}{2}+r}(v) \end{pmatrix}.
$$

The new variables $\hat{u}$ and $\hat{v}$ have the same weight, i.e., $w_{\hat{u}} = w_{\hat{v}} = w_u$. Without causing confusion we assign the same symbols ξ and ζ for the symbolic representation of the ring generated by $\hat{u}$, $\hat{v}$ and their derivatives. The exponents of symbols can be half-integer, which corresponds to half-differentiation in x-space.

Proposition 7 *In variables $\hat{u}$ and $\hat{v}$, system (5.54) takes the form*

$$
\begin{cases} \hat{u}_t = \hat{u}\xi_1^{n+\frac{1}{2}} + \frac{1}{2g} \sum_{s\geq 2} \sum_{l=0}^{s} \hat{u}^l \hat{v}^{s-l} \hat{a}_{l,s-l}(\xi_1,..,\xi_l,\zeta_1,..,\zeta_{s-l}) \\ \hat{v}_t = -\hat{v}\zeta_1^{n+\frac{1}{2}} - \frac{1}{2g} \sum_{s\geq 2} \sum_{l=0}^{s} \hat{u}^l \hat{v}^{s-l} \hat{a}_{l,s-l}(\xi_1,..,\xi_l,\zeta_1,..,\zeta_{s-l}) \end{cases},
$$

where $g = (\xi_1 + \cdots + \xi_l + \zeta_1 + \cdots + \zeta_{s-l})^{n+\frac{1}{2}-r}$ and $\hat{a}_{l,s-l}$ are defined in terms of $a_{i,s-i}(\xi_1,\ldots,\xi_i,\zeta_1,\ldots,\zeta_{s-i})$, $i = 0,\ldots,s$ as follows:

$$
\hat{a}_{l,s-l}(\xi_1,..,\xi_l,\zeta_1,..,\zeta_{s-l}) = \sum_{i=0}^{s} \sum_{p=max\{0,l-s+i\}}^{min\{i,l\}} C_i^p C_{s-i}^{l-p}
$$
$$
(-1)^{s-i-l+p} \langle\langle a_{i,s-i}(\xi_1,..,\xi_p,\zeta_1,..,\zeta_{i-p},\xi_{p+1},\ldots,\xi_l,\zeta_{i-p+1},..,\zeta_{s-l})
$$
$$
(\xi_{p+1}\cdots\xi_j\zeta_{i-p+1}\cdots\zeta_{s-l})^{n+\frac{1}{2}-r}\rangle S_l^\xi\rangle S_{s-l}^\zeta,
$$

where C_i^j are binomial coefficients defined by $C_i^j = \frac{i!}{j!(i-j)!}$.

In variables $\hat{u}$ and $\hat{v}$, the linear parts of symmetries (5.100) and (5.101) of system (5.46) are also diagonal matrices. We can now derive the symmetry conditions for the transformed forms as in Theorem 2 for the scalar case. This leads to the explicit recursive relations between symmetry (5.100) or (5.101) and system (5.46).

With explicit formulas at hand, we can prove the following theorem, crucial for global classification:

Theorem 17 *Assume the homogeneous system (5.46) ($p = 2n+1$) with $w_u > 0$ possesses a symmetry. Suppose there is another system of the same weight and of the same form*

$$
\begin{cases} u_t = v\zeta_1^r, \\ v_t = u\xi_1^{2n+1-r} + \sum_{k\geq 2} \sum_{i=0}^{k} u^i v^{k-i} b_{i,k-i}(\xi_1,..,\xi_k,\zeta_1,..,\zeta_{k-i}), \end{cases} \quad (5.102)
$$

whose quadratic terms equal to those of (5.46), that is, $b_{i,2-i}(x,y) = a_{i,2-i}(x,y)$, $i = 0,1,2$. Then if system (5.102) possesses a symmetry of the same order, then equations (5.102) and (5.46) are equal and share the same symmetry.

This can be viewed as another version of Theorem 4 for the systems case: the existence of infinitely many approximate symmetries of low degree together with one symmetry implies integrability.

Now we formulate the classification theorem:

Theorem 18 *If a homogeneous system (5.46) with a positive weight $w_u > 0$ of odd order ($p = 2n+1$) possesses a hierarchy of infinitely many higher symmetries, then it is one of the systems in the following list up to re-scaling $u \mapsto \alpha u, v \mapsto \beta v, t \mapsto \gamma t, x \mapsto \delta x$, where $\alpha, \beta, \gamma, \delta \in \mathbb{C}$:*

$$\begin{cases} u_t = v_1, \\ v_t = u_2 + 3uv_1 + vu_1 - 3u^2u_1, \end{cases}$$

$$\begin{cases} u_t = v_1, \\ v_t = (D_x + u)^{2n}(u) - v^2, \quad n = 1, 2, 3, \ldots. \end{cases}$$

These systems can be rewritten in the form of non-evolutionary equations if we introduce a new variable $u = w_x$:

$$w_{tt} = w_{xxx} + 3w_x w_{t,x} + w_{xx} w_t - 3w_x^2 w_{xx}.$$

$$w_{tt} = (\partial_x + w_x)^{2n}(w_x) - w_t^2.$$

The first equation is known to be integrable [53]. The second equation can be brought into linear form $f_{tt} = \partial_x^{2n+1} f$ by the Cole-Hopf transformation $w = \log(f)$. Its symmetries are given by

$$u_{\tau_m} = D_x(D_x + u)^{m-1} u, \quad m = 2, 3, \ldots$$

and

$$u_{\tau_m} = D_x(v + D_t)(D_x + u)^{m-1} u.$$

These symmetries correspond to the symmetries $f_{\tau_m} = f_m$ and $f_{\tau_m} = f_{t,m}$ of the equation $f_{tt} = f_{2n+1}$.

5.6 Symmetry structure of $(2 + 1)$–dimensional integrable equations

This section is devoted to the study of $(2 + 1)$–dimensional integrable equations. A famous example is the Kadomtsev-Petviashvili (KP) equation

$$u_t = u_{xxx} + 6uu_x + 3D_x^{-1}u_{yy}.$$

One of the main obstacles to extending the spectacular classification results of $(1+1)$-dimensional integrable equations to the $(2+1)$-dimensional case is that the equations themselves, their higher symmetries and conservation laws are non-local, i.e. they contain integral operators D_x^{-1} or D_y^{-1}. In 1998, Mikhailov and Yamilov, [45], introduced the concept of quasi-local functions based on the observation that the operators D_x^{-1} and D_y^{-1} never appear alone but always in pairs like $D_x^{-1} D_y$ and $D_y^{-1} D_x$ for all known integrable equations and their hierarchies of symmetries and conservation laws, which enabled them to extend the symmetry approach of testing integrability [12] to the $(2+1)$-dimensional case.

The results in this section are based on a recent paper [35]. We develop the symbolic representation method to derive the hierarchies of $(2+1)$-dimensional integrable equations from the scalar Lax operators and to study their properties globally. We prove that these hierarchies are indeed quasi-local as conjectured by Mikhailov and Yamilov in 1998, [45].

5.6.1 Quasi-local polynomials and Symbolic representation

The basic definitions and notations of the ring of (commutative) differential polynomials are similar to those in section 5.2.1. The derivatives of the dependent variable u with respect to its independent variables x and y are denoted by $u_{ij} = \partial_x^i \partial_y^j u$. For small values of i and j, we sometimes write the indices out explicitly, that is u_{xxy} and u instead of u_{21} and u_{00}.

A differential monomial takes the form $u_{i_1 j_1} u_{i_2 j_2} \cdots u_{i_n j_n}$. We call n the degree of the monomial. Let $\mathcal{R}^n$ denote the set of differential polynomials of degree n. The ring of differential polynomials is denoted by $\mathcal{R} = \oplus_{n \geq 1} \mathcal{R}^n$. It is a differential ring with total x-derivation and y-derivation

$$D_x = \sum_{i,j \geq 0} u_{i+1,j} \frac{\partial}{\partial u_{ij}} \quad \text{and} \quad D_y = \sum_{i,j \geq 0} u_{i,j+1} \frac{\partial}{\partial u_{ij}}.$$

Let us denote

$$\Theta = D_x^{-1} D_y, \qquad \Theta^{-1} = D_y^{-1} D_x.$$

The concept of **quasi-local (commutative) polynomials** $\mathcal{R}_\Theta$ was introduced in [45]. Its definition is similar to the ring extension of $\mathcal{R}_H$

as in section 5.4.1. We consider a sequence of extensions of $\mathcal{R}$ as follows:

$$\mathcal{R}_{\Theta^0} = \mathcal{R}, \quad \mathcal{R}_{\Theta^{n+1}} = \overline{\mathcal{R}_{\Theta^n} \bigcup \Theta(\mathcal{R}_{\Theta^n}) \bigcup \Theta^{-1}(\mathcal{R}_{\Theta^n})},$$

where the sets

$$\Theta(\mathcal{R}_{\Theta^n}) = \{\Theta(f); f \in \mathcal{R}_{\Theta^n}\}, \quad \Theta^{-1}(\mathcal{R}_{\Theta^n}) = \{\Theta^{-1}(f); f \in \mathcal{R}_{\Theta^n}\}$$

and the horizontal line denotes the ring closure. Each $\mathcal{R}_{\Theta^n}$ is a ring and the index n indicates the nesting depth of the operators $\Theta^{\pm 1}$. Clearly, we have

$$\mathcal{R}_{\Theta^0} \subset \mathcal{R}_{\Theta^1} \subset \mathcal{R}_{\Theta^2} \subset \cdots \subset \mathcal{R}_{\Theta^n} \subset \cdots \subset \mathcal{R}_{\Theta^\infty} = \mathcal{R}_\Theta.$$

To define the symbolic representation of $\mathcal{R}$, we replace u_{ij} by $\hat{u}\xi^i \eta^j$, where ξ and η are symbols (comparable to the definitions in section 5.2.3).

Definition 10 *The symbolic representation of a differential monomial is defined as*

$$u_{i_1,j_1} u_{i_2,j_2} \cdots u_{i_n,j_n} \longmapsto \frac{\hat{u}^n}{n!} \sum_{\sigma \in \mathcal{S}_n} \xi_{\sigma(1)}^{i_1} \eta_{\sigma(1)}^{j_1} \cdots \xi_{\sigma(n)}^{i_n} \eta_{\sigma(n)}^{j_n} .$$

The result of the action of operators $\Theta^{\pm 1}$ on a monomial

$$\hat{u}^k a(\xi_1, \ldots, \xi_k, \eta_1, \ldots, \eta_k)$$

in the symbolic representation is given by

$$\hat{u}^k a(\xi_1, \ldots, \xi_k, \eta_1, \ldots, \eta_k) \left(\frac{\eta_1 + \cdots + \eta_k}{\xi_1 + \cdots + \xi_k} \right)^{\pm 1} .$$

By induction, we can define the symbolic representation of any element in $\mathcal{R}_\Theta$, which is a rational function with its denominator being the products of the linear factors. The expression of the denominator uniquely determines how the operator $\Theta^{\pm 1}$ is nested. For example, the symbol of $u_x(\Theta u)\Theta^{-1}u \in \mathcal{R}_{\Theta^1}$ is

$$\frac{\hat{u}^3}{3}(\xi_1 \frac{\eta_2}{\xi_2} \frac{\xi_3}{\eta_3} + \xi_2 \frac{\eta_1}{\xi_1} \frac{\xi_3}{\eta_3} + \xi_3 \frac{\eta_2}{\xi_2} \frac{\xi_1}{\eta_1}) .$$

The symbolic representations of pseudo-differential operators in the $(2+1)$–dimensional case are similar to the case of one spatial variable in section [35]. However, we assign a special symbol X to the operator D_x in analogy to the symbol Y for the operator D_y.

The extension of the symbolic representation from one dependent variable to several dependent variables is straightforward. We need to assign new symbols for each of them such as assigning $\xi^{(1)}, \eta^{(1)}$ for u and $\xi^{(2)}, \eta^{(2)}$ for v and so on.

5.6.2 Lax formulation of $(2+1)$–dimensional integrable equations

We give a short description of the construction of $(2 + 1)$-dimensional integrable equations from a given scalar Lax operator based on the well-known Sato approach. For details on the Sato approach for the $(1 + 1)$-dimensional case, see the recent books [62, 63] and related references in them.

Let H be an m-th order pseudo-differential operator in two spatial variables of the form

$$H = -D_y + a_m D_x^m + a_{m-1} D_x^{m-1} + \cdots + a_0 + a_{-1} D_x^{-1} + \cdots, \quad m \geq 0,$$

where the coefficients a_k are functions of x, y. Let the commutator be the bracket on the set of pseudo-differential operators. Thus, the set of pseudo-differential operators forms a Lie algebra. For an integer $k < m$, we split into

$$H_{\geq k} = a_m D_x^m + a_{m-1} D_x^{m-1} + \cdots + a_k D_x^k$$
$$H_{<k} = H - H_{\geq k} = -D_y + a_{k-1} D_x^{k-1} + \cdots$$

This operator algebra decomposes as a direct sum of two subalgebras in both commutative and noncommutative cases when $k \in \{0, 1\}$. Similarly to the (1+1)-dimensional case, such decompositions are naturally related with integrability and lead to admissible scalar Lax operators for the case of $(2 + 1)$ dimensions:

a. $k = 0:$ $n \geq 2,$
$$L = D_x^n + u^{(n-2)} D_x^{n-2} + u^{(n-3)} D_x^{n-3} + \cdots + u^{(0)} - D_y;$$
b. $k = 1:$ $n \geq 2,$
$$L = D_x^n + u^{(n-1)} D_x^{n-1} + u^{(n-2)} D_x^{n-2} + \cdots + u^{(0)} + D_x^{-1} u^{(-1)} - D_y;$$
c. $k = 1:$ $L = u^{(0)} + D_x^{-1} u^{(-1)} - D_y;$

where $u^{(i)}$ are functions of two spatial variables x, y. We often use $u, v, w, \cdots$ in the examples. For the KP equation, the Lax operator is the case **a** when $n = 2$, namely, $L = D_x^2 + u - D_y$.

Let $S = D_x + a_0 + a_{-1}D_x^{-1} + \cdots$. For any operator L listed in cases **a**, **b** and **c**, the relation

$$[S, \; L] := SL - LS = 0, \tag{5.103}$$

uniquely determines the operator S by taking the integration constants to be zeros. Furthermore, we have $[S^n, \; L] = 0$ for any $n \in \mathbb{N}$. For each choice of i, we introduce a different time variable t_i and define the Lax equation by

$$\frac{\partial L}{\partial t_i} = [S^i_{\geq k}, L], \tag{5.104}$$

where k is determined by the operator L as listed in cases **a**, **b** and **c**.

Theorem 19 *For the operator S uniquely determined by (5.103), the flows defined by Lax equations (5.104) commute, i.e., $\partial_{t_j}\partial_{t_i} L = \partial_{t_i}\partial_{t_j} L$.*

5.6.3 Lax formulation in symbolic representation

We put the formalism of section 5.6.2 into the symbolic form. The strategy is to do the calculation as much as possible without symmetrisation and only perform the symmetrisation at the last stage to get the uniqueness of the symbolic representation since the symmetrisation complicates the calculation dramatically.

Let us assign the symbols $\xi^{(i)}, \eta^{(i)}$ for dependent variable $u^{(i)}$. The symbolic representations of the admissible scalar Lax operators are

   a. $k = 0: \quad n \geq 2,$
        $L = X^n - Y + \hat{u}^{(n-2)}X^{n-2} + \hat{u}^{(n-3)}X^{n-3} + \cdots + \hat{u}^{(0)} \; ;$
   b. $k = 1: \quad n \geq 2,$
        $\hat{L} = X^n - Y + \hat{u}^{(n-1)}X^{n-1} + \cdots + \hat{u}^{(0)} + \hat{u}^{(-1)}\frac{1}{X+\xi^{(-1)}} \; ;$
   c. $k = 1: \quad \hat{L} = -Y + \hat{u}^{(0)} + \hat{u}^{(-1)}\frac{1}{X+\xi^{(-1)}};$

Here we only treat the case **a**. The study of the cases **b** and **c** can be found in [35].

It is convenient to consider formal series in the form

$$S = X + \sum_{i=0}^{n-2} \hat{u}^{(i)}a_1^{(i)}(\xi_1^{(i)}, \eta_1^{(i)}, X) \tag{5.105}$$

$$+ \; \sum_{i_1=0}^{n-2}\sum_{i_2=0}^{n-2} \hat{u}^{(i_1)}\hat{u}^{(i_2)}a_2^{(i_1 i_2)}(\xi_{j_1}^{(i_1)}, \eta_{j_1}^{(i_1)}, \xi_{j_2}^{(i_2)}, \eta_{j_2}^{(i_2)}, X) + \cdots,$$

where $n \geq 2$ and a_i are functions of their specific arguments, the superindex $i_s \in \{0, 1, 2, \cdots, n-2\}$ and the subindex j_k is defined by the number of i_k in the list of $[i_1, i_2, \cdots, i_l]$. This implies that $j_1 = 1$ and

$j_k \geq 1$, $k = 1, 2, \cdots$. For example, when $i_1 = i_2$, the arguments of the function $a_2^{i_1 i_1}$ are $\xi_1^{(i_1)}$, $\eta_1^{(i_1)}$, $\xi_2^{(i_1)}$, $\eta_2^{(i_1)}$ and X.

It is easy to check that

$$[X^n - Y, \; \phi(\xi_{j_1}^{(i_1)}, \eta_{j_1}^{(i_1)}, \cdots, \xi_{j_l}^{(i_l)}, \eta_{j_l}^{(i_l)}, X)] =$$
$$= \; N_l(\xi_{j_1}^{(i_1)}, \eta_{j_1}^{(i_1)}, \cdots, \xi_{j_l}^{(i_l)}, \eta_{j_l}^{(i_l)}, X)\phi,$$

where the polynomial N_l is defined by

$$N_l(\xi_1, \eta_1, \xi_2, \eta_2, \ldots, \xi_l, \eta_l; X) = (\sum_{i=1}^{l} \xi_i + X)^n - X^n - \sum_{i=1}^{l} \eta_i \; .$$

Proposition 8 *For any operator L in case* **a**, *if the formal series (5.105) satisfies the relation $[S, \; L] = 0$ (cf. (5.103)), we have for $l \geq 1$,*

$$a_l^{(i_1 i_2 \cdots i_l)} = a_l(\xi_{j_1}^{(i_1)}, \eta_{j_1}^{(i_1)}, \cdots, \xi_{j_l}^{(i_l)}, \eta_{j_l}^{(i_l)}, X)$$
$$= \prod_{r=1}^{l}(X + \sum_{s=r+1}^{l} \xi_{j_s}^{(i_s)})^{i_r} b_l(\xi_{j_1}^{(i_1)}, \eta_{j_1}^{(i_1)}, \ldots, \xi_{j_l}^{(i_l)}, \eta_{j_l}^{(i_l)}, X).$$

The function b_l, $l \geq 1$, is defined by

$$b_l(\xi_{j_1}^{(i_1)}, \eta_{j_1}^{(i_1)}, \cdots, \xi_{j_l}^{(i_l)}, \eta_{j_l}^{(i_l)}, X) = \frac{c_l(\xi_{j_1}^{(i_1)}, \eta_{j_1}^{(i_1)}, \ldots, \xi_{j_l}^{(i_l)}, \eta_{j_l}^{(i_l)}, X)}{N_l(\xi_{j_1}^{(i_1)}, \eta_{j_1}^{(i_1)}, \ldots, \xi_{j_l}^{(i_l)}, \eta_{j_l}^{(i_l)}, X)}$$

with

$$c_l(\xi_{j_1}^{(i_1)}, \eta_{j_1}^{(i_1)}, \cdots, \xi_{j_l}^{(i_l)}, \eta_{j_l}^{(i_l)}, X)$$
$$= b_{l-1}(\xi_{j_1}^{(i_1)}, \eta_{j_1}^{(i_1)} \cdots, \xi_{j_{l-1}}^{(i_{l-1})}, \eta_{j_{l-1}}^{(i_{l-1})}, X + \xi_{j_l}^{(i_l)})$$
$$- b_{l-1}(\xi_{j_2}^{(i_2)}, \eta_{j_2}^{(i_2)}, \cdots, \xi_{j_l}^{(i_l)}, \eta_{j_l}^{(i_l)}, X), \quad l > 1$$

and the initial function $c_1(\xi_1^{(i)}, \eta_1^{(i)}, X) = \xi_1^{(i)}$.

To construct the hierarchy of the Lax equations we need to expand the coefficients of the operator (5.105) at $X \to \infty$ and truncate at the required degree. When $n \geq 2$, the expansion of $N_l(\xi_1, \eta_1, \xi_2, \eta_2, \ldots, \xi_l, \eta_l; X)^{-1}$ at $X \to \infty$ is of the form

$$N_l(\xi_1, \eta_1, \xi_2, \eta_2, \ldots, \xi_l, \eta_l; X)^{-1}$$
$$= \frac{1}{nX^{n-1}(\sum_{i=1}^{l} \xi_i)} \sum_{j \geq 0} \left\{ \frac{\sum_{i=1}^{l} \eta_i}{nX^{n-1}(\sum_{i=1}^{l} \xi_i)} \right.$$
$$\left. - \frac{1}{n} \sum_{k=0}^{n-2} C_n^k X^{k+1-n} (\sum_{i=0}^{l} \xi_i)^{n-k-1} \right\}^j, \quad n \geq 2.$$

Therefore, if we want to prove that the coefficients of operators (5.105), i.e. the functions a_l, are quasi-local defined similarly as in definition 7, we need to show that the functions c_l can be split into the sum of the image of D_x and the image of D_y. It is clear that $a_1^{(i)}(\xi_1^{(i)}, \eta_1^{(i)}, X)$ are quasi-local since we have $c_1(\xi_1^{(i)}, \eta_1^{(i)}, X) = \xi_1^{(i)}$. When $l > 1$, we have

Proposition 9 *The functions* $c_l(\xi_{j_1}^{(i_1)}, \eta_{j_1}^{(i_1)}, \cdots, \xi_{j_l}^{(i_l)}, \eta_{j_l}^{(i_l)}, X)$ *for* $l > 1$
vanish after substitution

$$\xi_{j_1}^{(i_1)} = -\xi_{j_2}^{(i_2)} - \cdots - \xi_{j_{l-1}}^{(i_{l-1})} \quad and \quad \eta_{j_1}^{(i_1)} = -\eta_{j_2}^{(i_2)} - \cdots - \eta_{j_{l-1}}^{(i_{l-1})}.$$

In fact, this proposition does not lead to our intended conclusion that b_l and thus a_l are quasi-local since the objects are rational, not polynomial. For example, the expression $u^2(\frac{\eta_2}{\xi_2} - \frac{\eta_1}{\xi_1})$ representing $u\Theta u - (\Theta u)u$ satisfies the above proposition. However, we cannot write $u\Theta u - (\Theta u)u = D_x f_1 + D_y f_2$, where both f_1 and f_2 are in $\mathcal{R}_\Theta$.

Notice that the formulae in Proposition 8 are without symmetrisation. Combining these expressions, we can obtain the formulae of high degree terms of the operator S when dependent variables are commuting. Every term in such S is quasi-local. This implies that every term in S^n is quasi-local. From Theorem 19 follows

Theorem 20 *The hierarchies of commutative (2 + 1)-integrable equations with scalar Lax operators are quasi-local.*

The above setting up is valid for the noncommutative case except Theorem 20. However, the extension of the concept of quasi-locality to the noncommutative case is rather complicated. D_x and D_y are the only derivations for the commutative differential ring. The extension simply enables us to apply D_x^{-1} and D_y^{-1} on the derivations. We know that the commutators are also derivations for a noncommutative associative algebra and we need to take them into consideration. There are some further discussions on this topic in [35].

Summary and discussion

In this article we have reviewed some recent developments in the symmetry approach in the symbolic representation. In particular we have discussed two different methods. One method is based on the study of conditions for the existence of a formal recursion operator, another one is based on the explicit analysis of approximate symmetries of low degrees.

The formal recursion method is a very efficient tool for testing the integrability of a given PDE. This method is based only on the fact of existence of an infinite hierarchy of higher symmetries and is not sensitive to possible lacunas in the hierarchy of symmetries. It is extendable to wide classes of equations, including certain types of non-local equations. The method is rather simple and convenient for classification of integrable systems of a fixed order. We have illustrated its power in applications to classification of integrable generalisations of Boussinesq, Benjamin-Ono and Camassa-Holm type equations.

Explicit analysis of approximate symmetries relies on the structure of dispersion relations (linear terms) of systems. We have seen that the structure of symmetries of a given equation is parametrised by their dispersion relations and the analysis of existence of approximate symmetries is based on divisibility properties of special polynomials determined by the dispersion laws. Such divisibility properties are often obtained via algebraic geometry and number theoretic methods. It gives detailed information on the structure of the hierarchy of higher symmetries. So far it is the only method which has proved to be suitable for a global classification. Here we mean a classification of integrable equations of all orders. In the frame of this method it has been demonstrated that any integrable homogeneous evolutionary equation

$$u_t = u_n + f(u_{n-1}, \ldots, u) \qquad n \geq 2,$$

where $w(u) \geq 0$, is a symmetry (a member of a hierarchy) of one of the equations of order $2, 3$ or 5 presented in Theorem 9.

A system of equations is a considerably more complicated object. Here the only "global" result available is a classification of integrable homogeneous Boussinesq type equations of odd order (Section 5.5.2). In the theory of integrable systems the description of admissible structures of linear terms (the dispersion laws) for equations and their symmetries is an important and as yet unsolved problem. It is the so-called *spectrum* or *dispersion* problem. Let us consider a system of PDEs

$$\mathbf{u}_t = A\mathbf{u}_n + \mathbf{F}(\mathbf{u}_{n-1}, \ldots, \mathbf{u}), \quad \mathbf{u} = (u_1, \ldots, u_N)^T, \quad n \geq 2 \qquad (5.106)$$

where A is a constant $N \times N$ matrix. Its higher symmetry, if it exists, is of the form

$$\mathbf{u}_{\tau_m} = B(m)\mathbf{u}_m + \mathbf{G}(\mathbf{u}_{m-1}, \ldots, \mathbf{u}), \qquad m \geq 2$$

where $B(m)$ is a constant matrix commuting with A. Let us denote by $\lambda_1, \ldots, \lambda_N$ the eigenvalues of the matrix A and assume that $\lambda_1 \neq 0$, then the set $S_A = \{\lambda_2/\lambda_1, \ldots, \lambda_N/\lambda_1\}$ is called the *spectrum* of system (5.106). Similarly define $S_{B(m)} = \{\mu_2(m)/\mu_1(m), \ldots, \mu_N(m)/\mu_1(m)\}$, the spectrum of symmetry, where $\mu_1(m), \ldots, \mu_N(m)$ are eigenvalues of the matrix $B(m)$. The spectrum is invariant with respect to any re-scaling. In many respects it reflects properties of symmetries, conservation laws and solutions of the system. For example, the existence of higher conservation laws for a system of two even order equations implies that $S_A = \{-1\}$ (see [20]).

The integrable system [57]

$$\begin{cases} u_t = (5 - 3\sqrt{5})u_3 - 2uu_1 + (3 - \sqrt{5})vu_1 + 2uv_1 + (1 + \sqrt{5})vv_1, \\ v_t = (5 + 3\sqrt{5})v_3 + (1 - \sqrt{5})uu_1 + 2vu_1 + (3 + \sqrt{5})uv_1 - 2vv_1, \end{cases}$$

has spectrum $S_A = \{\lambda_2/\lambda_1\} = \{-\frac{1}{2}(7 + 3\sqrt{5})\}$. It possesses an infinite dimensional algebra of symmetries of orders $m \equiv 1, 3, 7, 9 \mod 10$. The ratio of parameters $\mu_2(m)/\mu_1(m)$ is given by

$$\frac{\mu_2(m)}{\mu_1(m)} = \frac{\left(1 + \exp\left(\frac{2\pi i}{5}\right)\right)^m}{1 + \exp\left(\frac{2m\pi i}{5}\right)}.$$

The symmetry approach in symbolic representation can be applied to the study of possible spectra and classification of integrable systems. For systems of two homogeneous differential polynomial equations of second order the problem has been solved in [23]. Recently we have been working on this problem for systems of two equations of odd order. The results obtained will be published elsewhere. Here we present two rather non-trivial examples of integrable systems, which we believe are new.

The following system

$$\begin{cases} u_t & = (9 - 5\sqrt{3})u_5 + D_x\left\{2(9 - 5\sqrt{3})uu_2 + (-12 + 7\sqrt{3})u_1^2\right\} \\ & \quad +2(3 - \sqrt{3})u_3v + 2(6 - \sqrt{3})u_2v_1 + 2(3 - 2\sqrt{3})u_1v_2 \\ & \quad -6(1 + \sqrt{3})uv_3 + D_x\left\{2(33 + 19\sqrt{3})vv_2 + (21 + 12\sqrt{3})v_1^2\right\} \\ & \quad +\frac{4}{5}(-12 + 7\sqrt{3})u^2u_1 + \frac{8}{5}(3 - 2\sqrt{3})(vuu_1 + u^2v_1) \\ & \quad +\frac{4}{5}(24 + 13\sqrt{3})v^2u_1 + \frac{8}{5}(36 + 20\sqrt{3})uvv_1 - \frac{8}{5}(45 + 26\sqrt{3})v^2v_1, \\[2mm] v_t & = (9 + 5\sqrt{3})v_5 + D_x\left\{2(33 - 19\sqrt{3})uu_2 + (21 - 12\sqrt{3})u_1^2\right\} \\ & \quad -6(1 - \sqrt{3})u_3v + 2(3 + 2\sqrt{3})u_2v_1 + 2(6 + \sqrt{3})u_1v_2 \\ & \quad +2(3 + \sqrt{3})uv_3 + D_x\left\{2(9 + 5\sqrt{3})vv_2 - (12 + 7\sqrt{3})v_1^2\right\} \\ & \quad -\frac{8}{5}(45 - 26\sqrt{3})u^2u_1 + \frac{8}{5}(36 - 20\sqrt{3})vuu_1 + \frac{4}{5}(24 - 13\sqrt{3})u^2v_1 \\ & \quad +\frac{8}{5}(3 + 2\sqrt{3})(v^2u_1 + uvv_1) - \frac{4}{5}(12 + 7\sqrt{3})v^2v_1 \end{cases}$$

possesses an infinite dimensional algebra of higher symmetries with

$$\frac{\mu_2(m)}{\mu_1(m)} = \frac{(1 + \exp(\frac{\pi i}{6}))^m}{1 + \exp(\frac{m\pi i}{6})}, \qquad m \equiv 1, 5, 7, 11 \mod 12.$$

The system

$$\begin{cases} u_t &= -\frac{5}{3}u_5 - 10vv_3 - 15v_1v_2 + 10uu_3 + 25u_1u_2 - 6v^2v_1 \\ &\quad +6v^2u_1 + 12uvv_1 - 12u^2u_1, \\ v_t &= 15v_5 + 30v_1v_2 - 30v_3u - 45v_2u_1 - 35v_1u_2 - 10vu_3 \\ &\quad -6v^2v_1 + 6v^2u_1 + 12u^2v_1 + 12vuu_1. \end{cases}$$

possesses symmetries of orders $m \equiv 1, 5 \mod 6$ with

$$\frac{\mu_2(m)}{\mu_1(m)} = \frac{(1 + \exp(\frac{\pi i}{3}))^m}{1 + \exp(\frac{m\pi i}{3})}.$$

There is a reduction $v = 0$ to the Kaup-Kupershmidt equation.

Bibliography

[1] Zakharov, V.E., Manakov, S.V., Novikov, S.P., and Pitaevskii, L.P. (1980). *Teoriya solitonov. Metod obratnoi zadachi (Solitons Theory. The Inverse Transform Method)* (Nauka, Moscow), in Russian.

[2] Ablowitz, M. and Segur, H. (1981). *Solitons and the Inverse Scattering Transform* (SIAM, Philadelphia).

[3] Newell, A. (1985). *Solitons in Mathematics and Physics* (SIAM, Philadelphia).

[4] Ablowitz, M.J. and Clarkson, P.A. (1991). *Solitons, nonlinear evolution equations and inverse scattering*, volume 149 (Cambridge University Press, Cambridge).

[5] Olver, P.J. (1993). *Applications of Lie groups to differential equations*, second edition, volume 107 of *Graduate Texts in Mathematics* (Springer-Verlag, New York).

[6] Bakirov, I. (1991). On the symmetries of some system of evolution equations (Akad. Nauk SSSR Ural. Otdel. Bashkir. Nauchn. Tsentr, Ufa).

[7] Beukers, F., Sanders, J.A., and Wang, J.P. (1998). One symmetry does not imply integrability, *J. Differential Equations* **146**(1), 251–260.

[8] van der Kamp, P.H. and Sanders, J.A. (2002). Almost integrable evolution equations, *Selecta Math. (N.S.)* **8**(4), 705–719.

[9] Mikhailov, A., Novikov, V., and Wang, J.P. (2005). Partially integrable nonlinear equations with one high symmetry, *J. Phys. A* **38**, L337–L341.

[10] Sokolov, V.V. and Shabat, A.B. (1984). Classification of integrable evolution equations, in *Mathematical physics reviews, Vol. 4*, volume 4 of *Soviet Sci. Rev. Sect. C: Math. Phys. Rev.*, pages 221–280 (Harwood Academic Publ., Chur).

[11] Mikhailov, A.V., Shabat, A.B., and Yamilov, R.I. (1987). A symmetric approach to the classification of nonlinear equations. Complete lists of integrable systems, *Uspekhi Mat. Nauk* **42**(4(256)), 3–53.

[12] Mikhailov, A.V., Shabat, A.B., and Sokolov, V.V. (1991). The symmetry approach to classification of integrable equations, in *What is integrability?*, Springer Ser. Nonlinear Dynamics, pages 115–184 (Springer, Berlin).

[13] Shabat, A.B. and Mikhailov, A.V. (1993). Symmetries—test of integrability, in *Important developments in soliton theory*, pages 355–374 (Springer, Berlin).

[14] Mikhailov, A.V., Shabat, A.B., and Yamilov, R.I. (1988). Extension of the module of invertible transformations. Classification of integrable systems, *Comm. Math. Phys.* **115**(1), 1–19.

[15] Svinolupov, S.I. (1985). Second order equations with symmetries, *Uspekhi Mat. Nauk* **40**(5), 263.

[16] Svinolupov, S.I. and Sokolov, V.V. (1982). Evolution equations with nontrivial conservation laws, *Funktsional. Anal. i Prilozhen.* **16**(4), 86–87.

[17] Drinfel'd, V., Svinolupov, S., and Sokolov, V. (1985). Classification of fifth-order evolution equations having an infinite series of conservation laws, *Dokl. Akad. Nauk Ukrain. SSR Ser. A* **10**, 8–10.

[18] Yamilov, R.I. (1983). On classification of discrete evolution equations, *Uspekhi Mat. Nauk* **38**, 155–156.

[19] Levi, D. and Yamilov, R. (1997). Conditions for the existence of higher symmetries of evolutionary equations on the lattice, *J. Math. Phys.* **38**(12), 6648–6674.

[20] Mikhailov, A.V. and Shabat, A.B. (1986). Conditions for integrability of systems of two equations of the form $u_t = A(u)u_{xx} + F(u, u_x)$. II, *Teoret. Mat. Fiz.* **66**(1), 47–65.

[21] Mikhailov, A.V. and Shabat, A.B. (1985). Conditions for integrability of systems of two equations of the form $u_t = A(u)u_{xx} + F(u, u_x)$. I, *Teoret. Mat. Fiz.* **62**(2), 163–185.

[22] Svinolupov, S.I. (1989). On the analogues of the Burgers equation, *Phys. Lett. A* **135**(1), 32–36.

[23] Sanders, J.A. and Wang, J.P. (2004). On the integrability of systems of second order evolution equations with two components, *J. Differential Equations* **203**(1), 1–27.

[24] Gel'fand, I.M. and Dikii, L.A. (1975). Asymptotic properties of the resolvent of Sturm-Liouville equations, and the algebra of Korteweg-de Vries equations, *Uspehi Mat. Nauk* **30**(5(185)), 67–100, English translation: *Russian Math. Surveys*, 30 (1975), no. 5, 77–113.

[25] Wang, J.P. (1998). *Symmetries and Conservation Laws of Evolution Equations*, Vrije Universiteit/Thomas Stieltjes Institute Ph.D. thesis, Amsterdam.

[26] Sanders, J.A. and Wang, J.P. (1998). On the integrability of homogeneous scalar evolution equations, *J. Differential Equations* **147**(2), 410–434.

[27] Sanders, J.A. and Wang, J.P. (2000). On the integrability of non-polynomial scalar evolution equations, *J. Differential Equations* **166**(1), 132–150.

[28] Olver, P.J. and Wang, J.P. (2000). Classification of integrable one-component systems on associative algebras, *Proc. London Math. Soc. (3)* **81**(3), 566–586.

[29] Mikhailov, A.V. and Novikov, V.S. (2002). Perturbative symmetry approach, *J. Phys. A* **35**(22), 4775–4790.

[30] Mikhailov, A., Novikov, V., and Wang, J.P. (2007). On classification of integrable non-evolutionary equations, *Stud. Appl. Math.* **118**, 419–457.

[31] Novikov, V. and Wang, J.P. (2007). Symmetry structure of integrable nonevolutionary equations, *Stud. Appl. Math.* **119**(4), 393–428.

[32] Mikhailov, A. and Novikov, V. (2003). Classification of integrable Benjamin-Ono-type equations, *Moscow Mathematical Journal* **3**(4), 1293–1305.

[33] Beukers, F., Sanders, J.A., and Wang, J.P. (2001). On integrability of

systems of evolution equations, *J. Differential Equations* **172**(2), 396–408.

[34] van der Kamp, P.H. (2003). *Symmetries of Evolution Equations: a Diophantine Approach*, Vrije Universiteit Ph.D. thesis, Amsterdam.

[35] Wang, J.P. (2006). On the structure of (2 + 1)–dimensional commutative and noncommutative integrable equations., *J. Math. Phys.* **47**(11), 113508.

[36] Fokas, A.S. (1980). A symmetry approach to exactly solvable evolution equations, *J. Math. Phys.* **21**(6), 1318–1325.

[37] Sokolov, V.V. and Wolf, T. (2001). Classification of integrable polynomial vector evolution equations, *J. Phys. A* **34**(49), 11139–11148.

[38] Beukers, F., Private communication.

[39] Fokas, A. (1987). Symmetries and integrability, *Studies in Applied Mathematics* **77**, 253–299.

[40] Cauchy, A. and Liouville, J. (1839). Rapport sur un mémoire de M. Lamé relatif au dernier théoréme de Fermat, *C. R. Acad. Sci. Paris* **9**, 359–363.

[41] Beukers, F. (1997). On a sequence of polynomials, *J. Pure Appl. Algebra* **117/118**, 97–103, Algorithms for algebra (Eindhoven, 1996).

[42] Mirimanoff, D. (1903). Sur l'équation $(x + 1)^l - x^l - 1 = 0$., *Nouv. Ann. Math.* **3**, 385–397.

[43] Tzermias, P. (2007). On Cauchy-Liouville-Mirimanoff polynomials, *Canad.Math. Bull.* **50**(2), 313–320.

[44] Sanders, J.A. and Wang, J.P. (1998). Combining Maple and Form to decide on integrability questions, *Comput. Phys. Comm.* **115**(2-3), 447–459.

[45] Mikhailov, A.V. and Yamilov, R.I. (1998). Towards classification of (2+1)-dimensional integrable equations. Integrability conditions. I, *J. Phys. A* **31**(31), 6707–6715.

[46] Camassa, R. and Holm, D. (1993). An integrable shallow water equation with peaked solutions, *Phys. Rev. Lett.* **71**, 1661–1664.

[47] Degasperis, A. and Procesi, M. (1999). Asymptotic integrability, in *SPT 1999: Symmetry and perturbation theory*, ed. A. Degasperis and G. Gaeta, pages 23–37 (World Sci. Publishing).

[48] Joseph, R. (1977). Solitary waves in a finite depth fluid, *Journal of Physics. A: Mathematical and General* **10**, L225.

[49] Ablowitz, M., Fokas, A., Satsuma, J., and Segur, H. (1982). On the periodic intermediate long wave equation, *Journal of Physics A: Mathematical and General* **15**, 781.

[50] Hone, A. and Novikov, V. (2004). On a functional equation related to the intermediate long wave equation, *Journal of Physics. A: Mathematical and General* **37**, L399–L406.

[51] Hone, A. and Wang, J.P. (2003). Prolongation algebras and Hamiltonian operators for peakon equations, *Inverse Problems* **19**(1), 129–145.

[52] Zakharov, V. (1974). On stochastization of one-dimensional chains of nonlinear oscillators, *Sov.Phys. JETP* **38**, 108–110.

[53] Hernández Heredero, R., Shabat, A.B., and Sokolov, V.V. (2003). A new class of linearizable equations, *J. Phys. A* **36**(47), L605–L614.

[54] Sato, M. and Sato, Y. (1982). Soliton equations as dynamical systems on infinite dimensional Grassman manifolds, in *Lect. Notes in Num. Appl. Anal.*, volume 5, pages 259–271).

[55] Jimbo, M. and Miwa, T. (1983). Solitons and infinite dimensional Lie algebras, in *Publ. RIMS, Kyoto*, volume 19, pages 943–1001).

[56] Drinfel'd, V.G. and Sokolov, V.V. (1984). Lie algebras and equations of Korteweg– de Vries type, in *Current problems in mathematics, Vol. 24*, Itogi Nauki i Tekhniki, pages 81–180 (Akad. Nauk SSSR Vsesoyuz. Inst. Nauchn. i Tekhn. Inform., Moscow).

[57] Drinfel'd, V.G. and Sokolov, V.V. (1981). New evolution equations having an (*L*, *A*) pair, in *Partial differential equations*, volume 81 of *Trudy Sem. S. L. Soboleva, No. 2*, pages 5–9 (Akad. Nauk SSSR Sibirsk. Otdel. Inst. Mat., Novosibirsk).

[58] Shabat, A. (2005). Universal Solitonic Hierarchy, *J. Nonlinear Math. Phys.* **12**, 614–624.

[59] Hone, A., Novikov, V., and Verhoeven, C. (2006). An integrable hierarchy with a perturbed Henon-Heiles system, *Inverse Problems* **22**, 2001–2020.

[60] Hone, A., Novikov, V., and Verhoeven, C. (2008). An extended Henon-Heiles system, *to appear in Physics Letters A* .

[61] Sergeev, A. (2006). Zero curvature representation for a new fifth-order integrable system, *Fundam. Prikl. Mat* **12**(7), 227–229, arXiv:nlin.SI/0604064.

[62] Blaszak, M. and Szum, A. (2001). Lie algebraic approach to the construction of (2 + 1)– dimensional lattice–field and field integrable Hamiltonian equations, *J. Math. Phys.* **42**(1), 225–259.

[63] Kupershmidt, B. (2000). *KP or mKP: Noncommutative Mathematics of Lagrangian, Hamiltonian, and Integrable Systems*, volume 78 of *Mathematical Surveys and Monographs* (Amer. Math. Soc., Providence, RI).

6

Searching for integrable (P)DEs using algebraic conditions: two examples

Jarmo Hietarinta

Department of Physics, University of Turku,
FIN-20014 Turku, Finland

Abstract

Integrable dynamical systems are both rare and ubiquitous. It is therefore a worthy pursuit to find new integrable systems. In this contribution we give two examples of such search projects and discuss the computational algebraic problems this leads to.

6.1 Definitions of integrability

We will be discussing here the problem of searching for integrable systems. But at the very beginning it should be made clear that there is no *unique* definition of "integrability". In broad terms we can say that an equation is integrable if its *solutions behave nicely* (cf. Painlevé property). However, this is not an operational definition, because we do not normally know all the solutions. Thus we have to look for certain properties of the equation, namely such properties that in some way imply "niceness" of its solutions.†

One important observation that has been made about integrable systems is that there is always some interesting underlying mathematics. In a way this implies that among all equations the integrable ones are of measure zero, somewhat like prime numbers are among all numbers.

Since integrable dynamical system are rare it is of interest to *search for and classify* them. This is one of the important themes of research into integrable systems.

Different classes of dynamical systems have different properties that

---

† Note, however, that the existence of a closed form explicit solution is not equivalent to integrability: For example the logistic map $y_{n+1} = 4y_n(1 - y_n)$ has the explicit closed form solution $y_n = \frac{1}{2}[1 - \cos(2^n c)]$, which shows sensitive dependence on the initial value c and is therefore chaotic.

can be associated with integrability. In principle any property that guarantees regular behavior is of interest. The property can be directly associated with the equation:

- for sets of ordinary differential equations: sufficient number of conserved quantities
- for evolution equations: linearization through a Lax pair
- for almost all systems: symmetries
- 2D-lattice partial difference equations: consistent extension to 3D.

or with some calculable property of its solutions:

- for almost all systems: the absence of nasty singularities (Painlevé property)
- for ordinary difference equations: low growth of complexity (Nevanlinna theory)
- for soliton systems: the existence of multi-soliton solutions

Any of the above properties can be taken as the *key property* and a search can be organized around that chosen property. One just needs to choose a suitable class of dynamical systems on which the property can be applied. Indeed, for each property there are natural assumptions which restrict the class of equations under study. In practice applying "the method" on "the class" produces a large over-determined set of equations, which must then be solved. Usually this set of equations can at best be solved by computer algebra systems, but it should be stated that even fairly simple over-determined sets of equations are still unsolvable by modern computer methods.

In the following we will take a more detailed look at two aspects of integrability and the algebraic problems that are met in searching for integrable systems within those definitions.

6.2 Liouville integrability

6.2.1 Definitions

Liouville integrability is the suitable definition of integrability for Hamiltonian systems with a finite number of degrees of freedom. The dynamical variables consist of N coordinates $q_i(t)$ and N momenta $p_i(t)$, which are assumed to be C^∞-functions of time t, and the dynamics (time evolution) is given by a Hamiltonian $H = H(p, q)$ as follows:

$$\left\{ \begin{array}{l} \dot{q}_i = \{H, q_i\}, \\ \dot{p}_i = \{H, p_i\}, \end{array} \right. \quad \text{where} \quad \{A, B\} := \sum_{j=1}^{N} \left(\frac{\partial A}{\partial p_j} \frac{\partial B}{\partial q_j} - \frac{\partial A}{\partial q_j} \frac{\partial B}{\partial p_j} \right).$$

$\{\bullet, \bullet\}$ is called the *Poisson bracket*.

Definition 1 *A Hamiltonian $H(p,q)$ is said to be* **integrable** *if there are N functions $I_k(p,q)$ (H one of them) such that the I_k*
  (i) *are functionally independent,*
  (ii) *are in involution, i.e., $\{I_n, I_m\} = 0$, $\forall n, m$,*
  (iii) *are sufficiently regular.*

The usefulness of this definition follows from the *Liouville-Arnold theorem*, which states that a Liouville integrable system can indeed be integrated by quadratures [1].

Since $\{H, I_k\} = 0$ the I_k are called "constants of motion" or "conserved quantities". If there are M of them, then the different levels of integrability are:

- partially integrable if $M < N$,
- integrable, if $M = N$,
- superintegrable, if $M > N$ (in which case the I_k all commute with H, but they cannot all mutually commute).

Many examples of Liouville integrable systems are known, but an exhaustive classification exists only when $N = 2, 3$ and the p-degree of I_k is 2.

6.2.2 Searching for Liouville integrable systems

For simplicity let us take $N = 2$ and a Hamiltonian in the "standard form":

$$H = \tfrac{1}{2}(p_x^2 + p_y^2) + V(x, y). \tag{6.1}$$

Here the potential V is unknown, i.e., we would like to know for which V the system is integrable. Since $N = 2$ we need, by definition, just one sufficiently regular function $I(p_x, p_y, x, y)$ (the "second invariant"), which is functionally independent of H, but Poisson-commutes with it.

To simplify the search problem, and to guarantee regularity of I, we take I to be a polynomial in p. A search can then be organized by the degree of p in I. Since H is invariant under $p_i \rightarrow -p_i$ we must have $I(-p_x, -p_y, x, y) = \pm I(p_x, p_y, x, y)$, which simplifies the problem somewhat.

In practice one first generates a candidate I of degree K:

$$I := \sum_{k=0}^{[K/2]} \sum_{n=0}^{K-2k} c(x,y)_{K-2k,n} \, p_x^n p_y^{K-2k-n}$$

Next one calculates $\{H, I\} = 0$, collects coefficients of monomials $p_x^i p_y^j$, and solves the resulting differential equations.

Example: Take degree $K = 1$ in which case the ansatz will be $I = c_{11}p_x + c_{10}p_y$. Equations from $\{H, I\} = 0$ are

$$p_x^2 : \quad \partial_x c_{11} = 0.$$
$$p_y p_x : \quad \partial_x c_{10} + \partial_y c_{11} = 0.$$
$$p_y^2 : \quad \partial_y c_{10} = 0.$$
$$1 : \quad c_{10}\partial_y V + c_{11}\partial_x V = 0.$$

These are easy to solve: we get

$$c_{11} = \alpha y + \beta, \quad c_{10} = -\alpha x + \gamma, \quad V = v(\tfrac{1}{2}\alpha(x^2 + y^2) + \beta y - \gamma x).$$

Next, in order to *classify* the result we need to identify the canonical forms:

(i) If $\alpha \neq 0$ we can translate β and γ out to get $V = v(\sqrt{x^2 + y^2})$.
(ii) If $\alpha = 0$, we can rotate the potential into $V = v(x)$, or $V = v(x \pm iy)$.

For any degree K the leading (in p) terms of I are easy to solve, the coefficients are polynomials in x, y. The other equations are more difficult. For an ansatz of even degree K there will be $\sum_{k=0}^{K/2-1}(2k + 1) = (K/2)^2$ further unknown functions in I, one in H, namely V, and $\sum_{k=1}^{K/2} 2k = (K/2)((K/2) + 1)$ equations. This is an overdetermined set of differential equations if $K > 2$.

For $K = 2$ the complete solution is known: Equations of motion are separable in known orthogonal coordinate systems [2]. (Later some additional complex potentials were found [3].) Higher values of K correspond to non-separable cases. The complete solution is not known for any $K > 2$, although many isolated examples are known. For further details see the review [3].

6.2.3 Superintegrability

Superintegrability means that in addition to integrability (N mutually commuting quantities) there are still more quantities that commute with the Hamiltonian. The maximal number of such additional independent quantities commuting with the Hamiltonian is $N - 1$.

For example with the above Hamiltonian (6.1) ($N = 2$), for integrability we would need a second invariant I_2, and for superintegrability *two* constants of motion I_2, I_3, both commuting with H, but necessarily $\{I_2, I_3\} \neq 0$. Of course the I_i should be functionally independent and regular enough (for example polynomials in p).

Several superintegrable examples are known, usually I_2 is also quadratic, implying separability, the other may be higher.

Example: $V = x^2 + 4y^2 + d/x^2$ and $V = x^2 + 9y^2$ have second order I_2 (because they are separable) and also a third order I_3.

6.2.4 Quantum integrability

The classical definition of Liouville integrability carries easily over to quantum mechanics: Just replace $p_i \to i\,\tilde{}\,\partial_{q_i}$, so that the I_k become differential operators, and the Poisson bracket is replaced by the commutator.

Definition 2 *In quantum mechanics a system is said to be* **integrable** *if there exists a set of N independent mutually commuting differential operators, one of which is the Schrödinger operator.*

For $N = 2$ the Schrödinger operator would be

$$H := -\frac{\tilde{}^2}{2}(\partial_x^2 + \partial_y^2) + V(x, y),$$

and for quantum integrability we would need a second commuting quantity, which now would be a differential operator.

The equations ensuing from a search problem based on this are quite similar to the ones in the classical case, only slightly more complicated. Many classical systems have quantum integrable counterparts, with possibly $\tilde{}^2$ corrections [4].

In quantum mechanics there is the interesting additional possibility that it even makes sense to have *more than* N mutually commuting differential operators, although only N of them can be algebraically independent. In classical mechanics there is no use for algebraically dependent constants of motion, but in quantum mechanics, where we deal with differential operators, such relationships can be nontrivial. This leads to the concept of *algebraic integrability* [5].

As an example consider the Hamiltonian ($N = 3$)

$$\begin{aligned} H = &-\frac{\tilde{}^2}{2}\left(\mu_1 \partial_1^2 + \mu_2 \partial_2^2 + \mu_3 \partial_3^2\right) \\ &+ A\left[\mu_1 \mathcal{P}(q_2 - q_3) + \mu_2 \mathcal{P}(q_3 - q_1) + \mu_3 \mathcal{P}(q_1 - q_2)\right], \end{aligned}$$

where $\mathcal{P}$ is the Weierstrass elliptic function. The three known quantum integrable cases of this type are characterized as follows:

(i) The classical case: $\mu_1 = \mu_2 = \mu_3$, A free.
(ii) Elliptic case : $\mu_1 = \mu_2$, $A = -\,\tilde{}^2(\mu_1 + \mu_3)/\mu_1$ [5].
(iii) Hyperbolic case: $\mu_1 + \mu_2 + \mu_3 = 0$, $A = \tilde{}^2$ [8].

Note that in Cases 2 and 3 the strength of the potential is proportional to $\tilde{}^2$. This means they are purely quantum mechanical integrable potentials without classical counterparts. The four commuting quantities

I_n, for $n = 1, 2, 3, 4$, can be given as

$$I_n = \mu_1^{n-1}\partial_1^n + \mu_2^{n-1}\partial_2^n + \mu_3^{n-1}\partial_3^n + \text{ terms of degree } n - 2 \text{ and lower.}$$

The analysis of acceptable algebraic dependency is rather involved; in Case 1 above the dependency is trivial, but in the other two it brings in new information.

6.2.5 The computational problem

In this section the general problem is: *How to solve an overdetermined set of differential equations?* There are computer algebra programs designed just for this, for example CRACK [9].

The problem of solving sets of differential equations can be made into an algebraic problem by computing all differential consequences up to some fixed degree in the jet space. After a certain degree further differential consequences produce nothing new, in fact "completion algorithms" have been implemented on computer algebra systems, see e.g. [10]. The resulting set of equations can then be considered as a set of algebraic equations, by taking all partial derivatives of the unknown functions as independent variables. Since the system is highly overdetermined, and the number of integrable systems limited, the system of equations should be solvable. One can later on return to the original interpretation of partial derivatives and solve the final set of differential equations. For $K = 3$ the extension is not difficult to compute but it leads to complicated algebraic equations, and still has not been solved in full generality.

What is needed? A method to handle huge sets of overdetermined algebraic equations. If one tries to solve the full set by calculating its Gröbner basis the computers still tend to freeze due to the notorious intermediate expression swell. Since we are just trying to solve a set of polynomial equations, we just need the "factorized Gröbner basis" in which each factorization of an equation creates branches in the solution tree. This simplifies the size problem but does not solve it. One possible method of overcoming it is "incremental Gröbner basis computation". By this I mean the following algorithm:

Compute first a factorized Gröbner basis G for a small set of equations.

(i) choose the next equation E
(ii) for each $g \in G$ compute the factorized Gröbner base g_E of the set $\{E, g\}$

(iii) combine the lists $\{g_E | g \in G\}$ and eliminate sub-cases to create G_{new}.

(iv) if there are equations left, and no contradictions have been found, goto 1, else finish.

This would be efficient if the system is highly overdetermined and one can feed the new equations in some intelligent way.

6.3 Search for integrable soliton equations

Another search program that we will discuss here is the search for integrable PDE's. Again there are different signatures of integrability on which a search can be based. Let us compare two of them:

Symmetry [11]:

(i) dimension of space-time: 1+1
(ii) the ansatz is made for the leading part of the equation, which implies the dispersion relation (for example $u_t = u_{xxx} + f(u, u_x, u_{xx})$ implies curve $\omega = p^3$).
(iii) *nonlinearity is arbitrary*, to be found
(iv) the condition is on the existence of a sufficient number of symmetries of the equation

Three-soliton condition [12]:

(i) dimension of space-time arbitrary
(ii) *dispersion relation is arbitrary*, to be found
(iii) equations are assumed to be in the Hirota bilinear form, nonlinearity determined by dispersion relation
(iv) the condition is on the existence of multi-soliton solutions.

We see that the approaches are quite different, both in the integrability indicator and the suitable class of equations.

6.3.1 Hirota's bilinear method

At this point we would like to note that the variables in which an equation is given might not be the best for further analysis. Indeed, a fundamental ingredient in Hirota's approach is a dependent variable transformation. R. Hirota noted in 1971 [13] that since from ISTM we know that the soliton solutions of the Korteweg–de Vries (KdV) equation

$$u_t + 6uu_x + u_{xxx} = 0 \tag{6.2}$$

have the form

$$u = 2\partial_x^2 \log(\det M),$$

where the entries of M are of type $a + be^{px+\omega t}$, it would be useful to change variables from u to F by

$$u = 2\partial_x^2 \log F.$$

The new dependent variable F should be a good variable to use, since for soliton solutions it would be a polynomial in exponentials.

In practice it turns out that it is best to take the KdV equation into its potential form

$$w_{xxx} + 3w_x^2 + w_t = 0, \tag{6.3}$$

where $u = \partial_x w$. Now substituting

$$w = \alpha \partial_x \log F,$$

into (6.3) results with

$$F^2 \times (\text{something quadratic}) + 3\alpha(2 - \alpha)(2FF'' - F'^2)F'^2 = 0.$$

Thus we get a quadratic equation if we choose $\alpha = 2$:

$$F_{xxxx}F - 4F_{xxx}F_x + 3F_{xx}^2 + F_{xt}F - F_xF_t = 0.$$

This can be written as

$$(D_x^4 + D_xD_t)F \cdot F = 0,$$

where we have used the

Definition 3 Hirota's derivative operator D *is defined by*

$$\begin{aligned}
D_x^n f \cdot g &= (\partial_{x_1} - \partial_{x_2})^n f(x_1)g(x_2)\big|_{x_2=x_1=x} \\
&\equiv \partial_y^n f(x+y)g(x-y)\big|_{y=0}.
\end{aligned}$$

Note the crucial sign w.r.t Leibniz product rule.

Definition 4 *We say that an equation is in the* **Hirota bilinear form**, *if it is quadratic in the dependent variables and if all derivatives in the equation are given through Hirota's derivative operator.*

An important property of Hirota bilinear equations is their invariance under the gauge transformation

$$F \to e^\kappa F, \, G \to e^\kappa G : \quad P(D)(e^\kappa F \cdot e^\kappa G) = e^{2\kappa}P(D)F \cdot G, \quad \text{if } \kappa = \vec{c} \cdot \vec{x}. \tag{6.4}$$

This is the guiding principle in generalizing Hirota bilinear forms into trilinear forms, or into the discrete domain. Note also that

$$P(\vec{D})e^{\vec{a}\cdot\vec{x}} \cdot e^{\vec{b}\cdot\vec{x}} = P(\vec{a} - \vec{b})e^{(\vec{a}+\vec{b})\cdot\vec{x}}.$$

Bilinearization of the equation is a necessary first step in Hirota's analysis, but unfortunately this step is not algorithmic. In fact, the number of dependent or even independent variables cannot be fixed beforehand. Despite this almost all integrable equations have been written in bilinear form. In integrable cases the dependent variables are τ-functions, according to the Sato theory.

It should be noted that the existence of a bilinear form does not imply integrability. Once the bilinear form has been constructed one can try to find its multi-soliton solutions using a perturbative approach, but this will yield general N-soliton solutions only in integrable cases.

6.3.2 Soliton solutions

Let us consider the class of equations

$$P(\vec{D})F \cdot F = 0, \tag{6.5}$$

where P may be assumed even.

The background or trivial solution $u = 0$ corresponds to $F = 1$, which means that we must take $P(0) = 0$.

The one-soliton solution (1SS) is built on top of the background $F = 1$ perturbatively; the ansatz is:

$$F = 1 + e^{\eta}, \quad \text{where } \eta = \vec{x} \cdot \vec{p} + \eta^0.$$

It is easy to verify that this is a solution, provided that

$$P(\vec{p_i}) = 0, \tag{6.6}$$

which is the *dispersion relation*. This is the curve on which the solution lives. In the KdV case this solution corresponds to a soliton:

$$u = 2\partial_x^2(\log(F))$$

$$= \frac{2p^2 e^{\eta}}{(1 + e^{\eta})^2} = \frac{p^2/2}{\cosh(\frac{1}{2}\eta)^2}$$

Note that the dimension of the coordinate space is not limited.

The Ansatz for the two-soliton solution is a natural extension of the above:

$$F = 1 + e^{\eta_1} + e^{\eta_2} + A_{12}e^{\eta_1+\eta_2}, \quad \eta_i = \vec{x} \cdot \vec{p_i} + \eta_i^0, \tag{6.7}$$

The essential feature here is that the two-soliton solution is built from the 1SSs perturbatively, i.e., if we want a K-soliton solution the expansion should start as

$$F = 1 + \sum_{k=1}^{K} e^{\eta_k} + \dots$$

where each η_k is as in a 1SS.

After substituting (6.7) into (6.5) and collecting terms with different products of e^{η_k} one finds that all other terms cancel except $e^{\eta_1 + \eta_2}$ and from its coefficient one finds

$$A_{12} = -\frac{P(\vec{p}_1 - \vec{p}_2)}{P(\vec{p}_1 + \vec{p}_2)}. \tag{6.8}$$

Example, KdV: $\eta = px + \omega t + \eta_0$, and DR $\omega = -p^3$:

$$A_{12} = -\frac{(p_1 - p_2)^4 + (p_1 - p_2)(\omega_1 - \omega_2)}{(p_1 + p_2)^4 + (p_1 + p_2)(\omega_1 + \omega_2)} = \frac{(p_1 - p_2)^2}{(p_1 + p_2)^2}.$$

Thus the result is that *any* equation of type (6.5) has two-soliton solutions. This is a level of *partial integrability*: we can have elastic scattering of two solitons, for any dispersion relation, if the nonlinearity is suitable. However, the existence of three-soliton solutions is not automatic, on the contrary: it imposes severe restrictions on the polynomial P, and can be used as a search criterion.

6.3.3 The three-soliton condition

Definition 5 *If a Hirota bilinear equation has a 1SS given by*

$$F = 1 + \epsilon e^{\eta}, \eta_i = \vec{x} \cdot \vec{p}_i + \eta_i^0, P(\vec{p}_i) = 0,$$

and one can construct to that equation an N-soliton solution of the form

$$F = 1 + \epsilon \sum_{j=1}^{N} e^{\eta_j} + \text{(finite number of h.o. terms)}$$

without any further conditions on the parameters $\vec{p}_i$ of the individual solitons, then the equation is said to be **Hirota integrable**. *(Here ϵ is a formal expansion parameter.)*

In practice it seems that Hirota integrability is equivalent to other forms of integrability. In the above equation the condition "without any further conditions on the parameters" is essential, because all equations have multi-soliton solutions for some restricted set of parameters.

Let us now apply this principle to the three-soliton solution:

$$
\begin{aligned}
F = & 1 + e^{\eta_1} + e^{\eta_2} + e^{\eta_3} \\
& + A_{12} e^{\eta_1 + \eta_2} + A_{23} e^{\eta_2 + \eta_3} + A_{31} e^{\eta_3 + \eta_1} \\
& + A_{123} e^{\eta_1 + \eta_2 + \eta_3}
\end{aligned}
\tag{6.9}
$$

It seems that we have here one new parameter A_{123}. However we have the *separability condition*: If any one soliton goes far away, the rest should look like a $(N - 1)$-soliton solution. Here "going away" means that either $e^{\eta_k} \to 0$ or $e^{\eta_k} \to \infty$. When this is applied to (6.9) one finds

$$
A_{123} = A_{12} A_{23} A_{13}.
$$

Thus there is no freedom left: the parameters are restricted only by the dispersion relation (6.6); phase factors A_{ij} have been given already in (6.8). Therefore it turns out that the existence of a 3SS is a condition on the equation!

We can use the gauge transformation (6.4) to write the soliton solutions in a symmetric way. We get

$$
F_{1SS} = e^{-\frac{1}{2}\eta} + e^{\frac{1}{2}\eta},
\tag{6.10}
$$

$$
F_{2SS} = \frac{e^{\frac{1}{2}(-\eta_1 - \eta_2)}}{\sqrt{P(\vec{p}_1 + \vec{p}_2)}} + \frac{e^{\frac{1}{2}(-\eta_1 + \eta_2)}}{\sqrt{-P(\vec{p}_1 - \vec{p}_2)}} + \frac{e^{\frac{1}{2}(\eta_1 - \eta_2)}}{\sqrt{-P(\vec{p}_1 - \vec{p}_2)}} + \frac{e^{\frac{1}{2}(\eta_1 + \eta_2)}}{\sqrt{P(\vec{p}_1 + \vec{p}_2)}},
\tag{6.11}
$$

$$
F_{3SS} = \sum_{\sigma_i = \pm} \frac{e^{\frac{1}{2}(\sigma_1 \eta_1 + \sigma_2 \eta_2 + \sigma_3 \eta_3)}}{\sqrt{P(\sigma_1 \vec{p}_1 + \sigma_2 \vec{p}_2) P(\sigma_1 \vec{p}_1 + \sigma_3 \vec{p}_3) P(\sigma_2 \vec{p}_2 + \sigma_3 \vec{p}_3)}}.
\tag{6.12}
$$

Substituting (6.12) in (6.5) yields the "three-soliton-condition", which can be written as

$$
\sum_{\sigma_i = \pm} P(\sigma_1 \vec{p}_1 + \sigma_2 \vec{p}_2 + \sigma_3 \vec{p}_3) P(\sigma_1 \vec{p}_1 - \sigma_2 \vec{p}_2)
$$
$$
\times P(\sigma_2 \vec{p}_2 - \sigma_3 \vec{p}_3) P(\sigma_1 \vec{p}_1 - \sigma_3 \vec{p}_3) = 0.
$$

or

$$
\sum_{\sigma_i = \pm} \frac{P(\sigma_1 \vec{p}_1 + \sigma_2 \vec{p}_2 + \sigma_3 \vec{p}_3)}{P(\sigma_1 \vec{p}_1 + \sigma_2 \vec{p}_2) P(\sigma_2 \vec{p}_2 + \sigma_3 \vec{p}_3) P(\sigma_1 \vec{p}_1 + \sigma_3 \vec{p}_3)} = 0.
\tag{6.13}
$$

The search problem in the equation class (6.5) is then as follows [14]: **Find all polynomials P, such that (6.13) holds on the affine**

variety $\{(\vec{p}_1, \vec{p}_2, \vec{p}_3)|P(\vec{p}_i) = 0\}$. Naturally we are only interested in nontrivial results, which means that acceptable polynomials P should have a nonlinear irreducible factor.

Such a search program was carried out in [15]-[18], see also [19]. For the class (6.5) the complete result is as follows:

$$(D_x^4 - 4D_xD_t + 3D_y^2)F \cdot F = 0, \qquad (6.14)$$

$$(D_x^3D_t + aD_x^2 + D_tD_y)F \cdot F = 0, \qquad (6.15)$$

$$(D_x^4 - D_xD_t^3 + aD_x^2 + bD_xD_t + cD_t^2)F \cdot F = 0, \qquad (6.16)$$

$$(D_x^6 + 5D_x^3D_t - 5D_t^2 + D_xD_y)F \cdot F = 0. \qquad (6.17)$$

and their reductions. These equations also have 4SS and pass the Painlevé test. Of the above equations (6.16) was new, it is non-evolutionary and its Lax pair is still unknown.

6.3.4 Other types of equations

There are also other types of bilinear forms, sometimes pairs of bilinear equations etc. For them one can derive similar three-soliton conditions. Sometimes the existence of two-soliton solutions is not automatic.

Of the other types of equations we would like to mention here only the **nonlinear Schrödinger equation** given by

$$iu_t + u_{xx} + 2\epsilon|u|^2u = 0, \qquad (6.18)$$

where the function u is complex.

The substitution that bilinearizes (6.18) is

$$u = g/f, \quad g \text{ complex}, f \text{ real},$$

yielding

$$f\left[(iD_t + D_x^2)g \cdot f\right] - g\left[D_x^2 f \cdot f - \epsilon 2|g|^2\right] = 0,$$

For normal (bright) solitons we split this into

$$\begin{cases} (iD_t + D_x^2)g \cdot f &= 0, \\ D_x^2 f \cdot f &= \epsilon 2|g|^2. \end{cases} \qquad (6.19)$$

For the search point of view we define the "nlS-class" by

$$\begin{cases} B(D)\,G \cdot F &= 0, \\ A(D)\,F \cdot F &= \epsilon 2|G|^2, \end{cases} \qquad (6.20)$$

and start constructing multi-soliton solutions. As the vacuum soliton we take $f = 1, g = 0$. Next, using the formal expansion

$$f = 1 + \varepsilon f_1 + \varepsilon^2 f_2 + \ldots, \quad g = \varepsilon g_1 + \ldots$$

we get an ansatz for the 1SS

$$g = \varepsilon e^{\eta}, \; f = 1 + \varepsilon^2 a \, e^{\eta + \eta^*}, \; \eta = px + \omega t \text{ complex.} \tag{6.21}$$

which works if we take

$$\text{dispersion relation } B(\vec{p}) = B(-\vec{p}^*) = 0,$$
$$\text{phase factor } a = \frac{1}{A(\vec{p} + \vec{p}^*)}.$$

A search can be based on the existence of two and three-soliton solutions. This is because of the complex nature of the solution; in some way η and η^* should be counted separately.

The results of such a search are [18, 19]:

$$\begin{cases} (D_x^2 + iD_y + c) \, G \cdot F &= 0, \\ (a(D_x^4 - 3D_y^2) + D_xD_t) \, F \cdot F &= |G|^2, \end{cases}$$

$$\begin{cases} (i\alpha D_x^3 + 3cD_x^2 + i(bD_x - 2dD_t) + g) \, G \cdot F &= 0, \\ (\alpha D_x^3 D_t + aD_x^2 + (b + 3c^2)D_xD_t + dD_t^2) \, F \cdot F &= |G|^2, \end{cases}$$

$$\begin{cases} (i\alpha D_x^3 + 3D_xD_y - 2iD_t + c) \, G \cdot F &= 0, \\ (a(\alpha^2 D_x^4 - 3D_y^2 + 4\alpha D_xD_t) + bD_x^2) \, F \cdot F &= |G|^2. \end{cases}$$

The last equation combines the two most important $(2+1)$-dimensional equations, Davey-Stewartson and Kadomtsev-Petviashvili equations, and deserves further study.

6.4 Conclusions

We have here discussed the problem of searching for integrable systems. Integrability is associated with several properties and any one of them can in principle be used for a search project. However, the property that is natural and easily accessible varies from one class of equations to another.

Here we have discussed two search problems:

(i) Two-dimensional point-particle dynamics given by a standard type Hamiltonian (6.1), for which integrability is defined by the existence of a second invariant.

(ii) Soliton equations in Hirota bilinear form (6.5), for which integrability is associated with the existence of three-soliton solutions.

In each case the final problem is that of solving an overdetermined set of equations, for which new and more powerful methods are still needed.

Finally it should be noted that if we start with assumptions of high mathematical level the computations may be easier, but at the same time we may be unnecessarily restricting the set of results. In the examples discussed here the assumptions have been mathematically rather undemanding.

Bibliography

[1] Arnold, V.I. (1978). *Mathematical Methods of Classical Mechanics*, (Springer, Berlin) p. 271.

[2] Winternitz, P., Smorodinsky, J.A., Uhlir, M. and Fris, I. (1966). On symmetry groups in classical and quantum mechanics, *Yad. Fiz.* **4**, 625-635.

[3] Hietarinta, J. (1987). Direct methods in the search of the second invariant, *Phys. Reports* **147**, 87-154.

[4] Hietarinta, J. (1984). Classical versus quantum integrability, *J. Math. Phys.* **25**, 1833-1840.

[5] Burchnall, J.L. and Chaundy, T.W. (1928 and 1932). Commutative Ordinary Differential Operators, *Proc. Roy. Soc. (London)*, **118**, 557-583; ibid **134**, 471-485.

[6] Chalykh, O.A. and Veselov, A.P. (1990). Commutative Rings of Partial Differential Operators and Lie Algebras. *Comm. Math. Phys.* **126**, 597-611.

[7] Veselov, A.P., Feigin, M.V. and Chalykh, O.A. (1996). New integrable deformations of the Calogero-Moser quantum problem, *Russ. Math. Surveys,* **51**, 573-574.

[8] Hietarinta, J. (1998). Pure quantum integrability. *Phys. Lett. A* **246**, 97-104.

[9] Wolf, T. (2004). Applications of CRACK in the Classification of Integrable Systems, in *CRM Proceedings and Lecture Notes*, **37**, 283-300.

[10] Gerdt, V.P. and Blinkov, Yu.A. (1998). Minimal involutive bases, *Math. Comp. Simul.* **45**, 543-560.

[11] A. Mikhailov, article in this book, and references therein.

[12] Ito, M. (1980). An Extension of Nonlinear Evolution Equations of the K-dV (mK-dV) Type to Higher Orders, *J. Phys. Soc. Jpn.* **49**, 771-778.

[13] Hirota, R. (1971). Exact Solution of the Korteweg-de Vries Equation for Multiple Collisions of Solitons, *Phys. Rev. Lett.* **27**, 1192-1194.

[14] Hietarinta, J. (1991). Searching for integrable PDE's by testing Hirota's three-soliton condition, in *Proceedings of the 1991 International Symposium on Symbolic and Algebraic Computation, ISSAC'91*, ed. S.M. Watt (Association for Computing Machinery, New York), pp. 295-300.

[15] Hietarinta, J. (1987). search of bilinear equations passing Hirota's three-soliton condition: I. KdV-type bilinear equations. *J. Math. Phys.* **28**, 1732-1742.

[16] Hietarinta, J. (1987). A search of bilinear equations passing Hirota's three-soliton condition: II. mKdV-type bilinear equations. *J. Math. Phys.* **28**, 2094-2101.

[17] Hietarinta, J. (1987). A search of bilinear equations passing Hirota's three-soliton condition: III. sine-Gordon-type bilinear equations. *J. Math. Phys.* **28**, 2586-2592.

[18] Hietarinta, J. (1988). A search of bilinear equations passing Hirota's three-soliton condition: IV. Complex bilinear equations. *J. Math. Phys.* **29**, 628-635.

[19] Hietarinta, J. (1990). Recent results from the search for bilinear equations having three-soliton solutions, in *Nonlinear evolution equations: integrability and spectral methods*, eds. A. Degasperis, A.P. Fordy and M. Lakshmanan (Manchester U.P., Manchester), pp. 307-317.

7

Around Differential Galois Theory

Anand PillayThis work is supported by a Marie Curie Chair

University of Leeds, England

7.1 Introduction

I will give a short account of some of my talks at the very interesting mini-programme on the algebraic theory of differential equations. I have recently written several expository and research papers on the relevant topics, so I refer the reader to these papers for more details and restrict myself here to just conveying a few key points. In the first week I gave three talks, originally intended to consist of (I) model theory and differential algebra, (II) nonlinear differential Galois theory, and (III) differential algebra and diophantine geometry. As it turned out I only really covered (I) and (II) and I will discuss these below. References are [9] which gives an introduction to model theory and applications to differential algebra, and [8] which gives among other things an accessible account of a differential Galois theory going beyond both the Picard-Vessiot (linear) theory and Kolchin's strongly normal theory. Both these papers are written with an eye to the non logician. I also recommend the appendix to Hrushovski's paper [4] for another discussion of model theory and Galois theory.

In the second week of the programme I gave a talk on nonlinear generalizations of Grothendieck's conjecture on the arithmetic of differential equations. This Grothendieck conjecture (G), in a simple form, says that if $dy/dt = Ay$ is a linear differential equation (in vector/matrix form) over $\mathbb{Q}(t)$, then the equation has a fundamental system of solutions in the algebraic closure $\mathbb{Q}(t)^{alg}$ of $\mathbb{Q}(t)$ if and only if for almost all primes p the reduction mod p of the equation has a fundamental system of solutions in $\mathbb{F}_p(t)^{sep}$, the separable algebraic closure of $\mathbb{F}_p(t)$. Ekedahl, Shepherd-Barron and Taylor formulated a conjecture (F) relating the integrability of algebraic foliations on algebraic varieties defined over $\mathbb{Q}$

232

to the p-curvature of the reduction modulo p of the foliation for almost all p, and showed that (F) implies (G). I proved essentially that (G) is equivalent modulo (F) to a certain statement (D) whose conclusion belongs in a sense to differential algebraic geometry or the model theory of differential fields, and concerns the existence of definable functions (with arbitrary parameters) from suitable definable sets to the field of constants in differentially closed fields. The proof uses Kolchin's differential Galois theory (discussed and generalized in the rest of this paper) as well as results of J.-B. Bost. As my results are written up in detail in [10] (in a volume coming out of a 2005 program at the Isaac Newton Institute) I will refer interested readers to that paper rather than just repeat the content here.

7.2 Model theory and differential fields

Model theory and stability theory are quite abstract subjects, but many of their constructions and theorems are meaningful in more concrete contexts, although additional work may be required to find the appropriate translation. The theory of *internality* and *definable automorphism groups* originated with Zilber, was put in a general form by Hrushovski, and is still undergoing refinements, in connection with "definable groupoids". Poizat [12] first pointed out that this theory gives a model-theoretic explanation of the Picard Vessiot theory as well as Kolchin's more general strongly normal theory (both of which concern extensions of differential fields), and even suggests a further generalization. Details of this further generalization, at the level again of extensions of differential fields, were worked out in [7], but using heavily model-theoretic language. Another account concentrating on the differential equations giving rise to this generalized Galois theory, was given in [8]. The relevant equations are "logarithmic differential equations on algebraic D-groups". In fact in [2] Daniel Bertrand suggested that this general differential Galois theory may be relevant to proving Schanuel-type conjectures for nonconstant commutative algebraic groups, and this is confirmed to some extent in a current collaboration with Bertrand.

Key notions of model theory (and mathematical logic) include structure, formula, definable set, elementary extension and substructure, (first order) theory, complete theory, type. A rather free-ranging introduction for the general mathematician appears in [9], but there are of course many books on the subject, such as [3] and [6], which we recommend.

The important first order theory for our purposes is DCF_0, the theory of differentially closed fields of characteristic zero, in the vocabulary with symbols $+, -, \cdot, 0, 1, \partial$. DCF_0 consists of the axioms for differential fields of characteristic 0 together with axioms expressing, for a differential field F, that any finite system of differential polynomial equations over F with a solution in a differential field extension of F already has a solution in F. It is not completely trivial to express the latter in a first order manner. A differentially closed field (of characteristic zero) is precisely a model of DCF_0, that is a differential field in which the axioms are true. DCF_0 is complete with quantifier elimination. We typically work in a "sufficiently saturated" differentially closed field $\mathcal{U}$ (or $(\mathcal{U}, \partial)$ if we want to exhibit the derivation) namely a universal differential field in the sense of Kolchin. There is no harm in assuming that the field of constants $\mathcal{C}$ of $\mathcal{U}$ coincides with $\mathbb{C}$. Of course $\mathcal{U}$ (or any differentially closed field for that matter) is rather far from "natural" differential fields, such as $(\mathbb{C}(t), d/dt)$, $(\mathbb{Q}(t), d/dt)$, their finite extensions, and their Picard-Vessiot extensions. (However any countable differential field embeds in the differential field of germs of meromorphic functions at $0 \in \mathbb{C}$.) Differentially closed fields provide the appropriate notion of "geometric" for algebraic differential equations in the same way as algebraically closed fields do for polynomial equations. Namely to see all solutions of a system of differential polynomial equations one should consider points in a differentially closed field. This general point of view (like Kolchin's) is of course analogous to that of Weil's Foundations. Although possibly out of sync with a more scheme-theoretic approach, it is convenient for directly applying model-theoretic machinery.

The underlying objects of "differential algebraic geometry" are the common solution sets, $X \subseteq \mathcal{U}^n$, of finite systems of differential polynomial equations (with coefficients from $\mathcal{U}$), in n (differential) indeterminates. These are sometimes called Kolchin closed sets analogously to Zariski closed sets. Identifying an algebraic variety with its collection of $\mathcal{U}$-points we can and will also talk about Kolchin closed subsets of algebraic varieties. Up to finite Boolean combination the Kolchin closed sets are precisely the *definable sets* in $\mathcal{U}$ (by quantifier elimination). It makes sense to talk about a Kolchin closed set (or definable set) being defined over a differential subfield F of $\mathcal{U}$. For X defined over F, $X(F)$ denotes the set of points of X with coordinates from F.

The theory DCF_0 has an important model-theoretic property called ω-stability. I won't give the definition of ω-stability but in the case of

DCF_0 it amounts (using quantifier elimination) to the fact that for any countable differentially closed field F, the differential spectrum (set of prime differential ideals) of the differential polynomial ring $F\{x\}$ over F in one (differential) indeterminate is countable. One consequence of ω-stability is the existence and uniqueness of "prime models over sets" which in the case of DCF_0 amounts to the existence and uniqueness (up to isomorphism over F) of the differential closure F^{diff} of a differential field F.

7.3 Internality, differential galois theory , and logarithmic derivatives

As mentioned at the beginning of the last section the general theory of internality lies behind differential Galois theory. We fix an ω-stable theory T and let $\mathcal{U}$ be a saturated model of T of large cardinality in which we work. (So for $T = DCF_0$, $\mathcal{U}$ is a universal domain in the sense of Kolchin.) Let X, Y be $\emptyset$-definable sets. We say that Y is *internal* to X if there is a definable function f_c (defined with parameters c) from $X \times \ldots \times X$ onto Y. A typical example is where X is a group G, and there is a regular $\emptyset$-definable action of G on Y. In this case choose c to be any element of Y and let $f_c(g) = gc$.

The basic model-theoretic result (see Chapter 7, section 4 of [11] or the Appendix to [4] for general formulations and proofs) is:

Proposition 7.3.1 *Suppose X, Y are $\emptyset$-definable sets such that Y is internal to X. Then*
(i) We can choose f_c (witnessing internality) such that c is a tuple of elements of Y, in fact of elements from $Y(M_0)$ where M_0 is a prime model of T. We call c a "fundamental system of solutions".
(ii) Let $Gal(Y/X)$ be the group of permutations of Y induced by automorphisms of $\mathcal{U}$ which fix X pointwise. (Equivalently $Gal(Y/X)$ is the group of elementary permutations of Y over X.) Then there is a $\emptyset$-definable group G and a $\emptyset$-definable action of G on Y which is isomorphic (as an abstract group action) to the action of $Gal(Y/X)$ on Y.
(iii) The isomorphism in (ii) induces an isomorphism between $G(M_0)$ with its action on $X(M_0)$, and $Gal(Y(M_0)/X(M_0))$.
(iv) The definable group G from (ii) is also internal to X.

The general point is that when specialized to the theory $T = DCF_0$ and working over arbitrary differential fields, we obtain a rather comprehensive Galois theory for differential fields. Of course there are some delicate issues involved, first a move from "automorphism groups" of definable sets, as in Proposition 3.1, to automorphism groups of differential field extensions, but also a certain technical assumption of the "no new constants" kind. Anyway, we now take T to be DCF_0 and obtain the following:

Proposition 7.3.2 *Let F be a (small) differential field. Let X, Y be F-definable sets (namely defined with parameters from F). Assume that Y is internal to X, witnessed by c-definable function f_c with c from F^{diff} (by 3.1 (i)). Assume also*
(*) $X(F) = X(F^{diff})$.
Let $K = F\langle c\rangle$ (the differential subfield of $\mathcal{U}$ generated by F and c). Let G be the F-definable group given by Proposition 3.1 (for X and Y).
THEN there is a natural isomorphism i between $G(F^{diff})$ and $Aut_\partial(K/F)$, which establishes a Galois correspondence between F^{diff}-definable subgroups of G and differential fields lying between F and K.

Note that by $Aut_\partial(K/F)$ we mean the group of automorphisms of the *differential* field K which fix F pointwise.

We call an extension $F < K$ of differential fields obtained as in Proposition 3.2 a *differential Galois extension*.
Some remarks:
(I) There is a natural identification of $Gal(Y(F^{diff})/X(F^{diff}))$ with $Aut_\partial(K/F)$, which together with the isomorphism in 3.1(iii) gives the isomorphism i in 3.2.

(II) If $X = \mathcal{C}$ (the field of constants of $\mathcal{U}$) then condition (*) in Proposition 3.2 is equivalent to C_F being algebraically closed, and so $C_F = C_K$ as $K < F^{diff}$. The situation where $X = \mathcal{C}$ corresponds precisely to the extension $K > F$ in Proposition 3.2 being a *strongly normal* extension in the sense of Kolchin [5].

(III) If $X = \mathcal{C}$ and Y is the solution space of a *linear differential equation* over F, $\partial y = Ay$ (where y is a column vector of unknowns and A a square matrix over F) then K from Proposition 3.2 is called a *Picard-Vessiot extension* of F for the equation. Note that Y is a vector space over $\mathcal{C}$ and in fact the "fundamental system of solutions c" can be chosen to be a $\mathcal{C}$-basis (with coordinates in F^{diff}) of Y.

In cases (II) and (III) the differential Galois group of K over F is usually thought of as the group of C_F-rational points of some algebraic group over C_F. This follows from Proposition 3.1 (iv) (internality of G to X). We will discuss further below various incarnations of the Galois group.

We now briefly discuss the differential equations lying behind the differential field extensions of Proposition 3.2.

Note that in the Picard-Vessiot case (III) above, given a linear differential equation over F in vector form (*) $\partial y = Ay$, one is really interested in a fundamental system of solutions, which amounts to a nonsingular $n \times n$ matrix Z such that $\partial Z = AZ$. So we have replaced the original equation, where y is an unknown column vector, by the equation (**) $\partial z = Az$ where z is an unknown from GL_n. The differential rational map $z \to (\partial z)z^{-1}$ from GL_n to its Lie algebra is called the logarithmic derivative dln_{GL_n}, and the equation (**) becomes $dln_{GL_n}(z) = A$.

Kolchin ([5]) defines a logarithmic derivative dln_G for any algebraic group G over the constants. From the analytic point of view and working in the differential field K of meromorphic functions on some open set in $\mathbb{C}$, dln_G is obtained by applying the derivation $\partial = d/dt$ to the multivalued logarithm map from $G(K)$ to $LG(K)$ (where LG is the Lie algebra of G). From the differential algebraic point of view and working in the universal domain $\mathcal{U}$, dln_G is obtained by first applying ∂ to affine coordinates of $z \in G(\mathcal{U})$ to obtain a point in the tangent space to G at z, and then taking the corresponding element of $L(G)$.

Kolchin ([5]) proves

Proposition 7.3.3 *(i) Suppose C_F is algebraically closed, G is a connected algebraic group over F and $b \in LG(F)$. Then there is a solution $\alpha \in G(\mathcal{U})$ of $dln_G(-) = b$ such that $K = F(\alpha)$ is a strongly normal extension of F (i.e. a differential Galois extension as in Case (II) above). Moreover K is unique (as a differential field up to isomorphism over F). (ii) Conversely, if K is a strongly normal extension of F and F is algebraically closed, then there is a connected algebraic group G over C_F, and $b \in LG(F)$ such that K is generated by a solution of $dln_G(-) = b$.*

We now work towards stating a generalization of Proposition 3.3 to arbitrary differential Galois extensions.

Let G be an connected algebraic group over the differential field F (but not necessarily defined over the field of constants C_F of F). We need the notion of an algebraic ∂-group structure on G, due to Alex Buium [1]. We will explain this notion in the case where G is affine (for the general case replace coordinate ring by structure sheaf). Let $F[G]$ denote the coordinate ring of G. By an algebraic ∂-structure on G we mean an extension of the derivation ∂ on F to a derivation D on $F[G]$ which respects the group operation. "Respecting the group operation" means the following. Note that D extends canonically to a derivation on $F[G \times G] = F[G] \otimes_F F[G]$ which we also call D. For $f \in F[G]$ let f^* be the function on $G \times G$ taking (g, h) to $f(gh)$. Then the condition on D is that for any $f \in F[G]$, $D(f^*) = (Df)^*$. In the same way that a derivation of $F[G]$ which is 0 on F is, or corresponds to, a vector field on G, a derivation D extending ∂ corresponds to a section s of a certain "twisted" tangent bundle $T_\partial(G)$. $T_\partial(G)$ has the structure of a connected algebraic group over F which is also a torsor for $T(G)$. The compatibility of D with the group operation means precisely that s is a homomorphism. In any case we write an algebraic ∂-group as (G, D) or (G, s). Such an object really belongs to algebraic geometry or algebraic group theory. If G is defined over C_F then $T_\partial(G) = T(G)$ and we of course have the 0-section $G \to T(G)$, which we call s_0, giving the *trivial* algebraic ∂-group structure on G.

Before proceeding further it is worth noting the relationship of such objects to more familiar (to some people) things. For example, one can show rather easily that if (G, D) is an algebraic D-group, then D induces a derivation on the maximal ideal of the local ring of G at the identity, hence induces a connection, which we call D_V on the cotangent space V to G at the identity. If G happens to be the universal extension $\tilde{A}$ of an abelian variety A by a vector group, then V coincides with $H^1_{DR}(A)$ the first de Rham cohomology group of A, and the so-called Gauss-Manin connection on the latter turns out to be precisely D_V, where D is the *unique* algebraic ∂-group structure on $\tilde{A}$. So in this case, the unique ∂-group structure on $\tilde{A}$ can be viewed as a "lifting" of the Gauss-Manin connection.

An algebraic ∂-group (G, D) gives rise to objects and maps belonging to differential algebraic geometry: if $g \in G(\mathcal{U})$ we can apply the derivation ∂ to the coordinates of g and we obtain a point in the fibre of $T_\partial(G)$ above g which we call $\partial(g)$. But $s(g)$ is also in this fibre. Hence the difference $\partial(g) - s(g)$ is in the tangent space to G at g and hence yields a point in the Lie algebra of G which we call point $dln_{(G,D)}(g)$. So $dln_{(G,D)}$ is a differential rational map from $G(\mathcal{U})$ to $LG(\mathcal{U})$, and also a so-called crossed homomorphism. (When G is over the constants and $s = s_0$, the map coincides with Kolchin's dln_G described earlier). The kernel of $dln_{(G,D)}$ is denoted by $(G,D)^\partial$ or just G^∂ if D is understood, and is a "finite-dimensional differential algebraic group".

Note that if (G,D) is a connected algebraic ∂-group over F and $b \in LG(F)$, then letting Y be the solution space in $G(\mathcal{U})$ of the equation $dln_{(G,D)}(-) = b$, and taking $X = G^\partial$, then Y is internal to X. If moreover $G^\partial(F) = G^\partial(F^{diff})$ then we obtain a differential Galois extension K of F in the sense of 3.2 by letting g be a solution of the equation in $G(F^{diff})$ and putting $K = F(g)$. Conversely:

Proposition 7.3.4 *Let K be a differential Galois extension of F and assume that F is algebraically closed. Then there is a connected algebraic ∂-group (G,D) such that $G^\partial(F) = G^\partial(F^{diff})$, and there are $b \in LG(F)$ and $g \in G(K)$ such that $dln_{(G,D)}(g) = b$ and $K = F(g)$.*

Finally one may ask what the definable Galois group from 3.1 or 3.2 corresponds to. Bearing in mind the above few lines, let K be a differential Galois extension of F generated over F by a solution of an equation $dln_{(G,D)}(-) = b$ with $b \in LG(F)$ and with $G^\partial(F) = G^\partial(F^{diff})$. Then the definable Galois group from 3.1 and 3.2 is, up to F-definable isomorphism, of the form S^∂, where S is some ∂-subgroup, defined over F, of the algebraic ∂-group (G,D') where $D' = D + r_b - l_b$, r_b is the right invariant vector field on G determined by b and l_b the left invariant one. So $Aut_\partial(K/F)$ is canonically isomorphic to $S^\partial(F^{diff})$.

Another incarnation of the Galois group has the form H^∂ for H some ∂-subgroup of (G,D) defined over F. H^∂ and S^∂ will in general be only isomorphic over K. In the case where $G = GL_n$ and D is trivial, then H^∂ is the set of constant points of a linear algebraic group over C_F, and is what is commonly known as the differential Galois group of the corresponding Picard-Vessiot extension.

Bibliography

[1] Buium, A. (1992). *Differentially algebraic groups of finite dimension*, Lecture Notes in Mathematics 1506 (Springer, Berlin).

[2] Bertrand, D. (2008). Schanuel's conjecture for non-isoconstant elliptic curves over function fields, in *Model Theory and Applications to Algebra and Analysis*, ed. Chatzidakis, Macpherson, Pillay, and Wilkie (Cambridge University Press, Cambridge).

[3] Hodges, W.A. (1993). *Model Theory* (Cambridge University Press, Cambridge).

[4] Hrushovski, E. (2002). Computing the Galois group of a linear differential equation, *Banach Center Publ.* **58**, 97-138

[5] Kolchin, E. (1973). *Differential Algebra and Algebraic Groups* (Academic Press, New York).

[6] Marker, D. (2002). *Model theory - an introduction* (Springer-Verlag, Berlin).

[7] Pillay, A. (1998). Differential Galois theory I, *Illinois J. Math.* **42**, 678-699.

[8] Pillay, A. (2004). Algebraic D-groups and differential Galois theory, *Pacific J. Math.* **216**, 343-360.

[9] Pillay, A. (2008). Model theory and stability theory, with applications in differential algebra and algebraic geometry, in *Model Theory and Applications to Algebra and Analysis*, see above.

[10] Pillay, A. (2008). Differential algebra and generalizations of Grothendieck's conjecture on the arithmetic of linear differential equations, in *Model Theory and Applications...* as above.

[11] Pillay, A. (1996). *Geometric stability theory* (Oxford University Press, Oxford).

[12] Poizat, B. (2000). *A Course in model theory* (Springer-Verlag, Berlin).